Engineering Biomaterials for Neural Applications

Elisa López-Dolado • María Concepción Serrano

Editors

Engineering Biomaterials for Neural Applications

Targeting Traumatic Brain and Spinal Cord Injuries

Editors
Elisa López-Dolado
SESCAM
Hospital Nacional de Parapléjicos
Toledo, Spain

María Concepción Serrano
CSIC
Instituto de Ciencia de Materiales de
Madrid
Madrid, Spain

ISBN 978-3-030-81399-4   ISBN 978-3-030-81400-7   (eBook)
https://doi.org/10.1007/978-3-030-81400-7

This Springer imprint is published by the registered company Springer Nature Switzerland AG
The registered company address is: Gewerbestrasse 11, 6330 Cham, Switzerland

Scientific and technological research, from molecular to behavioural levels, has been carried out in many different places but they have not been developed in a really interdisciplinary way. Research should be based on the convergence of different interconnected scientific sectors, not in isolation, as was the case in the past.

Rita Levi-Montalcini

That is exactly what it is all about, meeting people who see things that you do not see. People who teach you to look with different eyes.

Mario Benedetti

Foreword

Our central nervous system defines how we function and perceive the world. When the brain or spine is injured, the consequences can be devastating to the individual and to society at large due to increased morbidity, productivity loss, and healthcare costs. Therefore, the restoration of cognitive and motor functions after damage to the central nervous system remains a major scientific, engineering, and medical challenge. In the quest to successfully address this challenge, it is now recognized that the development and implementation of engineered biomaterials will be an important part of the technological solution. The field of biomaterials science and engineering has significantly evolved within the past decade due to advances in the fields of cell and molecular biology, nanotechnology, materials science, and chemistry, which when in convergence can provide innovative solutions to medical problems. In Engineering Biomaterials for Neural Applications: Targeting Traumatic Brain and Spinal Cord Injuries, the editors and contributing authors do a thorough job at highlighting fundamental information on the anatomy and physiology of the nervous system, describing the consequences of trauma and disease on this system, and demonstrating how biomaterials can alter or monitor the course of disease. As a result, the reader will gain significant insight and appreciation for how biomaterials can be used to address clinical challenges due to trauma to nerves. The authors describe the biomaterial microenvironment and how the components and dynamics of that microenvironment are influenced by trauma or disease conditions. Throughout the book, they methodically review and highlight the importance of device and biomaterial biocompatibility, mechanical compliance, architecture, biodegradability, electrical activity, and form factor, all in the context of regenerating neural structures and restoring tissue functions. This thematic organization, which is consistent throughout the book, will help readers at all levels understand key topics and apply the learned material in their own research.

Advances in nanoscience and nanotechnology have been at the foundation of emerging areas of research and new fields such as regenerative engineering and synthetic biology. Similarly, nanotechnology has had a major impact in biomaterials research. In this regard, several chapters in this book demonstrate how nanomaterials, nanostructures, and nanoscale properties can be leveraged to promote neural regeneration and electrostimulation and discuss safety issues that must be addressed. Given the increase in cerebrovascular disease and brain injuries, the discussion of biomaterial- and cell-based therapeutic strategies that target stroke and traumatic brain injury is timely. The authors describe the role of devices and robotics in rehabilitation protocols for managing the consequences of traumatic brain injury as well as implantable devices for neurosurgery. Throughout the book, the authors emphasize function recovery and describe the role of biomaterials and devices in the process. Also appealing is the balance between regenerative/therapeutic and diagnostic strategies, giving the reader a great introduction into these exciting areas of biomaterials research. This book will be an excellent resource for graduate students and senior scientists pursuing research in biomaterials, neural prosthetics, artificial organs, and regenerative engineering as well as for medical students and professionals that are interested in applying biomaterials research to neural applications.

Professor Guillermo A. Ameer
Daniel Hale Williams Professor of Biomedical Engineering and Surgery
Director of the Center for Advanced Regenerative Engineering
Northwestern University

Preface

The nervous system is, undoubtedly, the most complex system within our body. Besides integrating all the information coming from the outside and inside media, it orchestrates individual and collective responses of cells, tissues, and organs. The enormous anatomical and structural complexity of this system, along with the superior functions that it dictates, has hidden this body component behind a mist of darkness and mystery for many centuries. Since breakthrough discoveries at the beginning of the twentieth century by neuroscientists such as Ramón y Cajal and Golgi, Nobel laureates in 1906 for their contribution to the understanding of the nervous tissue structure, neuroscience, neurology, and neurosurgery have remarkably advanced the knowledge on the physiopathology of this system.

Traumatic brain injury (TBI) and spinal cord injury (SCI) remain a challenge for scientists and clinicians at all stages, from their basic pathophysiology and evolution over time to their response to eventual therapeutics and consequences. Indeed, it has been generally accepted that, due to the complexity and heterogeneity of these pathologies among individuals, TBI and SCI can be divided in groups of subpathologies. Thus, it would be more appropriate to refer to them in plural rather than as singular and uniform diseases. Patients suffering from TBI and SCI have been largely benefitting from progress in the medical practice, although advances in their cure are dramatically limited to date. In this scenario, novel and more advanced biomaterials are emerging as a savior table, with enhanced biocompatibility features and promising versatility for addressing patients' heterogeneity, in line with the modern era of personalized medicine. The expectation is such that clinicians continue to invest efforts in clinical trials with biomaterial-based strategies.

Inspired by the enormous potential that this field is demonstrating, we compile in this book a selection of state-of-the-art topics related to engineered biomaterials primarily envisioned for the treatment of TBIs and SCIs. Chapters will first cover general aspects of the nervous system and their major traumatic pathological conditions. Specific advances in the use of biomaterials in this context, at both the nano- and macroscales, will be then presented. Final progress on novel orthotic and robotic therapies and implantable neural devices, as well as some auspicious diagnostic tools for the assessment of their morphological and functional effects, will be finally exposed and discussed.

There are five major features that differentiate this book from others previously published. First, its main focus is placed on relevant neural pathologies with unfortunately increasing incidence and prevalence worldwide. Second, it provides a new interdisciplinary perspective by experts in topics highly diverse but all related to neural applications. Third, it complementarily integrates clinical information on TBIs and SCIs with knowledge derived from basic and translational approaches. Fourth, it aims to generate a broad discussion with applicability in other neural diseases that could be, or are already being, targeted by similar material-based approaches. And fifth, this book is a short and concise compilation of the enormous progress that the field is experiencing in the last years.

Based on its multidisciplinary focus, this book is envisioned to reach a wide audience of readers, from basic scientists attracted by the need of therapeutics for neural repair to clinicians working routinely with patients suffering from neural diseases, as well as experts with more diverse backgrounds such as physicists, mathematicians, and computer bio-scientists, who are increasingly involved and imbricated in neuroscience research at all levels. Future professionals at a training stage of their careers can also benefit from the content of this book and, hopefully, find inspiration from it.

Also, we do wish this book serves patients affected by TBIs and SCIs, for whom the effort to compose it has been mainly placed, providing them an updated compilation of the scientific and clinical progress carried out by interdisciplinary experts all around the world in an attempt to comprehend these neural pathologies and approach their effective cure.

Every published book is the fruit of a considerable collective effort. And this book, without exception, certainly is. Without the contribution of a large number of experts, it could not have been written. We should not end these lines without expressing our most sincere gratitude to each particular co-author that has contributed to the elaboration of this book for their time and knowledge reflected in the final piece of their chapter. We also acknowledge the publisher's team, who conceived the idea of the book and accompanied us along the path of its elaboration. Also, to all those anonymous supporters, even when not expressively named in the final book, for their advice, silent contribution, and inspiration during the writing process. Finally, to you, the reader, the origin and goal of all editorial adventures, we thank you for having decided to take part in it.

A book is like a painting or a symphony. It is not only the colors or the notes that matter, but how they interact. What we see and hear, what we read, is the result of that engagement. We deeply hope that the content of each chapter will reach its full value at the end of the reading and that the messages resting on the pages will be a source of knowledge and inspiration.

To André Espinha, this book would not have been possible without his selfless and devoted assistance. To my parents, who have guided me with love to where I am now (M. C. Serrano). To my husband, my best support. To my nephews and siblings, my best motivation (E. López-Dolado).

Madrid, Spain   María Concepción Serrano

Toledo, Spain   Elisa López-Dolado
March 2021

Contents

Contributors

Melchor Álvarez-Mon Diseases of Immune System and Oncology Department, Hospital Universitario Príncipe de Asturias, Madrid, Spain

School of Medicine, Medicine and Medical Specialties Department, Universidad de Alcalá (UAH), Madrid, Spain

Instituto Ramón y Cajal de Investigación Sanitaria (IRYCIS), Carretera Colmenar Viejo, Madrid, Spain

Virginia Aranda Brain Damage Unit, Hospital Beata María, Madrid, Spain

Ángel Arévalo-Martín Laboratory of Neuroinflammation, Hospital Nacional de Parapléjicos, SESCAM, Toledo, Spain

Alejandro Babin Contreras Center for Prosthetics & Orthotics PRIM, Madrid, Spain

Laura Ballerini International School for Advanced Studies (SISSA/ISAS), Trieste, Italy

Ignacio Calvo-Arenillas Facultad de Fisioterapia y Enfermería, Universidad de Salamanca, Salamanca, Spain

Ana Calvo-Vera Facultad de Fisioterapia y Enfermería, Universidad de Salamanca, Salamanca, Spain

Nieves Casañ-Pastor Institut de Ciencia de Materials de Barcelona (ICMAB), Consejo Superior de Investigaciones Científicas (CSIC), Barcelona, Spain

Mónica Cicuéndez Departamento de Bioquímica y Biología Molecular, Facultad de Ciencias Químicas, Universidad Complutense de Madrid, Instituto de Investigación Sanitaria del Hospital Clínico San Carlos and CIBER de Bioingeniería, Biomateriales y Nanomedicina, CIBER-BBN Madrid, Spain

Teresa Criado Brain Damage Unit, Hospital Beata María Ana, Madrid, Spain

David Díaz Diseases of Immune System and Oncology Department, Hospital Universitario Príncipe de Asturias, Madrid, Spain

School of Medicine, Medicine and Medical Specialties Department, Universidad de Alcalá (UAH), Madrid, Spain

Ana Domínguez-Bajo Instituto de Ciencia de Materiales de Madrid (ICMM), Consejo Superior de Investigaciones Científicas (CSIC), Madrid, Spain

Nayra Fernández-Pinedo Children Rehabilitation Unit, Hospital Beata María Ana, Madrid, Spain

Lino Ferreira Center for Neuroscience and Cell Biology (CNC), University of Coimbra, Coimbra, Portugal

Faculty of Medicine, University of Coimbra, Coimbra, Portugal

Fernando García-García Department of Radiology, Hospital Nacional de Parapléjicos, SESCAM, Toledo, Spain

Daniel García-Ovejero Laboratory of Neuroinflammation, Hospital Nacional de Parapléjicos SESCAM, Toledo, Spain

Jesús García Ovejero Instituto de Ciencia de Materiales de Madrid (ICMM), Consejo Superior de Investigaciones Científicas (CSIC), Madrid, Spain

Antonio García Peris Department of Radiology, Hospital Nacional de Parapléjicos, SESCAM, Toledo, Spain

Ángel Gil-Agudo Biomechanics and Technical Aids Laboratory, Hospital Nacional de Parapléjicos, SESCAM, Toledo, Spain

Rehabilitation Department, Hospital Nacional de Parapléjicos, SESCAM, Toledo, Spain

Department of Medicine and Medical Specialities, School of Medicine, Universidad de Alcalá, Madrid, Spain

André F. Girão TEMA, Department of Mechanical Engineering, University of Aveiro, Aveiro, Portugal

Instituto de Ciencia de Materiales de Madrid (ICMM), Consejo Superior de Investigaciones Científicas (CSIC), Madrid, Spain

Jesús María González Instituto de Ciencia de Materiales de Madrid (ICMM), Consejo Superior de Investigaciones Científicas (CSIC), Madrid, Spain

Ankor González-Mayorga Laboratory of Interfaces for Neural Repair, Hospital Nacional de Parapléjicos, SESCAM, Toledo, Spain

Elisa López-Dolado Laboratory of Interfaces for Neural Repair, Hospital Nacional de Parapléjicos, SESCAM, Toledo, Spain

Rehabilitation Department, Hospital Nacional de Parapléjicos, SESCAM, Toledo, Spain

Research Unit of "Design and development of biomaterials for neural regeneration", Hospital Nacional de Parapléjicos, Joint Research Unit with CSIC, Toledo, Spain

Raquel Madroñero-Mariscal Fundación del Lesionado Medular, Madrid, Spain

Hospital Universitario Infanta Leonor, SERMAS, Madrid, Spain

Laboratory of Interfaces for Neural Repair, Hospital Nacional de Parapléjicos, SESCAM, Toledo, Spain

Paula A. A. P. Marques TEMA, Department of Mechanical Engineering, University of Aveiro, Aveiro, Portugal

Joana Mestre Veiga Center for Prosthetics & Orthotics PRIM, Madrid, Spain

María del Puerto Morales Instituto de Ciencia de Materiales de Madrid (ICMM), Consejo Superior de Investigaciones Científicas (CSIC), Madrid, Spain

João Peça Center for Neuroscience and Cell Biology (CNC), University of Coimbra, Coimbra, Portugal

Faculty of Medicine, University of Coimbra, Coimbra, Portugal

Sónia L. C. Pinho Center for Neuroscience and Cell Biology (CNC), University of Coimbra, Coimbra, Portugal

José Luis Polo Department of Electrical Engineering, Escuela de Ingeniería Industrial y Aeroespacial, Universidad de Castilla-La Mancha, Toledo, Spain

María Teresa Portolés Departamento de Bioquímica y Biología Molecular, Facultad de Ciencias Químicas, Universidad Complutense de Madrid, Instituto de Investigación Sanitaria del Hospital Clínico San Carlos and CIBER de Bioingeniería, Biomateriales y Nanomedicina, CIBER-BBN Madrid, Spain

Ann M. Rajnicek School of Medicine, Medical Sciences and Nutrition, University of Aberdeen, Scotland, UK

Catarina Rebelo Center for Neuroscience and Cell Biology (CNC), University of Coimbra, Coimbra, Portugal

Faculty of Medicine, University of Coimbra, Coimbra, Portugal

Tiago Reis Center for Neuroscience and Cell Biology (CNC), University of Coimbra, Coimbra, Portugal

Faculty of Medicine, University of Coimbra, Coimbra, Portugal

Ana de los Reyes Guzmán Biomechanics and Technical Aids Laboratory, Hospital Nacional de Parapléjicos, SESCAM, Toledo, Spain

Andrea Riendas Brain Damage Unit, Hospital Beata María Ana, Madrid, Spain

Artur Filipe Rodrigues Center for Neuroscience and Cell Biology (CNC), University of Coimbra, Coimbra, Portugal

Ángel Rodríguez de Lope Department of Neurosurgery, Complejo Hospitalario de Toledo, SESCAM, Toledo, Spain

João Sargento-Freitas Faculty of Medicine, University of Coimbra, Coimbra, Portugal

Department of Neurology, Centro Hospitalar e Universitário de Coimbra, Coimbra, Portugal

María Concepción Serrano Instituto de Ciencia de Materiales de Madrid (ICMM), Consejo Superior de Investigaciones Científicas (CSIC), Madrid, Spain

Pedro A. Serrano Brain Damage Unit, Hospital Beata María Ana, Madrid, Spain

Occupational Therapy Deparment, Faculty of Health Sciences, Centro Superior de Estudios Universitarios La Salle, Universidad Autónoma de Madrid, Madrid, Spain

Occupational Thinks Research Group, Institute of Neuroscience and Sciences of the Movement (INCIMOV), Centro Superior de Estudios Universitarios La Salle, Universidad Autónoma de Madrid, Madrid, Spain

Universidad de Castilla La Mancha, Talavera de la Reina, Toledo, Spain

Facultad de Fisioterapia y Enfermería, Universidad de Salamanca, Salamanca, Spain

Miriam M. Sevilla Brain Damage Unit, Hospital Beata María Ana, Madrid, Spain

Anabel Sorolla Harry Perkins Institute of Medical Research, QEII Medical Centre, Nedlands

Centre for Medical Research, The University of Western Australia, Perth, WA, Australia

Joana Sousa TEMA, Department of Mechanical Engineering, University of Aveiro, Aveiro, Portugal

João André Sousa Faculty of Medicine, University of Coimbra, Coimbra, Portugal

Department of Neurology, Centro Hospitalar e Universitário de Coimbra, Coimbra, Portugal

Cristina Suñol Institut de Investigacions Biomédiques de Barcelona (IIBB), Consejo Superior de Investigaciones Científicas (CSIC), Barcelona, Spain

Institut d'Investigacions Biomèdiques August Pi i Sunyer (IDIBAPS), Barcelona, Spain

Sadaf Usmani International School for Advanced Studies (SISSA/ISAS), Trieste, Italy

Sabino Veintemillas-Verdaguer Instituto de Ciencia de Materiales de Madrid (ICMM), Consejo Superior de Investigaciones Científicas (CSIC), Madrid, Spain

Edina Wang Harry Perkins Institute of Medical Research, QEII Medical Centre, Nedlands

Centre for Medical Research, The University of Western Australia, Perth, WA, Australia

Cristina Zafra Universidad de Castilla La Mancha, Talavera de la Reina, Spain

Acronyms

2D	Bi-dimensional (two-dimension)
3D	Tri-dimensional (three-dimension)
ABD	Acquired brain damage
AC	Alternating current
ADL	Activities of daily living
AFM	Atomic force microscopy
AIS	American Spinal Injury Association Impairment Scale
AON	Antisense oligonucleotide
APP	β-amyloid precursor protein
ASA	Anterior spinal artery
ATH	Atherosclerosis
BBB	Blood-brain barrier
BCI	Brain-computer interface
BDNF	Brain-derived neurotrophic factor
BDU	Brain damage unit at the HBMA
BOLD	Blood-oxygen-level dependent imaging
BSCB	Blood-spinal cord barrier
CAR	Chimeric antigenic receptor
Cas9	Caspasa 9
CDNF	Ciliary-derived neurotrophic factor
ChR	Channelrhodopsin
CIMT	Healthy side constrained-induced restriction movement therapy
CNS	Central nervous system
CNT	Carbon nanotube
Covid-19	Coronavirus disease 2019
CPG	Central pattern generator
CPPS	Cervical proprio-spinal premotoneuronal system
CRISPR	Clustered regularly interspaced short palindromic repeats
CSC	Charge storage capacity
CSF	Cerebrospinal fluid
CT	Computed axial tomography (see also "computerized tomography")

CTAB	Cetyltrimethylammonium bromide
DAPI	4′,6-diamidino-2-phenylindole
DBS	Deep brain stimulation
DC	Direct current
DCX	Doublecortin
dECM	Decellularized extracellular matrix
DEFO	Dynamic Elastomeric Fabric Orthosis
DH	Dorsal horn
DNA	Deoxyribonucleic acid
DOC	Disorder of consciousness
DPI	Days post-injury
DR	Dorsal root
DRG	Dorsal root ganglion
DTI	Diffusion tensor imaging
DWI	Diffusion-weighted imaging
ECG	Electrocardiogram
ECM	Extracellular matrix
EDX	Energy dispersive X-ray spectroscopy
EEC	Electrical equivalent circuit
EEG	Electroencephalography
EES	Epidural electrical stimulation
EF	Electric field
EGF	Epidermal growth factor
EGFR	Epidermal growth factor receptor
EIT	Electrical impedance tomography
EMA	European Medicines Agency
EMG	Electromyogram
EPC	Endothelial progenitor cell
EPR	Enhanced permeability and retention effect
ES	Electrical stimulation
ETI	Electrode-tissue interface
EV	Extracellular vesicle
FDA	Food and Drug Administration of the United States of America
FAST	Face drooping, arm/leg weakness, and speech difficulties triad
FES	Functional electrical stimulation
FGF-2	Fibroblast growth factor 2
FPGA	Field programmable gates array
FU	Forced use technique
GAP43	Growth-associated protein 43
GBT	Gut barrier translocation
GDNF	Glial-derived neurotrophic factor
GFAP	Glial fibrillary acidic protein
GFP	Green fluorescent protein
GM	Grey matter
GO	Graphene oxide

HA	Hyaluronic acid
Hb	Hemoglobin
HBMA	Hospital Beata María Ana
HbO_2	Oxygenated hemoglobin
HbR	Deoxygenated hemoglobin
hDI	Human dental stem cell
HDL	High-density lipoprotein
HER2	Human epidermal growth factor receptor 2
HFT	Human figure test
HI	Head injury
HIV	Human immunodeficiency virus
hMI	Human bone marrow-derived stem cell
hNuA	Human nucleus antigen
I-FABP	Intestinal fatty acid-binding protein
IgG	Immunoglobulin protein isotype G
IGF-1	Insulin-like growth factor 1
IgM	Immunoglobulin protein isotype M
IL-1	Interleukin 1
IL-1β	Interleukin 1β
IL-6	Interleukin 6
IP	Interphalangeal
IrOx	Iridium oxide
IS-BRAIN	Interventional synergistic model based on exer-gamed robotics to augment intensity in neurorehabilitation
ITO	Indium tin oxide
LBP	Lipopolysaccharide-binding protein
LDL	Low-density lipoprotein
LDLR	Low-density lipoprotein receptor
LFP	Local field potential
LL	Lower limb
LMV	Large multilamellar vesicle
LPS	Lipopolysaccharide
LRP-1	Low-density lipoprotein-related protein-1
MBP	Myelin basic protein
MCP	Metacarpophalangeal
MEMRI	Manganese enhanced magnetic resonance imaging
mGluR5	Metabotropic glutamate receptor 5
MF	Microfiber
MP	Methylprednisolone
MPC	Methacryloyloxyethyl phosphorylcholine
MPZ	Myelin protein zero
MRI	Magnetic resonance imaging
mRNA	Messenger ribonucleic acid
MSC	Mesenchymal stem cell
MWCNT	Multiwalled carbon nanotube

μCPP	Microscale continuous projection printing method
μCT	Micro-computed tomography
NaOL	Sodium oleate
NIR	Near-infrared
NF-200	Neurofilament 200
NGF	Nerve growth factor
NK	Natural killer
NMDEA	N-methyldiethanolamine
NP	Nanoparticle
NPC	Neural progenitor cell
NR	Nerve root
NSC	Neural stem cell
NT-3	Neurotrophin-3
OT	Occupational therapist
PAI	Photoacoustic imaging
PAMAM	Poly(amidoamine)
PAN	Polyacrylonitrile
PANI	Polyaniline
PANMA	Poly(acrylonitrile-co-methacrylate)
PC-12	Pheochromocytoma-derived cell line
PCL	Polycaprolactone
PDGFRβ	Platelet-derived growth factor receptor β
PDMS	Poly(dimethylsiloxane)
PEDOT	Poly(3,4-ethylenedioxythiophene)
PEG	Polyethylene glycol
PEI	Polyethylenimine
PEPE	Polyether-co-polyester
PFC	Perfluorocarbon
PFCE	Perfluoro-15-crown-5-ether
PGMA	Poly(glycidyl methacrylate)
PHEMA	Poly(2-hydroxyethyl methacrylate)
PHN	Progressive hemorrhagic necrosis
PLA	Poly(lactic acid)
PLA-DA	Poly(lactic acid) diacrylate
PLGA	Poly(lactic-co-glycolic acid)
PMMA	Polymethylmethacrylate
PMLA	poly(β-L-malic acid)
PNIPAM	Poly(N-isopropylacrylamide)
PNS	Peripheral nervous system
PPI	Poly(propylene imine)
PPy	Polypyrrole
PSA	Polysialic acid
PSS	Poly(styrenesulfonate)
Pt	Platinum
PT	Physiotherapist

PTA   Post-traumatic amnesia
PTFE   Poly(tetrafluoroethylene)
PVC   Polyvinylchloride
PVDF   Polyvinylidene fluoride
Q Electric charge
RAT   Robotics-assisted therapy
RGD   Arginine-glycine-aspartic acid (Arg-Gly-Asp) tripeptide adhesive sequence
rGO   Reduced graphene oxide
ROS   Reactive oxygen species
RTV   Room-temperature-vulcanizing (for silicones)
RVG   Rabies virus glycoprotein
SH-SY5Y   Human neuroblastoma cell line
SCI   Spinal cord injury
SCIP   Spinal cord injured patient
SEA   Series elastic actuator
SEM   Scanning electron microscopy
sEMG   Surface electromyography
SiNW   Silicon nanowire
siRNA   Small interfering ribonucleic acid
SMART   Sensory modality assessment and rehabilitation technique
SMI-32   Neurofilament H
soCoChR   Soma-targeted opsin
SVZ   Subventricular zone
TBI   Traumatic brain injury
tDCS   Transcranial direct current stimulation
TE   Tissue engineering
TEM   Transmission electron microscopy
TGF-β Transforming growth factor β
Ti   Titanium
TiO_2 Titanium oxide
Tf   Transferrin receptor
TLR   Toll-like receptor
TNF-α Tumor necrosis factor α
TMS   Transcranial magnetic stimulation
TOPO   Trioctylphosphine oxide
TOMR   Task-oriented motor re-learning
tPA   Tissue plasminogen activator
TUNEL   Terminal deoxynucleotidyl transferase dUTP nick end labeling
UCNP   Up-conversion nanoparticle
UL   Upper limb
UV   Ultraviolet
VEGF   Vascular endothelial growth factor
VH   Ventral horn
VMF   Ventral medial fissure

VR	Virtual reality
VSA	Variable stiffness actuator
WD	Wallerian degeneration
WDO	Wrist-driven orthosis
WHO	World Health Organization
WM	White matter

Chapter 1
General Aspects of Traumatic Neural Diseases and Requirements of Central Nervous System Implantable Biomaterials as Diagnostic and Therapeutic Tools

Ana Domínguez-Bajo, Ankor González-Mayorga, Elisa López-Dolado, and María Concepción Serrano

Abstract As living organisms, we have the ability to detect and respond to environmental stimuli for adaptation and survival purposes. The nervous system, responsible for orchestrating these functions, is a complex set of specialized structures functioning in a coordinated manner. Such system can be disrupted by external and/or internal conditions leading to pathology. Specifically, disorders of the nervous system are important causes of death and disability around the world. In this chapter, we will first introduce major concepts covering the central nervous system and traumatic brain and spinal cord injuries, as the target of the therapeutic strategies that are the focus of this book. We will later present an historical overview of biomaterials and tissue engineering approaches for neural applications. Finally, major requirements for implantable neuro-engineered materials will be covered. The aspects included in this introductory chapter are thought to assist the reader on the understanding of more specific topics enclosed in following chapters.

A. Domínguez-Bajo · M. C. Serrano (✉)
Instituto de Ciencia de Materiales de Madrid (ICMM), Consejo Superior de Investigaciones Científicas (CSIC), Madrid, Spain
e-mail: ana.dominguez@csic.es; mc.terradas@csic.es

A. González-Mayorga
Laboratory of Interfaces for Neural Repair, Hospital Nacional de Parapléjicos, SESCAM, Toledo, Spain
e-mail: fgmayorga@sescam.jccm.es

E. López-Dolado
Laboratory of Interfaces for Neural Repair, Hospital Nacional de Parapléjicos, SESCAM, Toledo, Spain

Rehabilitation Department, Hospital Nacional de Parapléjicos, SESCAM, Toledo, Spain

Research Unit of "Design and development of biomaterials for neural regeneration", Hospital Nacional de Parapléjicos, Joint Research Unit with CSIC, Toledo, Spain
e-mail: elopez@sescam.jccm.es

Keywords Biocompatibility · Biodegradability · Biomaterials · Central nervous system · Mechanical compliance · Requirement · Tissue engineering · Traumatic neural diseases

1.1   Introduction to the Central Nervous System

As living organisms, we have the ability to detect and respond to a wide range of stimuli arising from our environment for adaptation and survival purposes. The key element responsible for orchestrating these functions is the nervous system, composed of peripheral and central components acting together in a highly coordinated manner. Specifically, the nervous system is a complex set of specialized structures whose mission is to control and regulate the function of all the components of our body, coordinating their actions and the relation of the whole organism with the external ambient. Indeed, it is organized to detect changes in both the internal and external environments, evaluate such information and respond through variations in our whole body behaviour, including changes in movement, organs' activity and hormone levels. The nervous system is divided into two major components: the peripheral nervous system (PNS) and the central nervous system (CNS). The PNS is composed of all the nerves starting at the CNS and extending throughout the body. Nerves are bundles of peripheral nerve fibres that form pathways from the CNS to the rest of the body and vice versa. On its turn, the CNS is formed by the brain and the spinal cord [1]. The brain is protected by the skull and refers to the cerebrum, the cerebellum and the brainstem. The spinal cord is the part of the CNS located inside the vertebral canal and connects to the brain through the occipital hole of the skull (Fig. 1.1).

The CNS receives, integrates and correlates different types of sensory and motor information. After integrating such information, an adequate response is executed [1]. In what follows, major components of the CNS will be described.

1.1.1   The Brain

The brain is constituted by three main regions: the cerebrum, the cerebellum and the brainstem. The cerebrum has multiple functions, but, generally speaking, it is responsible for processing the information that comes from the senses, as well as controlling movement, emotions, memory, cognition and learning. Additionally, it is the centre of intellectual functions [1]. Anatomically, the cerebrum can be divided into two different parts: the telencephalon and the diencephalon. The telencephalon corresponds to the two cerebral hemispheres (right and left), which are communicated by nerve fibres on the corpus callosum (callosal commissure). Brain neural components can be divided into white and grey matter areas. While the white matter (mainly composed of neuronal axons) transmits information between

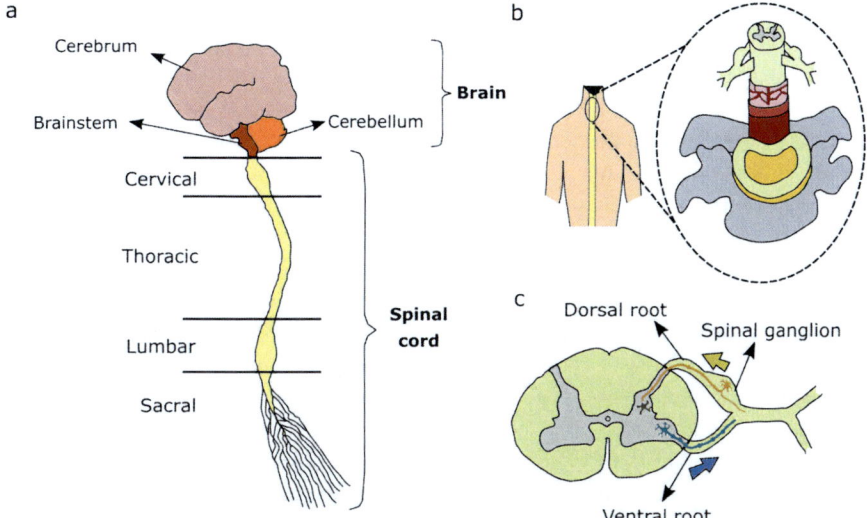

Fig. 1.1 Overview of the central nervous system anatomy [2]. (**a**) Scheme of its major components and regions. (**b**) Scheme of the spinal cord surrounded by the meninges and protected by a backbone vertebra. (**c**) Spinal nerve scheme. Adapted from Ana Domínguez-Bajo's PhD dissertation 2020 [2]

the brain and the rest of the body, the function of the grey matter (which contains neural cell bodies) is related to the processing of the information. Regarding functioning, the left hemisphere is responsible for movement and sensory perception from the right part of the body, logical reasoning, linguistic intelligence and mathematical ability. On its turn, the right hemisphere is responsible for movements and perception of the left side, vision in three dimensions, creativity and imagination [1]. Four lobes are identified in the cortex of each hemisphere: Frontal, occipital, temporal and parietal. The frontal lobe controls voluntary movements and it is related to the ability of planning, intelligence and personality. The occipital lobe integrates visual information. The temporal lobe deals with auditory information, memory and emotions. Finally, the parietal lobe integrates tactile information of the whole body and intervenes in balance [3].

The diencephalon is composed of the following different anatomical parts: Thalamus, hypothalamus and epithalamus. The thalamus receives sensations collected by other parts of the CNS and distributes them to diverse regions in the cerebral cortex. The main function of the hypothalamus is to regulate the balance of the body and the basic needs such as the intake of food and drink and the instinct of reproduction, also controlling the endocrine system [1]. The epithalamus belongs to the limbic system, thus related to emotions and intuition. It also contains the pineal gland, which regulates the states of sleep and wakefulness.

The cerebellum is attached to the brainstem by means of three pairs of fibre bundles or cerebellar peduncles [1]. In its upper or lower part, the cerebellum is shaped like a butterfly, the "wings" being the cerebellar hemispheres and the "body", the vermis. Each cerebellar hemisphere consists of lobes, separated by fissures. The cerebellum has an outer layer of grey matter, named as the cerebellar cortex, and grey matter nuclei located deep in the white matter. Its main function is the coordination of movements. Specifically, the cerebellum evaluates how the movements that initiate the motor areas of the brain are executed. In case they are not performed harmoniously and smoothly, the cerebellum detects it and sends feedback impulses to the motor areas so that they correct the error and modify the movements. In addition, the cerebellum participates in the regulation of posture and balance.

The brainstem consists of three parts: midbrain, pons and medulla oblongata. Ten of the twelve cranial nerves, which are responsible for the innervation of structures located in the head, leave the brainstem. They are the equivalent to the spinal nerves in the spinal cord. The midbrain is the upper structure of the brainstem and connects the pons and the cerebellum with the diencephalon. It consists of the cerebral peduncles, which are fibre systems that conduct impulses to and from the cerebral cortex. The pons is located between the medulla oblongata and the midbrain, inferior to the cerebellum. It is divided into a ventral part and a dorsal, tegmental, part separated at most levels by the medial lemniscus. The ventral pons consists of the precerebellar pontine nuclei and the brachium pontis (formed by efferent fibres of the pontine nuclei). The pontine tegmentum contains cranial nerve nuclei and several subnuclei of the reticular formation, including the parabrachial nuclei, caudal parts of the pedunculopontine tegmental nucleus, the laterodorsal tegmental nucleus, the locus coeruleus complex and the raphe nuclei [4]. The middle cerebellar peduncles are an intricate fibre system connecting the medulla oblongata to the cerebral hemispheres. The medulla oblongata, located between the spinal cord and the pons, is actually a pyramid-shaped extension of the spinal cord. The origin of the reticular formation, an important network of nerve cells, is a fundamental part of this structure. The impulses between the spinal cord and the brain are conducted through the medulla oblongata by major pathways of both ascending and descending nerve fibres. Control centres for heart, vasoconstrictor, and respiratory functions are also located here, as well as other reflex activities, including vomiting. Their survival relevance is such that injuries to these structures cause immediate death [5].

1.1.2   The Spinal Cord

The spinal cord is the part of the CNS located inside the vertebral foramen. It contains 31 spinal segments, which are the origin of the spinal nerves that communicate the spinal cord with the different parts of the organism. It has two fundamental functions: (1) To control many reflex acts and (2) to communicate the body with the brain through sensory ascending and motor descending pathways.

Like the rest of the CNS, the spinal cord is constituted by grey matter, in this case located in the central area, and white matter, located in the outermost area [1].

Grey matter neurons have variable characteristics regarding length and size of the dendrite and axon distribution. For instance, they can be multipolar and classified as Golgi I and II neurons. These neurons have either long or short axons. Those with long ones either exit the spinal cord through the ventral spinal roots to form ventral fascicles or connect long distances within this organ (for example the cervical part with the lumbar part). On the other hand, neurons with short axons make connections with cells located nearby at the grey matter. Some neurons are intrasegmental, distributed within a single segment; others extend through several segments, so they have an intersegmental distribution. Commonly, neuronal somas are clustered together, often in large numbers, responding to a common function. Sometimes, these groups can be subdivided into smaller ones, which would imply a particular function. Depending on their relative location in the spinal cord, there are anterior, posterior and lateral nuclei. The anterior horns mainly contain voluntary motor neurons in its upper part and visceral motor neurons in the base. In the posterior horns, exteroceptive sensitivity is found in the head, proprioceptive sensitivity in the medial part and visceral sensitivity at the base.

Bror Rexed identified that dorsal horn neurons were arranged in flattened sheets (i.e. laminae) parallel to the longitudinal axis of the spinal cord. Each plate can be distinguished from another by the neuronal morphology and the density of its junction, as also occurs in the architecture of the cortex. Then, each Rexed lamina contains neurons with different anatomical connections and functions. In the grey commissure or intermediate region and the posterior horn, the Rexed laminae are more similar to columns than flattened sheets. In any spinal segment, there are ten Rexed laminae, some of them corresponding to the nuclei of the spinal cord.

The white matter of the spinal cord is formed by the axons of sensory and motor neurons, the descending axons coming from supraspinal centres (brain and brainstem) and the central fibres of the sensory neurons of the spinal ganglion surrounding the grey matter. The millions of axons and fibres that form it are grouped into tracts and fascicles, which are named according to their origin and termination sites to easily identify the path and the direction of travelling of its information. Through the ascending tracts, information from the PNS and the spinal function itself is transmitted to the brain, while through the descending ones, the brain transmits its orders to the spinal cord. Regarding the distribution of the tracts, the long tracts are located peripherally and the short ones closer to the grey matter. These tracts and fascicles form three columns of white matter on either side of the midline of the spinal cord: dorsal, lateral, and ventral white columns. The dorsal white column contains somatic afferent fibres from the spinal nerves that are grouped into the graceful or thin and the cuneate or wedge-shaped fascicles, which ascend through the spinal cord without synapse until reaching the medulla oblongata. The lateral and ventral white columns are formed by ascending tracts that originate in the spinal cord itself and descending tracts that originate in different structures of the brain [5].

In addition, the spinal cord has an internal neural network formed by the propriospinal tracts, which connect the different spinal segments among them [6, 7]. The neural fibres that form these tracts follow a very rigorous organization: axons which travel through a specific area of the white matter project to a specific area contained in the grey matter. For example, fibres located at the dorsolateral part of the white matter project to the dorsolateral portion of the intermediate grey matter and to the motoneurons which innervate the distal limb muscles. Conversely, fibres located at the ventromedial area of the white matter project to the medial portion of lamina VII and the motoneurons which innervate the proximal muscles at the grey matter. Alongside, there are long propriospinal neurons able to communicate the cervical enlargement with the lumbar one through descending and ascending pathways. While descending cervico-lumbar projections are usually bilateral (ipsilateral and contralateral), ascending lumbo-cervical ones are, typically, only contralateral.

There are two neuronal circuits involved in the control of locomotion and forelimb motility: the cervical proprio-spinal premotoneuronal system (CPPS) and the central pattern generator (CPG). The somas of the neurons that form the CPPS are located mainly in C3 and C4 spinal segments and their axons project to the motoneurons of C7 and C8, as well as to the lateral reticular nucleus of the brainstem. Several studies in humans and different animal models have demonstrated that the CPPS receives excitatory inputs from corticospinal, rubrospinal, tectospinal and reticulospinal tracts [8–10], along with sensory inputs from muscles, joints, and skin [11], and inhibitory connections from intraspinal interneurons. Thanks to this, the CPPS could act as a relay between the tracts coming from supraspinal centres (e.g. cortical and subcortical tracts) and the motor nuclei of the spinal cord, thus modulating and updating the information from supraspinal centres before transmitting it to the motoneurons. Additionally, the CPPS sends a copy of the signals produced by the spinal motoneurons to the cerebellum through collaterals to the lateral reticular nucleus [12]. On its turn, the CPG, located in the lumbar enlargement, controls the flexion and extension patterns of the different joints during locomotion [13, 14]. It receives instructions from the sub-thalamic, midbrain and cerebellar locomotion regions [15] and information from the cerebral cortex, basal ganglia, brainstem and sensory systems.

1.1.3   The Meninges

The entire CNS is surrounded by three layers of connective tissue called meninges. The outmost and strongest membrane is the dura mater, formed by dense irregular connective tissue attached to the bone and receiving an intense blood supply. To follow, the arachnoid membrane is placed below the dura. It is formed by non-vascular connective tissue rich in collagen and elastic fibres forming a mesh. The subdural space is located between these two meningeal membranes. Finally, the pia mater is a very thin and transparent layer of connective tissue that is intimately adhered to the CNS that it covers. The subarachnoid space localizes between the arachnoid membrane and the pia mater and contains the cerebrospinal fluid (CSF) [16]. The meninges have three main functions: (1) to protect the underlying brain and spinal cord; (2) to support the vascular tree; and (3) to create a fluid-filled cavity, the subarachnoid space, vital for the proper function and survival of the CNS. The subarachnoid space is a widely used location for the placement of different implantable devices in human clinical practice (see Chap. 11) [17].

1.1.4 The Cerebrospinal Fluid and the Choroid Plexuses

The CSF is clear and colourless. It protects the brain and the spinal cord against chemical and physical injuries, besides transporting oxygen, glucose and other necessary chemicals from the blood to neurons and neuroglia. This fluid is produced in vascular structures located in the walls of the ventricles called choroid plexuses, which are capillary networks covered with ependymal cells that produce the CSF from the filtration of blood plasma. The CSF circulates continuously through the ventricular system, ependyma and subarachnoid space. Its production is an active process that requires energy expenditure by the choroid cells. Moreover, the CSF provides mechanical protection to the CNS by preventing the brain and the spinal cord to hit against the skull and the backbone [18]. Any situation involving an accumulation of CSF—either excessive production or obstructed drainage—can lead to serious neurological deficits, often requiring the surgical placement of valvular bypass devices [17].

1.1.5 Components at the Microscale: Neurons and Neuroglia

Early studies by the renowned neuroscientist Ramón y Cajal already identified neurons as discrete neural elements responsible for the functioning of our nervous system [19]. His work defined the origin of the so-called neuron doctrine, which marked the beginning of modern Neuroscience. Neural tissue consists of two main types of cells: neurons and neuroglia. Neurons are excitable cells that drive the electrical impulses that make all the functions of the nervous system possible, thus representing the basic functional and structural unit of the nervous system. It is estimated that the human brain contains over 86 ± 8 billion neurons [20]. Although with different shapes and sizes, all neurons have a basic and constant structure composed of soma, dendrites, axon initial segment and axon. Neurons are supported by neuroglia, a group of cells that are, in general, smaller than neurons and exceed them 5–10 times in number. The main glial cell types are astrocytes, oligodendrocytes, ependymal cells, Schwann cells and satellite cells [21]. Most CNS cells are of glial lineage, which explains the pathophysiological role played by the dysfunction of one or more of the glial cell types in diseases such as multiple sclerosis and CNS tumours.

1.2 Traumatic Neural Diseases: Traumatic Brain and Spinal Cord Injuries

The CNS, as a highly specialized system, can be disrupted by external and/or internal conditions leading to pathology. Disorders of the nervous system are important causes of death and disability around the world. As much as 1 in every 9 individuals dies of a disorder of the nervous system [22]. As the world population ages, the relative effects of disorders affecting this system are increasing. Within the most prevalent and dramatic CNS disorders, we could mention Alzheimer's disease and other dementias, epilepsy, Parkinson's disease, amyotrophic lateral sclerosis, multiple sclerosis, stroke and brain and spinal tumours, to cite a few. All these pathologies cause high burdens for the National Health Care systems, involving not only patients but also their relatives and society as a whole. At present, neurological disorders constitute 6.3% of the global burden of disease around the world. The World Health Organization (WHO), in their global public health project "Global Initiative on Neurology and Public Health" has defined all these neurological disorders as a public health challenge, predicting a 12% of the world population dying of neurological diseases by 2030 [23].

From the physio-pathological point of view, both traumatic brain injury (TBI) and spinal cord injury (SCI) can be classified as either penetrating or no penetrating injuries, compromising meninges and neural tissue integrity in a more severe way in the former. In TBI, the worst outcome primarily depends on whether the lesion was originated in either one or both sides of the brain (i.e. being the larger bilateral lesions those with the worst consequences). Conversely, in SCI, the upper located is the affected spinal segment, the worst are the consequences. In both TBI and SCI, events happening after the trauma can be classified as primary and secondary damage. Primary damage starts with the formation of hematomas, the disruption of blood vessels (blood–brain/spinal cord barriers), the initial activation of the immune response (at this point, mainly due to microglia activation), and the affectation and death of neural cells that comprise both brain and spinal tissue, including axonal shearing. Secondary damage appears after these initial events in a variable time frame. In TBI, one of the most studied consequences of cell damage is the subsequent cytotoxicity caused by an exaggerated release of the glutamate neurotransmitter. This causes additional impairments in information transmission in cells located around the lesion epicentre, even leading to the apparition of seizures in patients. In SCI, the ischemic injury due to major vessels deterioration has devastating consequences, such as severe haemorrhages and spinal cord swelling. These events lead to mechanical compression of the cord and worsening of the injury, causing the affectation of other spinal segments due to the extension of the damage (for a more meticulous explanation, see Chap. 2). In both TBI and SCI pathologies, cell membrane damage will drive to the release of different molecules. This will cause an imbalance in the normal ionic properties of the tissue, the liberation of more neurotransmitters and reactive oxygen species, and the augmentation of cell death, which will boost even more molecules release. Cells of the immune system such as macrophages, neutrophils and lymphocytes will be then attracted to the injury site, liberating immune mediators such as cytokines and driving to oedema formation (and, in the case of TBI, the increase of intracranial pressure). All these events, along with astrocyte recruitment and fibrous cells and proteins infiltration for the formation of the fibro-glial scar at the lesion site, leave a hostile scenario for neural regeneration. Afterwards, once the attenuation of the inflammatory response takes place, the injured tissue stabilizes, even with attempts of neural circuits remodelling. Both initial and secondary damages have different durations and consequences depending on the severity of the lesion. For further details on the physiopathology and clinical features of TBI [24–26] and traumatic SCI [27, 28], readers are referred elsewhere.

The therapeutic options for TBI are still limited. Specifically, TBI therapeutics under investigation include approaches to regulate neuroinflammation, aedema, and blood–brain barrier disintegration through pharmacotherapy and stem cell transplants [29]. As TBI pathology entails a progression from the primary injury to inflammation-mediated secondary cell death, limiting this inflammation to ameliorate a greater symptomology has emerged as an attractive treatment prospect. Reported studies to date have yielded some positive results, but superior outcomes are expected from combined therapies. At the clinical arena, the use of robotic-based rehabilitation strategies is extensively applied for functional recovery (for further details, see Chaps. 9 and 10). More recently, electrical stimulation is attracting much effort and proving therapeutic potential. At the moment, there are two active clinical trials where brain stimulation by using implanted electrodes is being tested to ameliorate and reverse cognitive deficits after TBI by inducing plastic changes in neural circuits (NCT02881151 [30], recruiting, phase not applicable; NCT03382626 [31], recruiting, phase not applicable).

Major advances and failures in the fields of neurosurgery, robotics, computational neuroscience and neuroengineering have populated SCI medicine with a multitude of wearable and implantable neurotechnologies to enable and augment function [32]. These treatments have historically been divided into restoration therapies, replacement strategies and rehabilitation procedures (Fig. 1.2). Within neuroengineering approaches, biomaterials are bringing so much promise that the first clinical trials have been initiated. At present, there is an on-going clinical trial on the safety and feasibility of a neural spinal scaffold made of poly(lactic-co-glycolic acid)-b-poly(L-lysine) for the treatment of subjects with American Spinal Injury Association Impairment Scale (AIS) grade A traumatic SCI at neurological level of injury of T2-T12 (NCT02138110) [33]. As early as 2016 and related to this clinical trial, a pioneer publication referring to the first human implantation of a bioresorbable polymer scaffold (InVivo Therapeutics) [34] for acute SCI came out (Fig. 1.3) [35]. Specifically, a 25-year-old man with a T11-12 fracture dislocation sustained in a motocross accident (AIS grade A) was treated with acute surgical decompression, spinal fixation with fusion and implantation of a 2 × 10 mm bioresorbable scaffold in the spinal cord parenchyma at T12. By 3 months, the

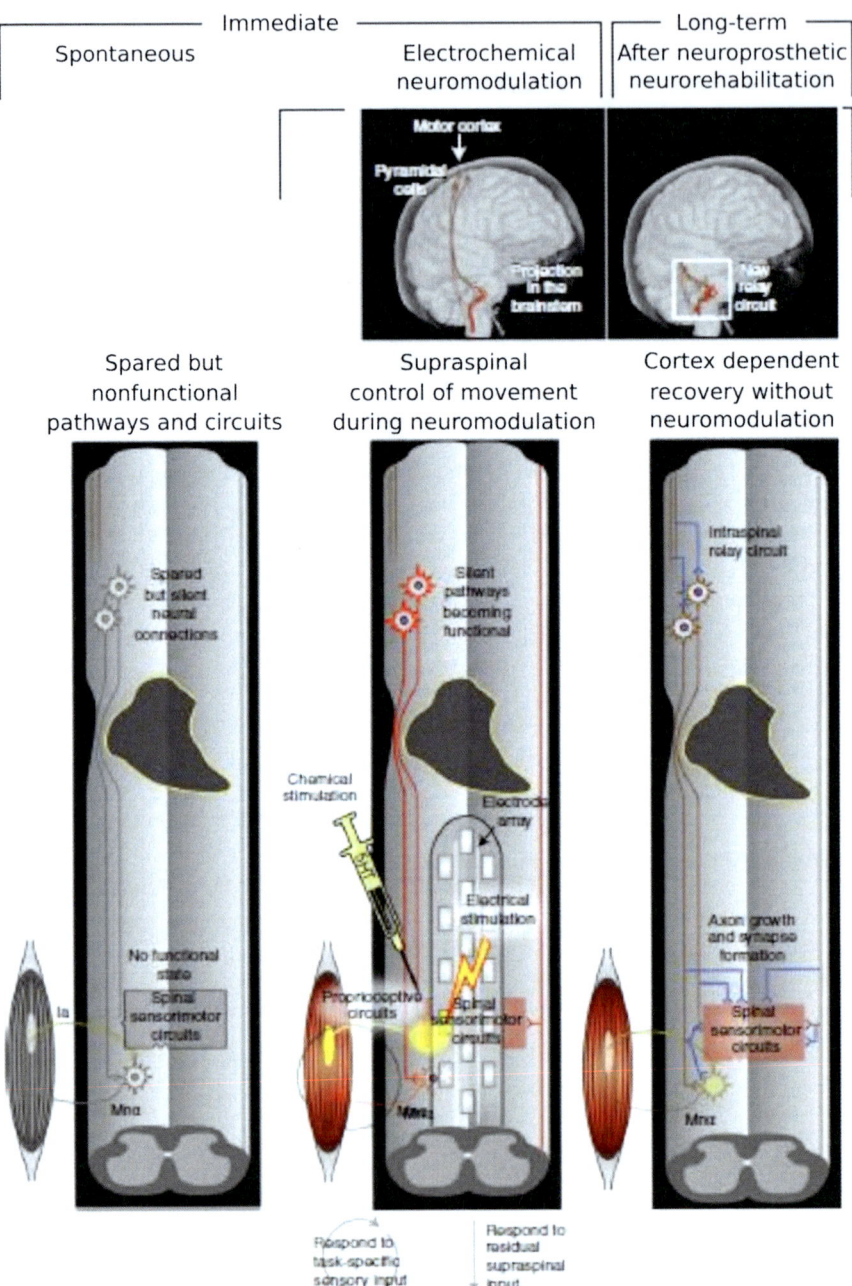

Fig. 1.2 Engineering strategies to enable immediate and long-term recovery of motor functions. After a severe but incomplete SCI, the sensorimotor circuits producing motor activity are spared, but they lack the essential source of modulation and excitation to be functional. Descending pathways are partly spared, but fail to activate these circuits for muscle contraction. The delivery of chemical and/or electrical neuromodulation therapies to the spinal cord below the injury immediately enables spinal sensorimotor circuits to respond to task-specific sensory and residual supraspinal inputs in order to produce functional movements. In conjunction

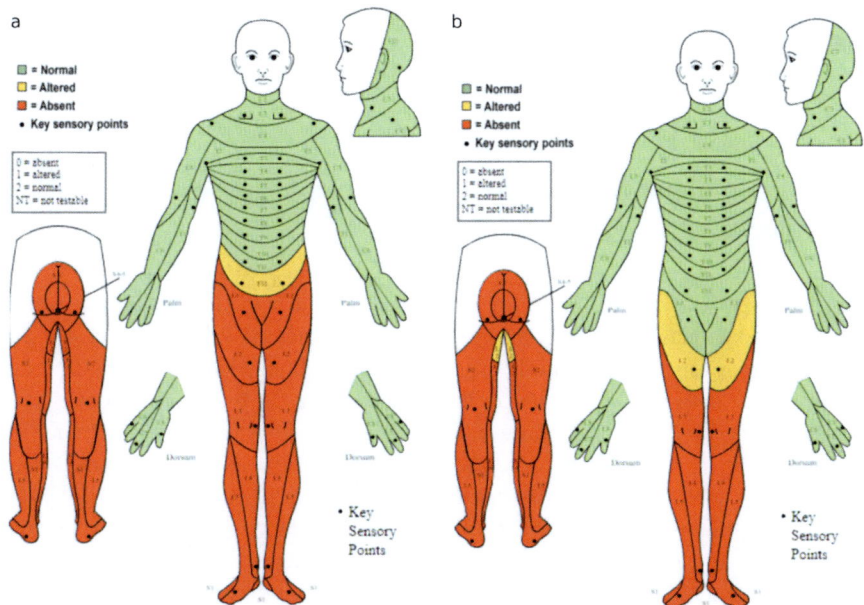

Fig. 1.3 Preoperative (**a**) and 3-month postoperative (**b**) score sheets adapted from the American Spinal Injury Association Impairment Scale (AIS): International Standards for Neurological Classification of Spinal Cord Injury (revised 2013; updated 2015) demonstrated improvement in the sensory examination of the patient. Preoperatively, he had no deep anal pressure sensation or voluntary anal contraction and a T11 sensory level with no motor function below that level, indicating a T11 AIS grade A complete SCI. At 3-month follow-up, he regained deep anal pressure sensation and voluntary anal contraction. In addition, he recovered normal sensation down through the L1 dermatomes bilaterally and some strength in the hip flexors and knee extensors, indicating conversion to an L1 AIS grade C incomplete injury. Orange indicates areas with no sensation, yellow indicates areas with diminished sensation and green indicates areas with normal sensation. Adapted from [35]

neurological examination of the patient improved to an L1 AIS grade C incomplete injury. At 6-month postoperative follow-up, there were neither procedural complications nor apparent safety issues related to the scaffold implantation. In China, the NeuroRegen Scaffold™, mainly composed of collagen and accompanied by either mesenchymal stem cells (NCT02688049 [36]) or bone marrow mononuclear cells (NCT02688062 [37]) transplantation, is being investigated for human patients with

Fig. 1.2 (continued) with neurorehabilitation, this intervention promotes an extensive reorganization of spared circuits and residual connections in the brainstem and spinal cord that mediates functional recovery. After training, motor control occurs without the need of the electrochemical neuromodulation therapies. Reprinted by permission from Springer Nature Customer Service Centre GmbH: Springer Nature, Nature Medicine [32], Copyright© (2019)

chronic SCI. In parallel, collagen scaffolds are being explored in another clinical trial (NCT02510365 [38]), along with comprehensive rehabilitation, in individuals with acute (≤21 days) complete SCI (C5-T12).

1.3 Introducing Biomaterials as Diagnostic and/or Therapeutic Tools

Generally speaking, biomaterials are defined as natural or synthetic substances that can be used either alone or combined with other materials, cells, or molecules to induce a regenerative desired response in any damaged organ or tissue of a living organism [39]. Depending on the biomedical goal, biomaterials could stay at the lesion site during the entire patient life (such as a hip prosthesis made of titanium) [40] or for a specific period (such as synthetic skin substitutes which will degrade with time to let the patient's own skin to regrow after a burn) [41]. Importantly, once implanted, the material–tissue interface will have a pivotal role in the proper integration and functioning of the biomaterial, as it is dynamic (i.e. it will be changing its features over biomaterial lifetime). Many types of biomaterials of natural and synthetic origin have been explored so far to offer alternative and/or complementary therapeutic approaches to patients with organ failure. For example, collagen [42, 43] and hyaluronic acid (HA) [44], two abundant biomolecules present in the extracellular matrix (ECM) of many tissues in the body, have been frequently used for the fabrication of biomedical implants. Even the biological structure that remains after the decellularization of a specific organ or tissue can be processed and used as a biomaterial in biomedical research (Fig. 1.4) [45, 46]. Natural biomaterials can be of animal source, such as chitosan, which is isolated from the cell walls of crustaceans, fungi and bacilli [47, 48]. Alternatively, examples of commonly used lab-made biomaterials for medical purposes are polycaprolactone (PCL) [42], polyethylene glycol [49] and polylactic acid [42].

The use of biomaterials for medical purposes can be framed within the inter-disciplinary field known as tissue engineering [50]. Concretely, scientists and physicians working in this field combine and apply the principles of basic biology, medicine, chemistry, material science, nanotechnology and bioengineering to construct biological substitutes for organs with different tissue complexity, such as the skin, trachea and spinal cord [51]. The definition of tissue engineering and regenerative medicine can be considered virtually the same, but it is also correct to say that tissue engineering is included within regenerative medicine [52]. Organ reparation by using the combination of biomaterials, cells and molecules is one of the most significant milestones in translational science in the last century, eventually improving the quality of life of millions of patients worldwide [53].

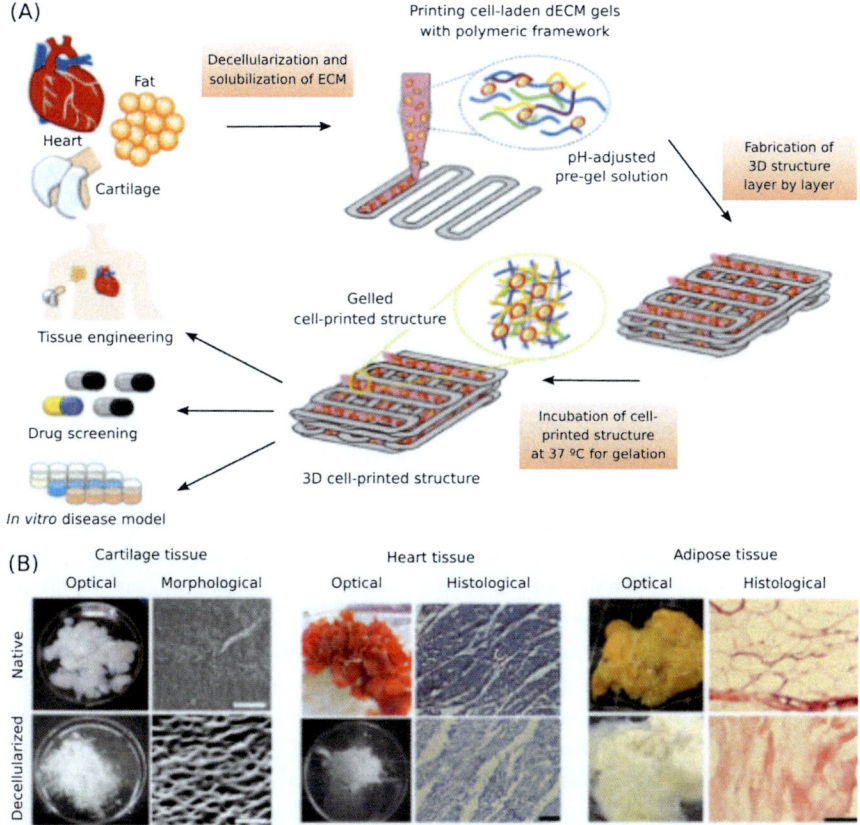

Fig. 1.4 3D bioprinting with decellularized extracellular matrix (dECM) bioinks of different tissue constructs. (**a**) dECM materials are obtained from various tissues via a multi-step decellularization process that combines physical, chemical and enzymatic treatments. The collected soluble dECM materials are mixed with stem cells and used as bioinks in a layer-by-layer bioprinting approach to fabricate tissue analogues. (**b**) Native tissues and bioprinted dECM constructs of the corresponding tissues show similar morphological and histological appearance. Reprinted by permission from Springer Nature Customer Service Centre GmbH: Springer Nature, Nature Communications [54], Copyright© (2014)

Although it is difficult to set the origin of tissue engineering, among its major early achievements, we can cite the first time that researchers were able to grow and maintain cells in culture in the laboratory [55] and the first complete organ transplant, a liver, between relative and non-relative subjects [56]. In the 1960–1970s decade, materials and cell biology approaches started to be combined in medical therapies, but it was during 1990s when Langer and Vacanti named this new field [50]. Regarding biomaterials use, metals, rubber and ceramics were the basis of the first materials generation employed for the fabrication of prosthesis that were placed at the lesion site. However, strictly speaking, these materials acted as physical substitutes, without inducing any change in the spared tissue [57, 58]. Following, these first materials were improved to prompt an initial regenerative response in the organism. For example, this resulted in new dental prostheses which were able to form an active interface with the native tissue, thus facilitating material integration. Another important advance was the creation of resorbable screws and sutures used in orthopaedic surgeries [58]. In the next period of biomaterials development (third generation), researchers realized that manipulating physicochemical features of biomaterials, such as giving to their surface some degree of roughness, modelling their architecture (from 2D to 3D), varying their degree of hardness and modifying surface chemical composition, materials could be able to elicit more diverse and beneficial effects once implanted. Importantly, the evoked responses were due to cellular and genetic adjustments that prompted both structural and functional changes to assist organ restoration [58, 59]. Finally, fourth-generation biomaterials are those that, once placed at the injured site, allow for additional functional properties such as the use of electrical stimulation to obtain desired tissue responses. In this last case, electrical impulses can regulate the activity of ion channels and pumps present in the cell membrane. Indeed, these materials differ from previous generation ones in providing real-time monitoring and recording of the cell responses generated. In this group, we could include electronic devices developed by nanotechnology tools to interface human tissues in vivo, such as cardiac pacemakers, cochlear implants and electrodes for deep brain stimulation [59, 60]. Biomaterials are also part of devices that do not substitute any tissue/organ but are implanted within our body for different specific purposes (Fig. 1.5) [61]. These devices allow researchers and clinicians to deliver drugs and molecules either spontaneously or in a controlled way, to act as sensors of relevant clinical parameters such as arterial, intraocular and intracranial pressures, and to maintain the normal position of an organ which underwent surgery, such as biocompatible meshes for hernia repair, among others.

Focused on neuroscience, neural tissue engineering (or neuroengineering) has emerged as a subfield of tissue engineering [62, 63]. As extensively explained above, the nervous system has an overly complex anatomical and functional structure. The information among its components is mainly transmitted through electric pulses (action potentials) and chemical signals (neurotransmitters). Moreover, information exchange with the nervous system occurs at different levels. For example, information from the brain and spinal cord is transported by peripheral nerves, and ultimately received and processed back to the CNS by a muscle, an organ or a gland. Thus, neural stimulation and recording could be done either at the muscle/organ level, at the beginning of the peripheral nerve, or directly at the CNS components [64]. Taking this into account and the fact that damage in the nervous system will comprise the loss of cellular and tissue components and then, the impairment of electric and chemical transmission, biomaterials used in nervous system diseases attempt to both regenerate nervous tissue and recover its functionality [65]. In the particular case of CNS injuries, several thousands of stimulators, including

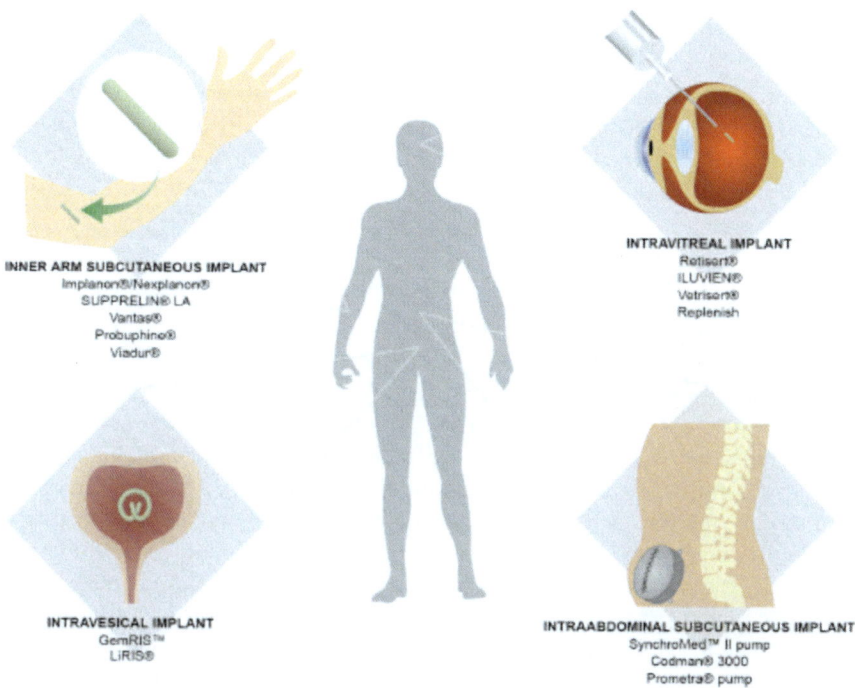

Fig. 1.5 Exemplary sites of implantation for US Food and Drug Administration (FDA)-approved implants and devices in clinical trials. Reprinted by permission from Springer Nature Customer Service Centre GmbH: Springer Nature, Biomedical Microdevices [61], Copyright© (2019)

deep brain and epidural stimulators, among others, have been implanted in TBI and SCI patients suffering pain, spasticity, loss of mobility and sensation in the extremities, and loss of bladder control, among others, since the 1960s. The most implanted neural interface is the above-mentioned cochlear implant, which have restored hearing of thousands of deaf patients for many years, followed by spinal cord stimulators [66]. Further details on these neuroimplants of current use in the clinics are provided in Chap. 11. Up-to-date research is also focused on the improvement of the physicochemical characteristics of biomaterials, from refining biomaterial selection to a better processing and biofunctionalization for improved anchorage of molecules (e.g. growth factors, drugs) and/or cells (e.g. stem cells, neural progenitor cells, Schwann cells) in order to get the most appropriate features for inducing neural tissue regeneration and functional recovery [65, 67–69]. Specific requirements for biomaterials with neural application are explained in the following section. An extensive revision on the state-of-the-art of biomaterials for neural repair in the context of TBI and SCI will be provided in the following chapters.

1.4   Requirements for CNS Implantable Biomaterials

Research along the last decades has proved that efforts to understand and tackle TBI and SCI have to be interdisciplinary due to the complexity of these pathologies. In words of the renowned neurologist Rita Levi-Montalcini, Nobel laureate for the discovery of the nerve growth factor in 1986:

> Scientific and technological research, from molecular to behavioural levels, has been carried out in many different places but they have not been developed in a really interdisciplinary way. Research should be based on the convergence of different interconnected scientific sectors, not in isolation, as was the case in the past.

The use of biomaterials and tissue engineering as therapeutic strategies to approach the repair of injured neural tissues is receiving expanding attention [70, 71]. Since early work in the 1970s by authors such as Hench and Ethridge [72], a big amount of articles has been published on the topic of biomaterials for neural applications, concurring with the emergence of the fourth generation of biomaterials, the so-called smart/biomimetic materials [73], previously mentioned. Within these materials, hydrogels represent an attractive option because they are structurally similar to the ECM of many tissues, can often be processed under relatively mild conditions and delivered in a minimally invasive manner (e.g. injection without tissue removal) [74], as well as fabricated by using 3D printing technologies to more precisely reproduce neural tissues architecture [75]. Additional advantages include soft mechanical properties, hydrophilic nature, biodegradability, biocompatibility and versatility to achieve high loads of bioactive molecules [76, 77]. Hydrogels were first placed into the CNS in the mid-1990s when poly-(2-hydroxyethyl methacrylate) (PHEMA) hydrogels containing Schwann cells were implanted into lesioned optic tracts in rats, with two-thirds of the scaffolds showing penetrating axons [78]. In a different study, collagen IV was combined with PHEMA and Schwann cells and implanted into lesioned optic tracts, resulting in increased neuron penetration [79]. More recently, amorphous non-fibrous hydrogels composed of HA have proved to promote angiogenesis and vascular remodelling, followed by subsequent neurite growth, in the brain after stroke [67]. Besides these exemplary strategies, multidisciplinary efforts including the use of biomaterials are providing encouraging results for restoring functions after SCI [80–82] and, more discretely, also TBI [83].

Knowledge from tissue engineering advances clearly indicates that successful approaches to repair damaged tissues and organs must achieve an optimal matching of their physicochemical and biological properties [84]. Implantable biomaterials for neural applications can be divided into three main groups: (1) Diagnostic devices to be shortly in contact with the tissue (i.e. commonly applied externally, so just minimal requirements are expected to be fulfilled), (2) Therapeutic neural interfaces including electrical sensors and stimulators aimed to be implanted for either short or long periods of time and (3) Therapeutic devices that are thought to integrate neural tissues chronically and somehow prompt inherent regenerative responses for neural repair. Major requirements for all these CNS implantable

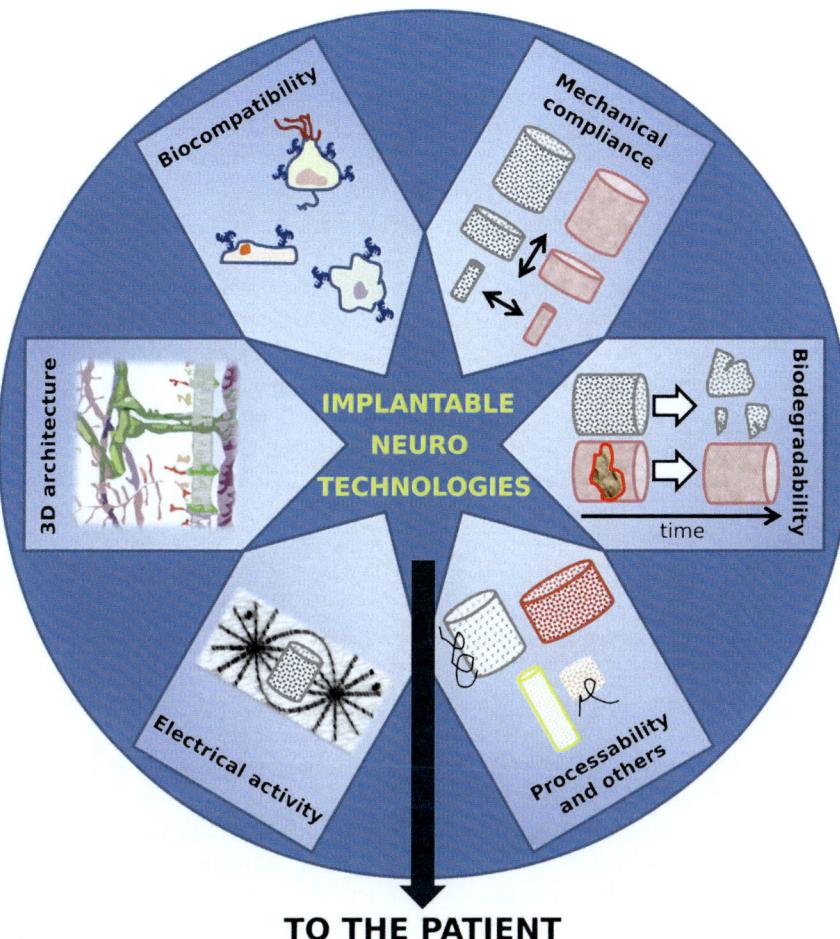

Fig. 1.6 Schematics of major requirements that neurotechnologies should fulfil to maximize performance after CNS implantation

biomaterials intend to guarantee a seamless interface with neural tissues and long-lasting performances. Although these implantable devices must comply with most of these requirements, prioritizing them depending on the final aim, biomaterial selection becomes stricter if the substrate need to be present for long periods of time, such as in chronic patients. In what follows, the most pivotal characteristics that implantable biomaterials and interfaces for neural repair must fulfil are described in more detail (Fig. 1.6).

1.4.1 Biocompatibility

As largely discussed above, the most critical feature for any element to be implanted in vivo, also applicable to those for neural applications, is biocompatibility. Besides general aspects, biomaterials interfacing neural tissues should be able to elicit biological compatibility with neural components including their peculiar ECM, as well as induce neural tissue remodelling and regeneration if pursued. In this sense, as many elements of neurons, neuroglia and neural ECM have nanoscale dimensions, the incorporation of components at the nanoscale in the biomaterial to be implanted offers a great promise to seamlessly integrate with neural tissues by simulating native features and functions [85]. Moreover, efforts to achieve biocompatible interfaces are mandatory to diminish local inflammation and posterior foreign body responses, major contributors to the failure of neural implants. To this goal, all chemical, physical and biological properties of implantable devices are vital factors for suppressing chronic tissue encapsulation [86, 87]. Biocompatibility features also include the absence of toxicity from degradation by-products that could eventually, or desirably if biodegradability is pursued, emerge from the biomaterial over time, as well as potential harmful residues and leachables that could remain in the biomaterial from its fabrication process.

There are diverse strategies to boost the biocompatibility of implantable devices for neural applications. One of the most extensively explored is the incorporation of adhesive moieties to prompt adhesion of selective cell types at the first stages of implantation. For instance, the fibronectin-derived Arginine–Glycine–Aspartic acid (RGD) peptide sequence has been shown to improve cell viability and migration in hydrogels [88, 89]. Importantly, neural stem cell adhesion and neurite outgrowth seems directly dependent on RGD sequence density in the hydrogel [90]. Alternatively, biocompatible polymers (e.g. silicone, poly(dimethylsiloxane)), some of them already available in medical grade, can be used to coat the implantable device and thus provide enhanced biocompatibility responses.

Another important approach to this regard is the functionalization with neural modulators, which could also add therapeutic features to the substrate. In this sense, a wide palette of neuroactive molecules ranging from proteins to nucleic acids and lipids has been identified. Particularly, neurotrophins are markedly relevant as they regulate development, maintenance and function of vertebrate nervous systems [91, 92]. Specifically, they are involved in the regulation of cell fate decisions, axon growth and dendrite pruning and outgrowth, the patterning of innervation, and the expression of proteins crucial for normal neuronal function, such as neurotransmitters and ion channels. In the mature nervous system, they control synaptic function and synaptic plasticity, while modulating neuronal survival. Precisely, neurotrophin 3 (NT-3) has been widely demonstrated as a beneficial growth factor in the injured spinal cord [93, 94]. In a more complex approach, Courtine and colleagues described a cocktail of growth facilitators boosting axon regeneration in complete rat and mouse SCI lesions (Fig. 1.7) [95, 96]. Factors investigated included: osteopontin, insulin-like growth factor 1 (IGF-1), ciliary-derived neurotrophic factor (CDNF),

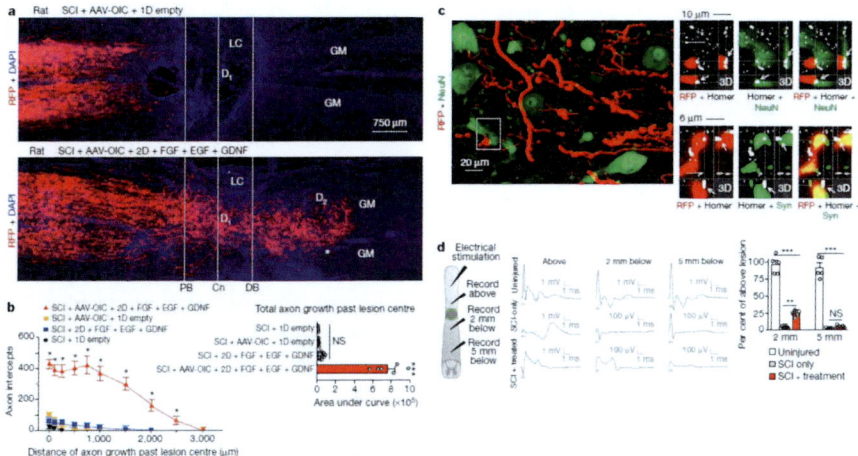

Fig. 1.7 Stimulated and chemoattracted proprio-spinal axon regrows robustly and conducts electrophysiological signals across anatomically complete SCI lesions in rats after combined delivery of AAV-OIC plus FGF, EGF and GDNF in two sequentially placed hydrogel depots. (**a**) RFP-labelled axons in composite tiled scans of horizontal sections. Dotted lines demarcate astrocyte proximal and distal borders around lesion core. Dashed line demarcates lesion centre (Cn). (**b**) Left, axon intercepts at specific distances past lesion centres (colour coding and n as in bar graph). Right, areas under axon intercept curves. *$P < 0.01$ versus all other groups, $P < 0.0001$ versus all other groups, one-way ANOVA with Bonferroni. (**c**) Detail images from the region indicated with an asterisk in (**a**). Left, RFP-labelled axons among NeuN$^+$ neurons in spared grey matter 2000 μm past the lesion centre. Top right, 3D detail of the outlined area in the left panel shows synapse-like contact of BDA-labelled terminal with the post-synaptic marker, homer, on a neuron. Bottom right, RFP-labelled terminal co-labelled with the presynaptic marker, synaptophysin (Syn) in synapse-like contact with the post-synaptic marker, homer. (**d**) Left, spinal cord stimulation and recording sites. Middle, representative individual electrophysiological traces after spinal cord stimulation. Right, peak-to-peak amplitude of the evoked potential at 2 and 5 mm below lesions relative to the potential above the lesion in SCI only or SCI + treatment (SCI + AAV-OIC + FGF + EGF + GDNF) or the equivalent distance in uninjured controls. **$P < 0.005$, ***$P < 0.0001$, one-way ANOVA with Bonferroni. For all bar graphs, data are mean ± standard error of the mean and dots shown mice per group. Reprinted by permission from Springer Nature Customer Service Centre GmbH: Springer Nature, Nature [96], Copyright© (2018)

fibroblast growth factor 2 (FGF-2), epidermal growth factor (EGF) and glial-derived neurotrophic factor (GDNF), delivered through the use of a diblock copolypeptide hydrogel with storage modulus between 75 and 100 Pa. Each of these factors was identified to play a particular role in the regeneration of the injured spinal tissue. In a different approach, a HA hydrogel modified with an anti-Nogo receptor antibody and loaded with poly(lactic-co-glycolic acid) (PLGA) microspheres containing brain-derived neurotrophic factor (BDNF) and vascular endothelial growth factor (VEGF) induced the formation of new blood vessels and regenerated nerve fibres at the injured T9-10 spinal cord [97]. Differently, Kadoya et al. described the effectiveness of the implantation of a fibrin matrix (25 mg/ml fibrinogen and 25 U/ml thrombin) loaded with rat multipotent neural progenitor cells (NPCs) and a growth factors cocktail to support graft survival [98]. This therapeutic biomaterial was tested in a T3 complete transection and a C4 corticospinal tract lesion. Further details on biomaterials for local and controlled drug delivery in SCI models can be found elsewhere [99].

1.4.2 Mechanical Compliance

Another pivotal requirement of implantable devices, and to date largely underestimated, is mechanical compliance with native and diseased neural tissues, which are naturally soft. For instance, mouse brain tissue characterizes by values as low as 200 Pa [100]. Atomic force microscopy (AFM) has recently evolved as a more precise technique than conventional rheology and mechanically testing instruments capable of providing precise mechanical measurements at the nano- and microscales of cells and tissues. For instance, AFM studies have demonstrated that the somata of pyramidal neurons in the cerebral cortex display elastic modulus between 480 Pa at 30 Hz and 970 Pa at 200 Hz, while values in astrocytes, even though being softer, reach 300 Pa and 520 Pa [101], respectively, and the somata of retinal neurons reach values as high as 1590 Pa at 200 Hz. Contrarily, fibroblasts from connective tissue have an elastic storage value dramatically higher ($\sim$5 kPa) [102]. More importantly, neural stem cells fate can be directed by the stiffness of the substrate (e.g. preferentially differentiating into neurons in stiffer substrates) [103]. Neurites of dorsal root ganglion cells grow longer on stiffer substrates [104], while spinal cord astrocytes adapt their morphology to the mechanical properties of the substrate [105].

When damaged, the mechanical properties of neural tissues significantly vary. Specifically, and contrary to scars in other tissues, the CNS tissue significantly softens after injury, both at the neocortex and the spinal cord, as demonstrated by AFM [106]. For instance, fresh spinal cord tissues have been reported to display Young's modulus values of 97 ± 69 Pa, for white matter regions, and 275 ± 99 Pa, for grey matter ones [107], in range with those previously reported for the spinal cord (177 and 420 Pa, respectively) and the rat cortical tissue (50–500 Pa) by other authors [106]. AFM stiffness maps of scar regions at the rat neocortex revealed a drastic drop in tissue elasticity by more than three-fold at 9 days after injury and a two-fold drop by day 22, with respect to the contralateral hemisphere [106]. Importantly, tissue softening was measured up to 1 mm away from the lesion site, with significant drops of 15% and 13% at 9 and 22 days, respectively, after injury. In the case of the spinal cord, although boundaries of the lesions are more irregular and diffuse than in the cortex, the stiffness of the lesioned spinal cord grey matter was also significantly lower than in control tissue. This effect was also significant outside the visible lesion. These neural scars are mainly composed of glial cells (e.g. astrocytes and NG2-glia) accompanied by some non-neural cells such as pericytes and meningeal cells [108], along with collagen and other ECM components. As the mechanical properties of the environment play a pivotal role in cell and tissue responses, it seems reasonable to hypothesize that the reduced ability of neurons to regrow in CNS injuries could be partially associated with changes in the mechanical properties at the lesion site.

Unfortunately, biomaterials described to date for neural applications are still far from matching the mechanical properties of native neural tissues. Materials such as methacrylate hydrogels (2.9 kPa perpendicular to pores and 6.7 kPa along pores) [109], cholesterol-modified methacrylate hydrogels (10.1 kPa) [110], protein-functionalized methacrylate scaffolds (263 kPa) [111] and PCL/gellan gum blends (79.7 kPa) [112] are some of the biomaterials reported for neural applications with softer mechanical properties. Alternatively, electrically active nanomaterials have been hybridized with hydrogels to decrease the mechanical mismatch between stiff electrodes and soft neural tissues [113, 114]. Even when reporting favourable regenerative responses such as axon growth, no data exist on the potential distress caused by their mechanical mismatch with neural tissues.

1.4.3 3D Architecture

All tissues and organs in the body display a tridimensional (3D) structure, so the brain and the spinal cord also encompass this characteristic. It seems reasonable then that engineered materials for neural repair are designed to reproduce the native ECM of native neural tissues, including their 3D architectural features. Moreover, it has been extensively proved that cells behave differently in 2D and 3D environments [115, 116], stressing the need of satisfying this requirement. Specifically, cells seem to behave more similarly to their native fate when cultured in 3D vs. 2D substrates.

In the past decade, 3D printing and bioprinting have emerged as useful, customizable and versatile techniques for the design and fabrication of a variety of biomaterials in a 3D configuration [117, 118]. They have been used to produce medical devices with patient-specific shapes, engineered tissues for in vivo regeneration, and in vitro tissue models for screening therapeutics. In the particular case of neural engineering, Koffler et al. recently reported the utility of a 3D-printed biomimetic hydrogel for SCI (Fig. 1.8) [75]. The authors used a microscale continuous projection printing method (μCPP) to create a complex CNS structure. μCPP can print scaffolds tailored to the dimensions of the rodent spinal cord in 1.6 s, and it is scalable to human spinal cord lesions. These μCPP 3D-printed scaffolds loaded with NPCs demonstrated ability to support axon regeneration and form new neural relays across sites of complete SCI in rodents.

Alternatively, fibre-like biomaterials, either in a 2D or 3D configuration, are being under investigation providing anisotropic guidance cues and then triggering white matter tract regeneration. Initial approaches in this sense were focused on materials such as agarose and PLGA. A pioneer study by Tuszynski and colleagues described the implantation of multi-channel agarose conduits containing bone marrow stromal cells engineered to release BDNF. In a rat cervical microaspiration

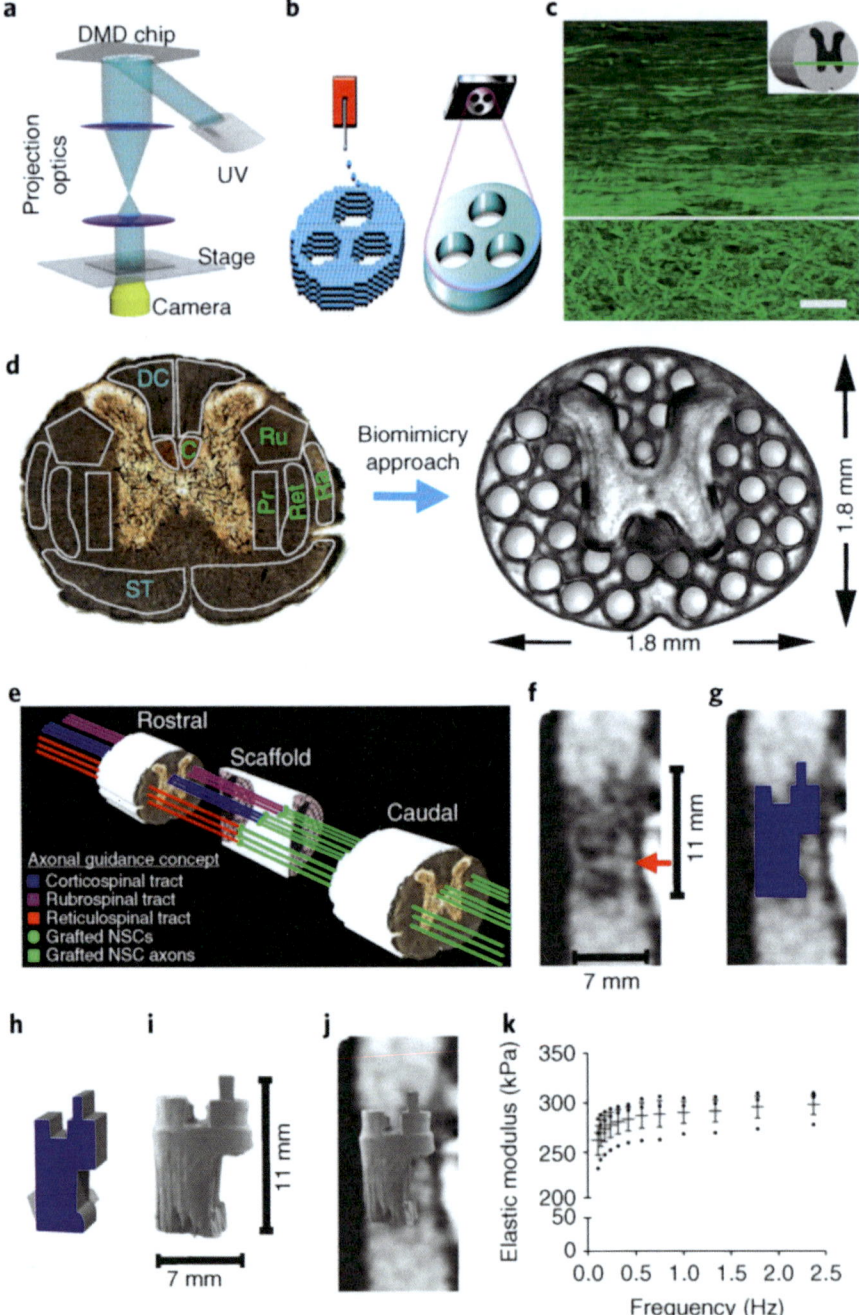

Fig. 1.8 (a) 3D-printer setup: an ultraviolet light source (365 nm wavelength); a computer for sliced image-flow generation and system synchronization; a digital micromirror device (DMD) for optical pattern generation; a set of projection optics; a stage for sample position control;

injury model, the combination of BDNF release and guidance conduits enabled neurite outgrowth in an organized, anisotropic manner [119].

At the microscale, architecture parameters governing the movement of cells and the diffusion of molecules (from large proteins such as growth factors to small molecules such as oxygen) become critical. Specifically, implantable biomaterials for neural regeneration must be able to serve as supportive and guidance structures that allow cell and ECM protein infiltration, as well as exchange of nutrients and waste products. In hydrogels, size at the microscale is defined by the mesh size, known as the distance between crosslinking points typically measured in angstroms (Å). Diffusion mechanisms are then significantly affected by the mesh size and, on its turn, largely influence degradation of the biomaterial. Once initiated, degradation will also start controlling diffusion of cells and molecules inside the biomaterial.

Fig. 1.8 (continued) and a CCD (charge-coupled device) imaging system for on-line monitoring of the fabrication process. (**b**) μCPP layerless 3D printing creates structures without discrete layers. In extrusion-based 3D-printing approaches (left), as the stage moves in the z axis and material is extruded, layers are created between the drops. μCPP printing (right) creates a structure with one continuous layer. (**c**) Heavy chain neurofilament (NF200) labelling of axons in intact T3 rat spinal cord. Rostral is to the left, and caudal is to the right of the image. The axons in the white matter (top of the panel) are highly organized into parallel arrays travelling from rostral to caudal. The axons in the grey matter (bottom of the panel) are not linear. The white line demarcates the interface between the white and grey matter. The inset schematic diagram indicates the orientation of the horizontal section. (**d**) Axonal projections in the spinal cord are linearly organized into regions (fascicles) containing axons of related function. Motor systems are shown in green and sensory systems are shown in blue. C, corticospinal tract; Ru, rubrospinal tract; Ra, raphespinal tract; Ret, reticulospinal tract; Pr, proprio-spinal tract; ST, spinothalamic tract; DC, dorsal column sensory axons. The scaffold mimics the linear organization of white matter. Channels are precisely printed in 3D space. (**e**) A schematic diagram explaining the axonal alignment and guidance hypothesis. A lesioned host axon (for example, a corticospinal tract (CST), rubrospinal or reticulospinal axon) regenerates into the scaffold and forms a synapse onto a NPC-derived axon inside a channel, and the NPC neuron in turn extends an axon out of the scaffold below the injury site (green lines) into the same white matter fasciculus below the lesion, guided by a linear microchannel. The scaffold maintains its 3D coordinates across the lesion, matching the natural host architecture. (**f**) A sagittal mid-cervical T1-weighted magnetic resonance image of a human clinically complete AIS A SCI. A sliver of spared host white matter is evident on the anterior aspect (right side) of the lesion (arrow). The size of the lesion is noted next to the image. (**g**) The traced outline of the cystic lesion cavity from (**f**). (**h**) A computer-aided design (CAD) 3D model of the scaffold to be 3D-printed, corresponding to the precise lesion shape. (**i**) The printed scaffold matches the size shown in (**f**). (**j**) A hypothetical fit of the printed 3D scaffold in a human contusion cavity. (**k**) Mechanical measurement of the scaffold elastic modulus using dynamic mechanical analysis ($n = 3$). The data presented are mean ± standard error of the mean. Scale bar: 50 μm (**c**). Reprinted by permission from Springer Nature Customer Service Centre GmbH: Springer Nature, Nature Medicine [75], Copyright© (2019)

1.4.4 Biodegradability

Among optional characteristics of implantable biomaterials and interfaces for neural regeneration, biodegradability can be cited as one of special interest. For those particular applications that mainly rely on tissue reparative features, the temporal presence of the biomaterial might be sufficient and, moreover, desirable as chronic implants could drive to secondary fibrotic responses due to suboptimal physico-chemical properties. Ideal biomaterial degradation profiles in vivo would be those matching the kinetics of the tissue repair induced so, by the end of the regeneration process, the biomaterial is no longer there. On their turn, biomaterials serving to diagnostic purposes are typically required only for a concrete period of time, so superior stability over time rather than biodegradability might be aimed instead. This feature is closely related to biocompatibility, as biodegradable biomaterials are obligated to assure that degradation by-products are also biocompatible. For further details on this requirement, please refer to previous sections.

To this regard, biodegradable polymers and hydrogels are particularly attractive as they typically contain liable bonds that can be cleaved and result in smaller chemical groups able to be cleared from the implant site. For instance, degradation can occur through the hydrolytic and enzymatic cleavage of bonds made during polymerization, the incorporation of degradable monomeric units with bonds cleaved by water and the spontaneous hydrolysis of ester bonds [76]. Hydrogels can be customized with enzyme specific cleavage sites for driving degradation by tissue specific enzymes.

In general terms, degradation can mainly occur via surface or bulk erosion [120, 121], being the process dependent on the amount of degradable units, crosslinking degree and access of water and/or enzymes to the inner degradable parts of the biomaterial, among other factors. Typically, surface degradation occurs when water and enzymes cannot penetrate the interior areas, thus forcing the surface or exterior bonds to cleave first. Conversely, bulk erosion is characterized by degradation that occurs homogeneously throughout the biomaterial, with interior and exterior bonds being cleaved simultaneously. Hydrogels predominantly experience bulk erosion due to their higher water content and fast diffusion of liquids.

1.4.5 Electrical Activity

In the recent decades, electrical conductivity and activity have been identified as eventual pivotal features of interest for implantable technologies in neural tissues based on their inherent electrical activity. Signalling in the CNS relies on the ability of nerve cells to respond to very small stimuli with rapid and large changes in the electrical potential difference across the cell membrane [122]. Importantly, this rapid signalling depends on ion channels, a class of integral membrane proteins optimally tuned to respond to physical and chemical signals. They recognize and select specific ions, open and close in response to specific electrical, mechanical and chemical signals, and conduct ions across the cell membrane. Ion transporters/pumps, although also present in nerve cells, do not participate in rapid signalling but in the establishment and maintenance of concentration gradients of physiologically relevant ions. Today, we know that information is carried within neurons and from neurons to their target cells by means of electrical and chemical signals. This information can travel long distances by the generation and propagation of action potentials, arising from sequential changes in the membrane permeability to K^+ and Na^+ ions and for which voltage-gated ion channels play a relevant role.

Progress in the understanding of the electrical functioning of the CNS has permitted the exploration of electrical stimulation (ES) as a promising therapy for SCI patients [123], being extensively used as a therapy to relapse cardiac arrhythmias or seizures, to alleviate pain and to improve strength of paralysed muscles. The application of ES by means of conductive materials to promote and/or facilitate neural regeneration continues to be a rarely explored, and hardly understood, therapy for neural repair, especially in the injured spinal cord. Approaches involving the use of this type of materials are being the focus of a significant amount of publications [124–126], although in most of the cases the therapy is exclusively explored in vitro. For instance, the application of a potential of $100\,mV\,mm^{-1}$ for 2 h induced neural differentiation in PC-12 cells cultured on biodegradable citrate-based polymers doped with aniline tetramers to gain electrical conductivity [127]. In a different work, Casañ-Pastor and colleagues explored a palette of diverse materials including poly(3,4-ethylenedioxythiophene) (PEDOT) and iridium-titanium oxide for promoting nerve growth in amphibian neurons [128]. Specifically, neurites on substrates containing iridium oxide grew faster and in random directions. ITO substrates, with and without PCL/PLGA, were also able to promote neuronal differentiation and neurite growth in neural stem/progenitor cells under ES (80 mV) [129]. More recently, Blesch and colleagues have proved that ES has a time-dependent influence on the regenerative capacity of sensory neurons and might enhance axonal regeneration in combinatorial approaches for SCI [130]. Specifically, in vivo ES protocols (20 Hz, 2× motor threshold, 0.2 ms, 1 h) applied on the intact sciatic nerve in adult rats induced a 2-fold increase in neurite length of sensory neurons in vitro 1 week later. Longer (7 h) and repeated (7 days, 1 h/each) ES protocols increased growth by 56–67% at 7 days. A deeper discussion on the application of electric fields to conductive materials and neural cultures and eventual harmful events (e.g. faradaic reactions) is covered in Chap. 5.

1.4.6 Processability and Surgical Requirements

Those biomaterials and interfaces envisioned to be implanted in the brain and the spinal cord for clinical use should be easily processed in a scalable, reproducible and cost-effective manner, so the resulting therapy becomes available for a majority, if not the totality, of the affected population in a short period of time. Moreover, surgical needs must be also taken into consideration, including the capacity to fabricate the engineered material in a diversity of sizes and shapes to accommodate individual patient needs. If suturing is needed, biomaterials have to sustain it without breaking or tearing. However, given the softness of neural tissues, the use of sutures might be rarely needed for their CNS implantation. Finally, these neurotechnologies should be stable during storage at adequate conditions before implantation and easily handable at the surgery room by clinicians and healthcare personnel.

1.4.7   Other Particular Requirements for Functioning

Besides general characteristics described above, certain biomaterials for neural applications might have additional requirements related to their functioning and performance. For instance, implantable neuroelectrodes should include communication with as many individual neurons as possible and a high degree of signal-to-noise ratio for either short (days and months) or chronic (years) implantation times [126]. On their turn, biomaterials envisioned for drug delivery should be able to sustain a therapeutic drug release according to the physiopathology phases of traumatic neural tissue damage.

1.5   Final Considerations

In this chapter, we have presented an overview of the CNS, major physio-pathological features of traumatic brain and spinal cord injuries and a general description of engineered materials for neural applications, both biomaterials and interfaces in the context of their specific biological target. Major requirements for implantable biomaterials aimed at neural regeneration are also summarized, from mandatory biocompatibility and mechanical compliance to optional biodegradation and electrical conductivity. The aspects covered by this introductory chapter are thought to assist the reader on the understanding of more specific topics enclosed in the following chapters. In what follows, diverse current subjects related to this main axis will be described and discussed, from biomaterials and conductive nanomaterials to sensors and robotic-based rehabilitation strategies for TBI and SCI. An exposition of current neuro-engineered technologies applied to date in the clinics will be finally provided.

Acknowledgements This work was funded by the Agencia Estatal de Investigación de España and the Fondo Europeo de Desarrollo Regional (Grant MAT2016-78857-R, AEI/FEDER, UE).

References

1. Kandel ER, Schwartz JH, Jessell TM et al. (2012) Principles of neural science, 5th edn. Appleton & Lange: McGraw Hill, p 338–343
2. PhD Dissertation, Ana Domínguez Bajo, Design and development of biomaterials for spinal cord injury repair, 21st December 2020, Universidad Complutense de Madrid, Spain
3. Rosdahl CB, Kowalski MT (2008) Textbook of basic nursing, 9th edn. Lippincott Williams & Wilkins, p 189
4. Haines DE, Mihailoff GA (2018) Fundamental neuroscience for basic and clinical applications 5th edn, Chapter 12—The Pons and Cerebellum. Elsevier, Amsterdam, p 172–182
5. Afifi AK, Bergman RA (2005) Functional neuroanatomy: text and atlas, 2nd edn. McGraw-Hill, New York
6. Chung K, Coggeshall RE (1983) Propriospinal fibers in the rat. J Comp Neurol 217:47–53
7. Chung K, Langford LA, Coggeshall RE (1987) Primary afferent and propriospinal fibers in the rat dorsal and dorsolateral funiculi. J Comp Neurol 263:68–75
8. Alstermark B, Sasaki S (1986) Integration in descending motor pathways controlling the forelimb in the cat. 15. Comparison of the projection from excitatory C3–C4 propriospinal neurons to different species of forelimb motoneurons. Exp Brain Res 63:543–556
9. Lundberg A (1999) Descending control of forelimb movements in the cat. Brain Res Bull 50:323–324
10. Pauvert V, Pierrot-Deseilligny E, Rothwell JC (1998) Role of spinal premotoneurones in mediating corticospinal input to forearm motoneurones in man. J Physiol 508:301–312
11. Baldissera F, Cavallari P, Fournier E et al. (1987) Evidence for mutual inhibition of opposite Ia interneurones in the human upper limb. Exp Brain Res 66:106–114
12. Alstermark B, Lundberg A, Norrsell U et al. (1981) Integration in descending motor pathways controlling the forelimb in the cat. 9. Differential behavioural defects after spinal cord lesions interrupting defined pathways from higher centres to motoneurones. Exp Brain Res 42:299–318
13. Brown TG (1911) The intrinsic factors in the act of progression in the mammal. Proc Roy Soc B 84:308–319
14. Grillner S, Zangger P (1984) The effect of dorsal root transection on the efferent motor pattern in the cat's hindlimb during locomotion. Acta Physiol Scand 120:393–405
15. Eidelberg E, Walden JG, Nguyen LH (1981) Locomotor control in macaque monkeys. Brain 104:647–663
16. Drake RL, Vogl W, Mitchell AWM (2010) Gray's Anatomy for students, 2nd edn. Churchill Livingstone, Elsevier, p 830–833
17. Haines DE (2013) Fundamental neuroscience for basic and clinical applications, 4th edn. Saunders, Elsevier
18. Shoykhet M, Clark RSB (2011) Structure, function, and development of the nervous system. In: Pediatric critical care, 4th edn. Mosby, Elsevier, p 783–804
19. Ramón y Cajal S (1904) Textura del sistema nervioso del hombre y los vertebrados. Imprenta y Librería de Nicolás Moya, Madrid
20. Azevedo FA, Carvalho LR, Grinberg LT et al. (2009) Equal numbers of neuronal and non-neuronal cells make the human brain an isometrically scaled-up primate brain. J Comparative Neurol 513:532–541
21. Mtui E, Gruener G, Fitzgerald MJT (2011) Clinical neuroanatomy and neuroscience, 6th edn. Saunders, Elsevier, p 38
22. Bergen DC, Silberberg D (2002) Nervous system disorders: a global epidemic. Arch Neurol 59:1194–1196
23. World Health Organization (2006) Neurological disorders, public health challenges. https://www.who.int/mental_health/neurology/neurological_disorders_report_web.pdf. Cited 27 Sep 2021

24. Capizzi A, Woo J, Verduzco-Gutierrez M (2020) Traumatic brain injury: an overview of epidemiology, pathophysiology, and medical management. Med Clin North Am 104:213–238
25. Ladak AA, Enam SA, Ibrahim MT (2019) A review of the molecular mechanisms of traumatic brain injury. World Neurosurg 131:126–132
26. Najem D, Rennie K, Ribecco-Lutkiewicz M et al. (2018) Traumatic brain injury: classification, models, and markers. Biochem Cell Biol 96:391–406
27. Ahuja CS, Wilson JR, Nori S et al. (2017) Traumatic spinal cord injury. Nat Rev Dis Primers 3:17018
28. Ahuja CS, Badhiwala JH, Fehlings MG (2020) "Time is spine": the importance of early intervention for traumatic spinal cord injury. Spinal Cord 58:1037–1039
29. Bonsack B, Heyck M, Kingsbury C et al. (2020) Fast-tracking regenerative medicine for traumatic brain injury. Neural Regen Res 15:1179–1190
30. https://clinicaltrials.gov/ct2/show/NCT02881151, Cited 27 Sep 2021
31. https://clinicaltrials.gov/ct2/show/NCT03382626, Cited 27 Sep 2021
32. Courtine G, Sofroniew MV (2019) Spinal cord repair: advances in biology and technology. Nat Med 25:898–908
33. https://clinicaltrials.gov/ct2/show/NCT02138110?term=NCT02138110&draw=2&rank=1, Cited 27 Sep 2021
34. https://www.invivotherapeutics.com, Cited 27 Sep 2021
35. Theodore N, Hlubek R, Danielson J et al. (2016) First human implantation of a bioresorbable polymer scaffold for acute traumatic spinal cord injury: A clinical pilot study for safety and feasibility. Neurosurgery 79:E305-12
36. https://clinicaltrials.gov/ct2/show/NCT02688049?term=NCT02688049&draw=2&rank=1, Cited 27 Sep 2021
37. https://clinicaltrials.gov/ct2/show/NCT02688062?term=NCT02688062&draw=2&rank=1, Cited 27 Sep 2021
38. https://clinicaltrials.gov/ct2/show/NCT02510365?term=NCT02510365&draw=2&rank=1, Cited 27 Sep 2021
39. Bergmann CP, Stumpf A (2013) Dental ceramics, Microstructure, properties and degradation. Topics in mining, metallurgy and materials engineering. Springer, Berlin
40. Kaur M, Singh K (2019) Review on titanium and titanium-based alloys as biomaterials for orthopaedic applications. Mater Sci Eng C Mater Biol Appl 102:844–862
41. Vijayavenkataraman S, Lu WF, Fuh JY (2016) 3D bioprinting of skin: a state-of-the-art review on modelling, materials, and processes. Biofabrication 8:032001
42. Kim MS, Kim JH, Min BH et al. (2011) Polymeric scaffolds for regenerative medicine. Polym Rev 51:23–52
43. Liu S, Xie YY, Wang B (2019) Role and prospects of regenerative biomaterials in the repair of spinal cord injury. Neural Regen Res 14:1352–1363
44. Pettikiriarachchi JTS, Parish CL, Shoichet MS et al. (2010) Biomaterials for brain tissue engineering. Aust J Chem 63:1143–1154
45. Zhang YS, Yue K, Aleman J et al. (2017) 3D Bioprinting for tissue and organ fabrication. Ann Biomed Eng 45:148–163
46. Song JJ, Ott HC (2011) Organ engineering based on decellularized matrix scaffolds. Trends Mol Med 17:424–432
47. Patel H, Bonde M, Srinivasan G (2011) Biodegradable polymer scaffold for tissue engineering. Trends Biomater Artif Organs 25:20–29
48. Ikeda T, Ikeda K, Yamamoto K et al. (2014) Fabrication and characteristics of chitosan sponge as a tissue engineering scaffold. Biomed Res Int 2014:786892
49. Bonnet M, Trimaille T, Brezun JM et al. (2020) Motor and sensitive recovery after injection of a physically cross-linked PNIPAAm-g-PEG hydrogel in rat hemisectioned spinal cord. Mater Sci Eng C Mater Biol Appl 107:110354
50. Langer R, Vacanti JP (1993) Tissue engineering. Science 260:920–926
51. Atala A, Kasper FK, Mikos AG (2012) Engineering complex tissues. Sci Transl Med 4:160rv12

52. Baptista PM, Atala A (2016) Regenerative medicine: the hurdles and hopes. In: Laurence J, translating regenerative medicine to the clinic. Academic Press, Elsevier, p 3–7
53. Lysaght MJ, Reyes J (2001) The growth of tissue engineering. Tissue Eng 7:485–493
54. Pati F, Jang J, Ha DH et al. (2014) Printing three-dimensional tissue analogues with decellularized extracellular matrix bioink. Nat Commun 5:3935
55. Harrison RG (1910) The outgrowth of the nerve fiber as a mode of protoplasmic movement. J Exp Zool 9:787–846
56. Atala A (2009) Engineering organs. Curr Opin Biotechnol 20:575–592
57. Lazurko C, Harden S, Suuronen EJ, Alarcon EI (2019) Biomaterials for organ and tissue repair. Front Young Minds 7:8
58. Bhat S, Kumar A (2013) Biomaterials and bioengineering tomorrow's healthcare. Biomatter 3:e24717
59. Ning C, Zhou L, Tan G (2016) Fourth-generation biomedical materials. Mater Today 19:2–3
60. Feiner R, Dvir T (2018) Tissue–electronics interfaces: from implantable devices to engineered tissues. Nat Rev Mater 3:17076
61. Pons-Faudoa FP, Ballerini A, Sakamoto J, Grattoni A (2019) Advanced implantable drug delivery technologies: transforming the clinical landscape of therapeutics for chronic diseases. Biomed Microdevices 21:47
62. Gu X (2015) Progress and perspectives of neural tissue engineering. Front Med 9:401–411
63. Prochazka A (2017) Neurophysiology and neural engineering: a review. J Neurophysiol 118:1292–1309
64. Grill WM, Norman SE, Bellamkonda RV (2009) Implanted neural interfaces: biochallenges and engineered solutions. Annu Rev Biomed Eng 11:1–24
65. Yang B, Zhang F, Cheng F et al. (2020) Strategies and prospects of effective neural circuits reconstruction after spinal cord injury. Cell Death Dis 11:439
66. Prochazka A, Mushahwar VK, McCreery DB (2001) Topical review neural prostheses. J Physiol 533:99–109
67. Nih LR, Gojgini S, Carmichael ST, Segura T (2018) Dual-function injectable angiogenic biomaterial for the repair of brain tissue following stroke. Nat Mater 17:642–651
68. Zhang M, Tang Z, Liu X et al. (2020) Electronic neural interfaces. Nat Electron 3:191–200
69. Tam RY, Fuehrmann T, Mitrousis N, Shoichet MS (2014) Regenerative therapies for central nervous system diseases: a biomaterials approach. Neuropsychopharmacol 39:169–188
70. Schmidt CE, Leach JB (2003) Neural tissue engineering: strategies for repair and regeneration. Annu Rev Biomed Eng 5:293–347
71. Gu X, Ding F, Yang Y, Liu J (2011) Construction of tissue engineered nerve grafts and their application in peripheral nerve regeneration. Prog Neurobiol 93:204–230
72. Hench LL, Ethridge EC (1975) Biomaterials - the interfacial problem. Adv Biomed Eng 5:35–150
73. Holzapfel BM, Reichert JC, Schantz UG et al. (2013) How smart do biomaterials need to be? A translational science and clinical point of view. Adv Drug Deliver Rev 65:581–603
74. Drury JL, Mooney DJ (2003) Hydrogels for tissue engineering: scaffold design variables and applications. Biomaterials 24:4337–4351
75. Koffler J, Zhu W, Qu X et al. (2019) Biomimetic 3D-printed scaffolds for spinal cord injury repair. Nat Med 25:263–269
76. Aurand ER, Lampe KJ, Bjugstad KB (2012) Defining and designing polymers and hydrogels for neural tissue engineering. Neurosci Res 72:199–213
77. Peppas NA, Hilt JZ, Khademhosseini A, Langer R (2006) Hydrogels in biology and medicine: from molecular principles to bionanotechnology. Adv Mater 18:1345–1360
78. Plant GW, Harvey AR, Chirila TV (1995) Axonal growth within poly (2-hydroxyethyl methacrylate) sponges infiltrated with Schwann cells and implanted into the lesioned rat optic tract. Brain Res 671:119–130
79. Plant GW, Woerly S, Harvey AR (1997). Hydrogels containing peptide or aminosugar sequences implanted into the rat brain: influence on cellular migration and axonal growth. Exp Neurol 143:287–299

80. Führmann T, Schoichet MS (2018) The role of biomaterials in overcoming barriers to regeneration in the central nervous system. Biomed Mater 13:050201
81. Führmann T, Anandakumaran PN, Shoichet MS (2017) Combinatorial therapies after spinal cord injury: how can biomaterials help? Adv Healthc Mater 6:1601130
82. Haggerty AE, Maldonado-Lasunción I, Oudega M (2018) Biomaterials for revascularization and immunomodulation after spinal cord injury. Biomed Mater 13:044105
83. Lacalle-Aurioles M, Camps CC, Zorca CE et al. (2020) Applying hiPSCs and biomaterials: towards an understanding and treatment of traumatic brain injury. Front Cell Neurosci 14:594304
84. Langer R, Tirrell D (2004) Designing materials for biology and medicine. Nature 428:487–492
85. Silva GA (2006) Neuroscience nanotechnology: progress, opportunities and challenges. Nat Rev Neurosci 7:65–74
86. Nicolelis MAL, Dimitrov D, Carmena JM et al. (2003) Chronic, multisite, multielectrode recordings in macaque monkeys. Proc Natl Acad Sci USA 100:11041–11046
87. Polikov VS, Tresco PA, Reichert WM (2005) Response of brain tissue to chronically implanted neural electrodes. J Neurosci Methods 148:1–18
88. Zhu J (2010) Bioactive modification of poly(ethylene glycol) hydrogels for tissue engineering. Biomaterials 31:4639–4656
89. Burdick JA, Anseth KS (2002) Photoencapsulation of osteoblasts in injectable RGD-modified PEG hydrogels for bone tissue engineering. Biomaterials 23:4315–4323
90. Schense JC, Hubbell JA (2000) Three-dimensional migration of neurites is mediated by adhesion site density and affinity. J Biol Chem 275:6813–6818
91. Huang EJ, Reichardt LF (2001) Neurotrophins: roles in neuronal development and function. Ann Rev Neurosci 24:677–736
92. Allen SJ, Dawbarn D (2006) Clinical relevance of the neurotrophins and their receptors. Clinical Sci (Lond) 110:175–191
93. Almutiri S, Berry M, Logan A, Ahmed Z (2018) Non-viral-mediated suppression of AMIGO3 promotes disinhibited NT3-mediated regeneration of spinal cord dorsal column axons. Sci Rep 8:10707
94. Alto LT, Havton LA, Conner JM et al. (2009) Chemotropic guidance facilitates axonal regeneration and synapse formation after spinal cord injury. Nat Neurosci 12:1106–1113
95. Asboth L, Friedli L, Beauparlant J et al. (2018) Cortico-reticulo-spinal circuit reorganization enables functional recovery after severe spinal cord contusion. Nat Neurosci 21:576–588
96. Anderson MA, O'Shea TM, Burda JE et al. (2018) Required growth facilitators propel axon regeneration across complete spinal cord injury. Nature 561:396–400
97. Wen Y, Yu S, Wu Y et al. (2016) Spinal cord injury repair by implantation of structured hyaluronic acid scaffold with PLGA microspheres in the rat. Cell Tissue Res 364:17–28
98. Kadoya K, Lu P, Nguyen K et al. (2016) Spinal cord reconstitution with homologous neural grafts enables robust corticospinal regeneration. Nat Med 22:479–487
99. Ziemba AM, Gilbert RJ (2017) Biomaterials for local, controlled drug delivery to the injured spinal cord. Front Pharmacol 8:245
100. Song B, Song J, Zhang S et al. (2012) Sustained local delivery of bioactive nerve growth factor in the central nervous system via tunable diblock copolypeptide hydrogel depots. Biomaterials 33:9105–9116
101. Lu YB, Franze K, Siefert G et al. (2006) Viscoelastic properties of individual glial cells and neurons in the CNS. Proc Natl Acad Sci USA 103:17759–17764
102. Garcia PD, Guerrero CR, Garcia R (2017) Time-resolved nanomechanics of a single cell under the depolymerization of the cytoskeleton. Nanoscale 9:12051–12059
103. Pathak MM, Nourse JL, Tran T et al. (2014) Stretch-activated ion channel Piezo1 directs lineage choice in human neural stem cells. Proc Natl Acad Sci USA 111:16148–16153
104. Koch D, Rosoff WJ, Jiang J et al. (2012) Strength in the periphery: growth cone biomechanics and substrate rigidity response in peripheral and central nervous system neurons. Biophys J 102:452–460

105. Moshayedi P, Costa LF, Christ A et al. (2010) Mechanosensitivity of astrocytes on opti-mized polyacrylamide gels analyzed by quantitative morphometry. J Phys Condens Matter 22:194114
106. Moeendarbary E, Weber IP, Sheridan GK et al. (2017) The soft mechanical signature of glial scars in the central nervous system. Nat Commun 8:14787
107. Domínguez-Bajo A, González-Mayorga A, Guerrero CR et al. (2019) Myelinated axons and functional blood vessels populate mechanically compliant rGO foams in chronic cervical hemisected rats. Biomaterials 192:461–474
108. Göritz C, Dias DO, Tomilin N et al. (2011) A pericyte origin of spinal cord scar tissue. Science 333:238–242
109. Kubinová S, Horák D, Hejčl A et al. (2015) SIKVAV-modified highly superporous PHEMA scaffolds with oriented pores for spinal cord injury repair. J Tissue Eng Regen Med 9:1298–1309
110. Kubinová S, Horák D, Hejčl A et al. (2011) Highly superporous cholesterol-modified poly(2-hydroxyethyl methacrylate) scaffolds for spinal cord injury repair. J Biomed Mater Res Part A 99A:618–629
111. Tsai EC, Dalton PD, Shoichet MS, Tator CH (2006) Matrix inclusion within synthetic hydrogel guidance channels improves specific supraspinal and local axonal regeneration after complete spinal cord transection. Biomaterials 27:519–533
112. Silva NA, Salgado AJ, Sousa RA et al. (2010) Development and characterization of a novel hybrid tissue engineering-based scaffold for spinal cord injury repair. Tissue Eng Part A 16:45–54
113. Abidian MR, Martin DC (2009) Multifunctional nanobiomaterials for neural interfaces. Adv Funct Mater 19:573–585
114. Green RA, Hassarati RT, Goding JA et al. (2012) Conductive hydrogels: mechanically robust hybrids for use as biomaterials. Macromol Biosci 12:494–501
115. Duval K, Grover H, Han LH et al. (2017) Modeling physiological events in 2D vs. 3D cell culture. Physiology (Bethesda) 32:266–277
116. Yamada KM, Sixt M (2019) Mechanisms of 3D cell migration. Nat Rev Mol Cell Biol 20:738–752
117. Li H, Fan W, Zhu X (2020) Three-dimensional printing: the potential technology widely used in medical fields. J Biomed Mater Res A 108:2217–2229
118. Li T, Chang J, Zhu Y, Wu C (2020) 3D Printing of bioinspired biomaterials for tissue regeneration. Adv Healthc Mater 9:e2000208
119. Stokols S, Sakamoto J, Breckon C et al. (2006) Templated agarose scaffolds support linear axonal regeneration. Tissue Eng 12:2777–2787
120. Göpferich A (1996) Mechanisms of polymer degradation and erosion. Biomaterials 17:103–114
121. von Burkersroda F, Schedl L, Göpferich A (2002) Why degradable polymers undergo surface erosion or bulk erosion. Biomaterials 23:4221–4231
122. Kandel ER, Schwartz JH, Jessell TM et al. (2013) Principles of neural science, 5th edn. McGraw HillNew York, p 100–306
123. Torregrosa T, Koppes RA (2016) Bioelectric medicine and devices for the treatment of spinal cord injury. Cells Tissues Organs 202:6–22
124. Guimard NK, Gomez N, Schmidt CE (2007) Conducting polymers in biomedical engineering. Prog Polym Sci 32:876–921
125. Balint R, Cassidy N, Cartmell SH (2014) Conductive polymers: towards a smart biomaterial for tissue engineering. Acta Biomater 10:2341–2353
126. Fattahi P, Yang G, Kim G, Abidian MR (2014) A review of organic and inorganic biomaterials for neural interfaces. Adv Mater 26:1846–1885
127. Shan D, Kothapalli SR, Ravnic DJ et al. (2018) Development of citrate-based dual-imaging enabled biodegradable electroactive polymers. Adv Funct Mater 28:1801787.

128. Rajnicek AM, Zhao Z, Moral-Vico J et al. (2018) Controlling nerve growth with an electric field induced indirectly in transparent conductive substrate materials. Adv Healthc Mater 7:e1800473

129. Lei KF, Lee IC, Liu YC, Wu YC (2014) Successful differentiation of neural stem/progenitor cells cultured on electrically adjustable indium tin oxide (ITO) surface. Langmuir 30:14241–14249

130. Goganau I, Sandner B, Weidner N et al. (2018) Depolarization and electrical stimulation enhance *in vitro* and *in vivo* sensory axon growth after spinal cord injury. Exp Neurol 300:247–258

Chapter 2
Characteristics of the Spinal Cord Injured Patient as a Host of Central Nervous System Implanted Biomaterials

Daniel García-Ovejero, Ángel Arévalo-Martín, David Díaz, and Melchor Álvarez-Mon

Abstract Spinal cord injury (SCI) is a systemic injury. The spinal damage provokes not only local responses but also a number of dysregulations in peripheral organs that, in turn, affect the spinal cord. Each human lesion is unique and thus biomaterial-based therapies should be individually tailored. Studies performed on human subjects and *postmortem* tissue samples are burdened by important limitations. Four categories of human SCI have been established based on the macroscopic/low magnification appearance of the injured cord: solid cord injuries, contusions/cavities, lacerations, and massive compressions. In addition, two different damages occur after SCI: the insult itself that leads to the primary damage and the subsequent cascade of pathological processes that further enlarge the lesion, defined as the secondary damage. Beside, there is increasing evidence that acute and chronic SCI are associated with a severe dysfunction of the immune system, which includes three main syndromic alterations: (i) systemic inflammation; (ii)

D. García-Ovejero (✉) · Á. Arévalo-Martín
Laboratory of Neuroinflammation, Hospital Nacional de Parapléjicos, (SESCAM), Toledo, Spain
e-mail: dgarciao@sescam.jccm.es; aarevalom@sescam.jccm.es

D. Díaz
Diseases of Immune System and Oncology Department, Hospital Universitario Príncipe de Asturias, Madrid, Spain

School of Medicine, Medicine and Medical Specialties Department, Universidad de Alcalá (UAH), Madrid, Spain
e-mail: david.diaz@uah.es

M. Álvarez-Mon (✉)
Diseases of Immune System and Oncology Department, Hospital Universitario Príncipe de Asturias, Madrid, Spain

School of Medicine, Medicine and Medical Specialties Department, Universidad de Alcalá (UAH), Madrid, Spain

Instituto Ramón y Cajal de Investigación Sanitaria (IRYCIS), Carretera Colmenar Viejo, Madrid, Spain
e-mail: melchor.alvarezdemon@uah.es

© The Author(s), under exclusive license to Springer Nature Switzerland AG 2022
E. López-Dolado, M. Concepción Serrano (eds.), *Engineering Biomaterials for Neural Applications*, https://doi.org/10.1007/978-3-030-81400-7_2

immunodeficiency; and (iii) autoimmune reactions. SCI dysfunction is mainly due to the alteration in the communication between the immune and neuroendocrine systems. Intestinal dysbiosis has been found in SCI patients, accompanied by loss of the integrity of the intestinal barrier. This increased intestinal permeability may play a role in the observed increase in bacterial translocation and, through it, in the disturbances of the innate and adaptive immune responses found.

Keywords Bacterial translocation · Human studies · Immunodeficiency · Inflammation · Intestinal dysbiosis · Primary damage · Secondary damage · Spinal cord injury

2.1 Introduction

In the context of this book, one of the applications of engineered biomaterials that may readily come to the mind of the reader is their use to repair spinal cord injuries (SCIs). This challenge has been considered a chimera for a long time, but advances in the field are shedding new hope, and it is not so clear now that the spinal cord will not be able to regenerate one day. In the current chapter, we aim to draw a reconnaissance map of the battlefield in which biomaterial researchers will develop their strategies. Much of the information currently available for SCI comes from experimental models using mammalian species (mostly rodents, monkeys, cats, and dogs), and those models will probably be the ones that many experts will face first when testing their devices in vivo. There are many good reviews on the state of the art of molecular, cellular, and functional mechanisms of SCI obtained from preclinical animal models [1–4, 102]. However, we want to compile here the information available on humans, the current knowledge of the terrain that biomaterials will find when approaching patients in the clinics. This has been summarized before in excellent reviews from the 1990s and early 2000s, but a lot of new data have been produced since then, not gathered yet in a single review. We hope this chapter will be of help.

We wanted to start by stating some general considerations. First, it is important to understand that SCI is a systemic injury, a syndrome. It represents the damage of a key center of coordination that provokes not only local responses, but also a number of dysregulations in peripheral organs that, in turn, affect the spinal cord. For this, we will summarize the cellular and molecular events observed inside the human spinal cord during the time course of the lesion, as well as the changes in the periphery that condition the response of the neural tissue to the damage and, thus, may either boost or limit regeneration. Second, it should be kept in mind that each human lesion is unique. In experimental models, performed under controlled conditions on homogeneous animal populations, interindividual variability is highly limited; but in humans, the acute insult can be, and usually is, extremely variable in nature, extension, and severity and also finds an extremely variable substrate (e.g. patients with different age, sex, health state). Consequently, biomaterial-based therapies should be individually tailored. A common pattern of response can be drawn, anyway, and we will show it here, but the intrinsic variability of human SCIs should always be kept in mind. Otherwise, interventions could end up in no significant functional improvements. Third, studies performed on human subjects and *postmortem* tissue samples are burdened by important limitations. One is the scarcity of human SCI tissue available, which usually prevents most studies of examining enough number of samples to fully cover the time course of the lesion for a specific feature. This forces them to establish a continuous view of a process with fragmentary datasets. Another one is the preservation of the tissue, that is usually good enough for anatomical and histochemical observations, but frequently not compatible with obtaining good immunostainings due to overfixation and the delicate nature of some antigens [5–8].

2.2 General Features of Human SCIs

2.2.1 Types of Human SCIs

As stated in the introduction, high interindividual variability can be found in human SCIs, but a general classification of cases is still helpful for describing and understanding human responses to SCI. Accordingly, four categories of human SCI were established by Bunge et al. in 1993 [9], extensively used ever since [6, 10–12]. They are based on the macroscopic/low magnification appearance of the injured cord and are defined as follows [9, 12]:

(i) **Solid cord injuries** (approximately 10% of the cases) classify spinal cords that look normal macroscopically—without softening, discoloration, or cavity formation, but show different features of damage and pathology under microscopic examination.

(ii) **Contusions/cavities** (around 50% of the cases) are lesions in which a blunt force deforms the cord without invading it (i.e. the continuity of the spinal cord surface is preserved, and no adhesions are found between this surface and the dura mater). This type of displacement usually affects more severely the innermost aspects of the cord in which evident areas of hemorrhage and necrosis are formed. After weeks/months, these areas end up forming cysts (fluid-filled cavities) that usually extend rostrally and caudally through the ventral aspects of the dorsal columns.

(iii) **Lacerations** (about 20% of the cases) involve a penetrating disruption of the spinal cord surface by bone fragments, knifes, or projectiles.

(iv) **Massive compressions** (about 20% of the cases) are very severe lesions in which the cord is macerated or pulpified to a varying degree and the epicenter of the lesion ends up in a cavitation full of connective tissue and nerve roots invasion. In that regard, tissue response may share similarities with contusions, but also with the fibrosis observed in laceration injuries. These lesions are also usually accompanied by vertebral fractures.

In types (i) and (ii), pial surface continuity is preserved; whereas in types (iii) and (iv), it is not. This causes a substantial connective tissue response after lacerations and massive compressions that participates in scarring at the lesion epicenter [9]. Both in animals and humans, responses may qualitative and quantitatively vary depending on the injury type (e.g. ischemia, contusion/compression, transection) [13, 14] and the nature of the lesion. Figure 2.1 shows two exemplary SCIs in human patients. Since contusion and compression are the most frequent among patients, these models are also the most frequently used in animals as clinically relevant [13, 14]. Therefore, most of the descriptions in the literature are referred to the responses elicited after contusion/cavitation and massive compression types, both for animals and humans, which may not fully apply to penetrating injuries (laceration). However, models like transections and hemisections (i.e. penetrating injuries) are indeed useful for specific approaches, such as those using growth supporting biomaterials [14, 15]. Therefore, we have tried to specify the particularities of all of these lesions when available.

2.2.2   Clinical Versus Anatomical Score of Human SCIs

Clinically, injuries at the spinal cord can be also classified as either complete or incomplete according to the degree of preservation of sensory and motor function below the level of injury (American Spinal Injury Association Impairment Scale, AIS grades A–E) [17]. However, this functional division does not necessarily reflect the anatomical state. Consequently, most of the patients classified as clinically complete still maintain some tissue preservation in the injury site [9, 11, 12]. Some authors define these lesions as "discomplete" (clinically complete but anatomically incomplete) [9, 11, 18, 19]. Discompleteness was first uncovered by the finding of electrophysiological transmission of signals across the lesion in patients who were clinically complete [18] and then confirmed by histological observations of anatomical continuity of the white matter across the lesion in clinically complete patients [11, 19]. This state of complete clinical but incomplete anatomical lesion is not anecdotal. In *postmortem* pathological studies, continuity of the central nervous system parenchyma can be found in around 60–70% of clinically complete cases [9, 11, 19].

Among incomplete lesions, severity is also variable. Studies using histology and magnetic resonance imaging (MRI) have shown a direct correlation of lesion severity with lesion size (longitude and diameter) and an inverse correlation with the amount of preservation of a tissue ventral bridge, both in humans and rats [20, 21]. Here, it may be relevant to highlight the importance of MRI to study the morphology and dynamics of SCI. MRI presents lower resolution and accuracy than histology for anatomical and morphological measurements, but it offers additional advantages like the possibility of performing longitudinal in vivo studies with the

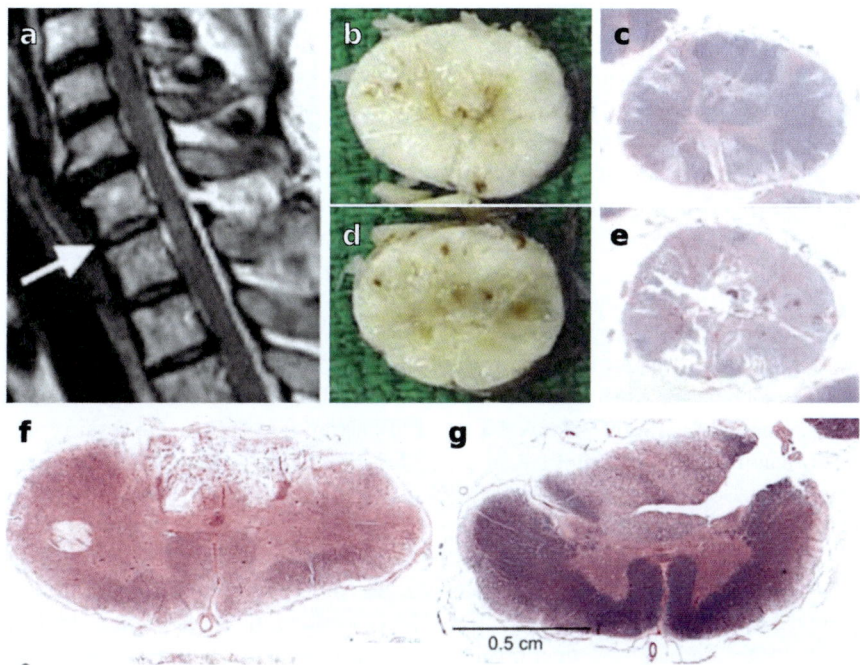

Fig. 2.1 Examples of two different human SCIs. Features shown by MRI, gross morphology, and histology. (**a–e**) 49 years old patient with a C7/T1 contusion, who died 15 days after the accident. (**a**) MRI shows T2 hyperintensity between C7/T1 vertebrae. (**b, d**) Macroscopic views of rostral (**b**) and thick epicenter (**d**) segments in which petechiae and hemorrhagic core can be observed. (**c, e**) Myelin staining shows the histological correlation. (**f, g**) 59 years old patient with a C5/C6 contusion/compression, who died 60 days after the accident. Myelin staining shows the histological correlation: (**f**) Areas of Wallerian degeneration in the ascending tracts of the rostral spinal segments and (**g**) a dorsal necrotic core that forms a cyst at the epicenter. Reproduced by permission from Springer Nature Customer Service Centre GmbH: Springer Nature, Spinal Cord [16], Copyright© 2007

same individual and avoiding fixation and *postmortem* artifacts. In addition, MRI enables measuring dynamic processes such as water mobility and liquid diffusion [22].

2.2.3 Time Course of Injury Progression in the Cord: Primary and Secondary Damage

Two different damages occur after SCI: the insult itself (e.g. contusion, compression, laceration) and the subsequent cascade of pathological processes that further enlarge the lesion and are collectively known as the secondary damage. The time course of gross damages is as follows. In victims that die at the scene, only rare petechiae (i.e. small bleedings) can be found in the gray matter, except for lacerations and massive compressions that show distortion of the tissue [6, 12, 23]. In a short time period (1–4 h after the injury), tissue fragmentation, congestion (i.e. hyperemia), petechiae and malorientation of fiber tracts can be found, but the true extent of injury, in terms of further necrosis, cannot be predicted at that moment [6, 12, 23]. From hours to 3 days (acute phase), tissue changes are notable and highly related to vascular impairment. Vasogenic and cytotoxic edema develop due to the accumulation of fluids into the extracellular space and inside the cells (mostly in astrocytes). Edema and petechial hemorrhages are found in the white matter tracts, while a hemorrhagic necrosis area surrounded by neutrophils is established mostly in the gray matter [6, 12]. Additionally, hemorrhage can be also found within meninges [24, 25]. All hemorrhages are, nevertheless, small, and major bleeding (i.e. hematoma) is rarely observed, except in lacerations. They come from rupture of postcapillary venules and sulcal arterioles either by mechanical disruption or from intravascular coagulation [12, 26]. In the preserved gray matter, pathological features can be observed in cells (e.g. eosinophilia, vacuolation, karyorrhexis) reflecting ischemic necrosis or chromatolysis. In the white matter, vacuolation and axonal swelling can be found at various degrees [6, 12].

In the following 5–10 days post injury (DPI), the necrotic areas contain abundant macrophages and many axonal profiles with swelling at the levels adjacent to the injury site [6, 11, 27]. The following weeks to months after injury, areas of necrosis are progressively eliminated and replaced by cysts (i.e. non-expanding cavities) or filled with abundant macrophages (Fig. 2.2). Moreover, edema is resolved, astrocyte reactivity increases, and revascularization begins [6, 12]. Reactive astrocytes intermingle their processes forming a glial scar, that is not as dense after human SCI, but it is accompanied by a "mesenchymal scar" [12] consisting of fibroblasts, meningeal cells, and a specific extracellular matrix mainly composed of a notable deposit of collagen, laminin, and inhibitory chondroitin sulfate proteoglycans [12, 28, 29], as further discussed below (Fig. 2.3). In segments adjacent to the lesion level, reactive astrocytes, macrophage accumulation, β-amyloid precursor protein (APP) immunopositive, and argyrophilic axonal swellings are common. Also, axons retract due to Wallerian degeneration (WD), which further induces glial and inflammatory responses in white matter areas. In cases of severe injuries (especially after massive compression), there is a strong infiltration of Schwann cells into the cavity months after damage. Interestingly, the borders of these cavities show a moderate astrocytosis, not a thick wall of processes, and the limit between fluid filling the cavity and the parenchyma is minimal (in contrast to what is seen in expanding cavities known as syrinxes) [9, 12, 20].

A summary of the main events described after human SCI during the time course of the lesion is shown in Table 2.1. We offer there the data published to date on cell death, axonal plasticity, local inflammation, astrocytosis, vascular changes, myelin, and stem/progenitor cells after human SCI. We gathered events in three-time frames (i.e. early, subacute, and chronic) and specified which events are observed at or next to the epicenter of the lesion in comparison to those in adjacent and distal regions.

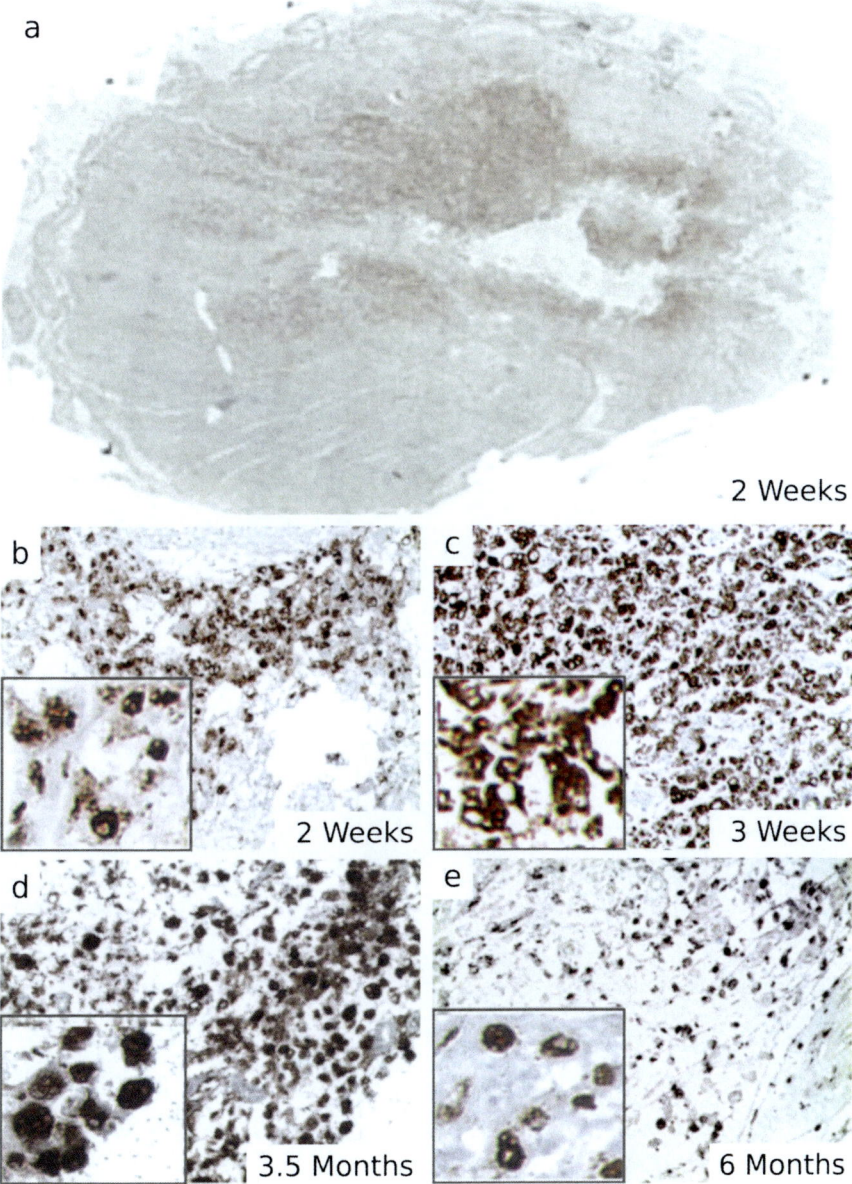

Fig. 2.2 Microglia/macrophages accumulate in the injury site from weeks to months after human SCI. (**a**) Panoramic view of the immunostaining against CD68 at the epicenter of a contusion 2 weeks after damage. Morphology of CD68$^+$ microglia/macrophages is shown in medium and high magnification (nested squares) at the injury site after 2 weeks (**b**), 3 weeks (**c**), 3.5 months (**d**), and 6 months (**e**). Reproduced by permission from Springer Nature Customer Service Centre GmbH: Springer Nature, Acta Neurophatologica [30]. Copyright© 2011

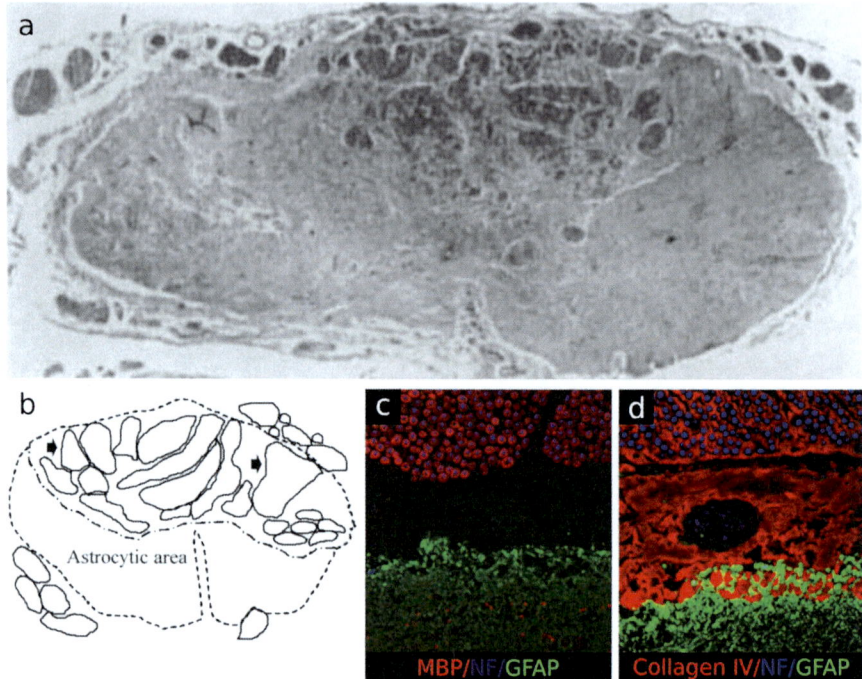

Fig. 2.3 Infiltration of Schwann cells in chronic severe injuries. (**a**) Histological section caudally adjacent to the epicenter of a chronic severe dorsal lesion (fracture dislocation, 17 years previously). A high density of Schwann cells myelinated peripheral axon bundles can be observed inside the lesion core. (**b**) Scheme depicting a severe lesion with partial tissue preservation, similar to that shown in (**a**). Two areas can be distinguished: Preserved spinal cord tissue is dominated by astrocytosis (astrocytic area) and is completely segregated from the lesion core, while the lesion core shows infiltration of axons and Schwann cells (schwannosis) and a distinct extracellular matrix (ECM). (**c**) The astrocytic area (glial fibrillary acidic protein, GFAP, staining, green) is segregated from Schwann cells and peripheral myelinated axons (myelin basic protein, MBP, staining, red; neurofilament staining, blue). (**d**) ECM composition differs between both areas. Only the lesion core includes collagen IV deposition. (**a**) Reproduced from [31]. Copyright© 1967 The Pathological Society of Great Britain and Ireland. Copyright© 2005 John Wiley and Sons. (**b–d**) Reproduced from [28] by permission of Oxford University Press

2.2.4 Other Considerations That May Be Relevant for the Use of Biomaterials

In some cases, the lesions, even though causing severe impairments, may be relatively small and limited to one spinal segment [12]. However, the most frequent scenario is that cavities and damage are usually not limited to the level of initial impact, so intra- and extramedullary circulatory impairments occur also in adjacent levels [11, 32]. The sum of circulatory impairments and narrowing of the spinal canal cause increased intramedullary pressure, which may induce the upward and

Table 2.1 Summary compilation of main events described after human SCI at different time frames and locations with respect to the lesion epicenter. DPI: days post injury; ECM: extracellular matrix; GM: Gray matter; PHN: Progressive hemorrhagic necrosis; WD: Wallerian degeneration; WM: white matter

Time frame	Lesion to 1 week		From 1 to 8 weeks		From 4 months to years	
Area of interest	Epicenter	Distal areas	Epicenter	Distal areas	Epicenter	Distal areas
Cell death	Neuronal necrosis [6, 39]	–	Spare apoptosis of neurons [30]	–	–	–
Cell death	First apoptotic features [6, 39]	–	Apoptosis of oligodendrocytes (CNPase$^+$), macrophages (CD68$^+$) and neutrophils (MPO$^+$) in the perilesional rim and WM [30, 39]	Apoptosis of oligodendrocytes in degenerating tracts, mostly above the lesion [30, 39]	None	None
Cell death	Astrocyte numbers drop [12, 28, 39] and microglia (Iba1$^+$TMEM119$^+$) cell numbers drop in the lesion core [27]	–	–	–	–	–

(continued)

Table 2.1 (continued)

Time frame	Lesion to 1 week		From 1 to 8 weeks		From 4 months to years	
Area of interest	Epicenter	Distal areas	Epicenter	Distal areas	Epicenter	Distal areas
Axonal plasticity and degeneration	Varicose enlargements and swollen axons in WM [11, 19, 27]	Varicose enlargements and swollen axons in WM, 1 segment above and below the lesion [11, 19]	Signs of regeneration failure (sterile endbulbs). Axonal spheroids [20]	First features of WD from 12 DPI [40]. Established WD after 7 weeks [41]	Sterile endbulbs and axonal dieback. Intraspinal plasticity [20]	Intraspinal plasticity. Aberrant plasticity of sensory fibers in the dorsal horn (CGRP$^+$) [42]
Axonal plasticity and degeneration	Retraction bulbs [11, 19]					
Local inflammation	Neutrophils associated with first vessels during first hours and entered the cord 1–3 DPI [6, 43]	Neutrophils found 1–3 segments rostral and caudal to the lesion [6, 43]	Neutrophil no longer detected beyond 2 weeks (in some studies, before 5 DPI). Presence of some MPO$^+$ cells (macrophages?) [6, 12, 30, 43]	–	–	–

| Local inflammation | Microglial activation. Upregulation of CD68 (1–3 DPI). Some macrophages found already in the cord [6, 43] | Activated microglia and macrophages, but rare foamy macrophages. CD68, MPO and gp91 phox expression, but in lower levels and associated with clear tissue damage. Begin to overexpress LCA and MHCII in areas of WD [6, 43, 44] | More activated microglia. Macrophages accumulation in necrotic core that progressively acquire foamy phenotype. Activated morphology and proinflammatory profiles in lesion surroundings. Microglial proliferation in the lesion rim [6, 16, 27, 30, 43] | Microglial/macrophage overexpression of CD68, LCA, and MHCII in areas of WD [44] | Loss of CD68 expression, but microglial/macrophage activation still clear in the lesion core (TMEM119- macrophages) and rim (TMEM119+ microglia) [6, 27, 30, 44] | MHCII expression in WD areas maintained in WM, lost in GM [44] |
| Local inflammation | Microglial cell numbers drop at the lesion core (Iba1+, TMEM119+) [27] | – | CD8+ lymphocytes in cavity margins in low numbers. CD4+ in lower numbers, same time course. Some reports numerous CD3+ lymphocytes. General absence of B cell infiltration with a few exceptions [6, 27, 30] | – | CD8+, CD4+, CD3+ lymphocytes found at first weeks-months [6, 30] | – |

(continued)

Table 2.1 (continued)

Time frame	Lesion to 1 week		From 1 to 8 weeks		From 4 months to years	
Area of interest	Epicenter	Distal areas	Epicenter	Distal areas	Epicenter	Distal areas
Local inflammation	Increase of some cytokines in spinal cord (IL1a, IL1b, IL-6, IL-8, IL-18, MIO1a, MIP1b, GROa) and decrease of IL-4 and IL-10. TGFB1 increases in neurons, astrocytes, and invading macrophages [6, 27, 45–47]	–	TGFB1 expression maintained. TGFB2 in macrophages and astrocytes from 24 DPI [47]. IL18 expression in microglia/macrophages [27]	–	TGFB2 maintained [47]. IL18 expression in microglia/macrophages [27]	–
Astrocytes	Decreased astrocyte numbers [12, 28, 48]	–	Astrocytes begin to react and hypertrophy occupying preserved tissue up to 24 DPI. Then extend processes to form glial scar [12, 28, 40, 48–50]	Astrocyte reactivity in WD areas. NG2 is found in macrophages and oligodendrocyte precursors [40, 48, 50]	Two segregated regions can be found: (1) astrocyte domains (preserved rim of tissue) that include preserved WM, axons, central myelin, and ECM (phosphacan, NG2), and (2) lesion core invaded by ECM (collagen, laminin, NG2, fibronectin, phosphacan), regenerated nerves and Schwann cells forming peripheral myelin with associated versican and neurocan. AQP4$^+$ and AQP4$^-$ astrocytes [2, 16, 24, 28, 29, 48, 50–54]	Segregation in two different regions not as frequent as in epicenter. When found, region with Schwannosis much less prominent than that of reactive astrocytes. In WD areas, peak of astrocyte reactivity at 4 months [49]

Vasculature/blood–spinal cord barrier	Disruption of small vessels, mostly in GM. Hemorrhage and PHN. Angiogenesis begins at 2 DPI [23]	Angiogenesis from 4 DPI (1–2 segments away from the injury site) [28, 48]	Recovery of vessel density (collagen IV, laminin, and fibronectin signals) [28]	Angiogenesis lasts until 2 weeks post injury in GM and WM [28]	Increasing numbers of blood vessels with irregular appearance and thickened walls [28, 48]	–
Myelin	Swollen, vacuolated, and faintly stained myelin sheaths (2–3 DPI). Oligodendrocytes show first apoptotic features [11, 19, 27, 55]	No noticeable changes	Oligodendrocytes with apoptotic features (up to 4 weeks) and high levels of activated caspase-3, -7, and -9 in the rim of spared tissue (only in WM) [30, 39]	Myelin degeneration in WD areas. Presence of apoptotic oligodendrocytes. At 24 DPI, myelin debris found and ring-like myelin profiles number reduced. Changes in myelin associated proteins after 5 weeks [30, 39–41, 56]	Progressive loss of myelin associated proteins. First, MAG (up to 4–10 months), then MBP, PLP, MOG, NOGO-A, that last beyond 3 years. Complete loss of them all at 10 years. Infiltration of Schwann cells and peripheral myelin around invading nerves. Peripheral myelination of a limited number of central axons at the margins of the cavity [11, 29, 31, 51, 52, 55–58]	–
Stem/precursor cells	No proliferation at ependymal region, but expression of nestin [8, 59]	No proliferation at ependymal region [8]	No proliferation at ependymal region [8]	No proliferation at ependymal region [8]	No proliferation at ependymal region [8]	No proliferation at ependymal region [8]

downward spreading of the lesion. This extension can range beyond the spinal cord and even reach the medulla oblongata [32] and involve necrosis in the ventral region of the dorsal columns that produce pencil-shape (en crayon) lesions [32, 33]. In these cases, the appearance of marginal spongiosis can be also observed in distal levels and a large number of veins and small arteries in the subarachnoid space contain organized thrombi surrounding the damaged segments [32].

In some specific clinical conditions (i.e. central cord syndrome), disturbances of MRI signals in the center of the cord lead to the assumption that a central hemorrhage may underlie posterior degeneration. However, histological analysis has shown that this condition is caused by diffuse white matter damage, especially in the lateral columns—occupied by the corticospinal tract—rather than gray matter necrosis [34].

Once the cavity is formed, it usually stabilizes. However, from months to years after injury, new cavities (i.e. syrinx) may develop. These are "active" cysts with pressure inside that forces them to expand rostro-caudally and may cause enormous damage with progressive loss of function. Expanding post-traumatic intramedullary cysts are uniformly associated with significant scarring in the subarachnoid space, causing a tethering of the pial surface of the contused cord to the surrounding thickened dura. In contrast with flaccid (i.e. non-expanding) cavities, where the borders are loose with no defined astrocytic boundary, syrinxes can have an outer border composed of multilayered reactive astrocytes [20]. This post-traumatic syringomyelia is hypothesized to be caused by the combination of two factors: (1) a damage or frailty in the nervous tissue, and (2) a disturbance in the normal circulation of the cerebrospinal fluid (CSF) along the subarachnoid space, caused by fibrous plugs that impede CSF flow and force it to enter the cord through perivascular spaces [35–38]. In some cases, patients have symptoms similar to post-traumatic syringomyelia, although MRI do not show evidence of cavity expansion but a tethering of the cord to the dural surface and an apparent cord expansion [20]. This is known as progressive non-cystic myelomalacia, and symptoms may be stopped by duroplasty and untethering.

2.3   Peripheral Immune Responses After SCI

There is increasing evidence that acute and chronic SCIs are associated with a severe dysfunction of the immune system [60, 61]. The immune system plays a dual role in the pathogenesis of SCI such that, besides the role of immune-mediated inflammatory mechanisms, SCI itself also leads to immune system dysfunction. A distinctive feature of SCI is the dynamic coexistence of acquired immunodeficiency and systemic inflammation [62–64]. The immunodeficiency develops within hours after injury and provokes an impaired response to pathogens that favors the increased incidence of infectious diseases among these patients [65–67]. The systemic inflammation seems to be differentiated in two stages: a first acute inflammatory response to spinal cord damage and a chronic response consequent to the persistent and inadequate stimulation of cells of the immune system [45, 62, 68–71]. Anyhow, the systemic inflammatory response has been proposed to contribute to the pathogenesis of the metabolic syndrome, increased cardiovascular risk and fatigue [72–77], all these being complications observed in SCI patients. Furthermore, both nervous and immune systems affect each other through bidirectional interactions. For instance, the immune-inflammatory response is involved in the pathogenesis of mental health conditions such as depression and anxiety [78], also frequently suffered by patients with SCI.

SCI associated immune system dysfunction may be separated into three main syndromic alterations: (i) systemic inflammation, due to a maintained and inadequate stimulation of immune system cells; (ii) immunodeficiency, secondary to an impaired function of these cells; and (iii) autoimmune reactions, due to a loss of self-tolerance of the immune system (for a review on this topic, readers are referred to Schwab et al. [60]). Finally, SCI is often associated with widespread visceral dysfunction mainly due to alterations in the communication between the immune and neuroendocrine systems [60].

2.3.1   Proinflammatory State After SCI

Acute SCI induces a severe immune-inflammatory response both at the lesion site and systemically. These responses do not resolve and they are chronically maintained with dynamic variations [6, 45, 62, 68–71]. Several studies have shown that chronic SCI patients have increased serum concentrations of proinflammatory cytokines including interleukin 1 (IL-1) and 6 (IL-6) and tumor necrosis factor alpha (TNF-α). Furthermore, abnormal serum levels of immunoregulatory cytokines, chemokines, and growth factors have been also reported [45, 62, 68–71, 79]. These findings appear to be independent of infections or pressure ulcers secondary to SCI, events that enhance the systemic immune-inflammatory response. The anomalous pattern of serum cytokines shown by SCI patients suggests that a low grade chronic inflammatory process is occurring, which has been proposed to be related to the pathogeny of obesity, diabetes [80], accelerated atherogenesis [81], and bone loss, complications also observed after SCI. Of note, this peripheral cytokine disturbance might also affect the progression of the injury, as well as interfere with implanted biomaterials. Circulating cytokines may be actively transported through the normal blood–brain barrier, but passively enter the central nervous areas where the barrier is absent or impaired. In this regard, enhanced cytokine levels have been associated with dysfunction of neural networks and axons, neurological damage, expanding trauma-induced axonopathy and myelinopathy [82].

2.3.2 Changes in the Innate and Adaptive Immune Responses Induced by SCI

SCI associates with several alterations both in innate and adaptive components of the immune system response. Monocytes play a critical role in the innate and adaptive immunity, as well as in the induction and regulation of the inflammatory response [83]. Decreased monocyte counts have been observed in acute SCI patients, but these counts seem to normalize at chronic stages. However, there is evidence of in vivo abnormal activation of circulating monocytes with a proinflammatory pattern of cytokine production. The increased TNF-α expression in monocytes from chronic SCI patients is ascribed to their $CD14^{+high}CD16^-$ subset [84]. Monocytes from SCI patients also show abnormal expression of activation receptors such as Toll-like receptors (TLRs). Increased TLR4 and TLR9 expression in monocytes has been found among patients with spinal cord damage. Furthermore, monocytes from patients with chronic SCI show a marked defect in phagocytosis [85]. Additionally, Natural killer (NK) cells play a relevant role in the defense against microorganisms, especially against virus infections. Chronic SCI is associated with a decreased NK cytotoxic activity in a SCI level-dependent manner [86, 87].

On their turn, T lymphocytes play a central role in the adaptive immune response. Helper T lymphocytes regulate the activity of the effector cytotoxic T lymphocytes and the proliferation and differentiation of B lymphocytes into antibody secreting cells. In addition, helper T lymphocytes modulate the function of cells of the innate response such as NK cells, monocytes, and other antigen presenting cells. A reduction in the absolute numbers of circulating T lymphocytes has been described in chronic SCI patients, mainly due to a decrease in the T helper cell subpopulation. Furthermore, a marked disbalance in T helper subsets is observed in these patients, who exhibit an expansion of both activated regulatory T cells and T helper 17 cells, with a concomitant increased production of suppressor cytokines such as transforming growth factor beta (TGF-β) and IL-10. An increased percentage of T helper 1 cells have been also found in some stages of T cell activation/differentiation in SCI patients. Circulating T lymphocytes responses after SCI are also impaired. Expansion of $CD28^-$ T lymphocytes is observed in those patients with expression of markers of exhaustion and aging [88]. Suggesting a premature immunosenescence, there is also evidence of diminution in the number of naïve cells [89]. Still, contradictory results about T lymphocyte counts and subset distributions and functions have been described, thus suggesting a dynamic pattern of the immune system dysfunction also observed in other inflammatory and autoimmune chronic diseases.

SCI patients show a redistribution of the T and B lymphocytes repertoire with an expansion of the clones reactive to both central nervous system and peripheral antigens. An increased frequency of T lymphocytes reactive to myelin proteins (e.g. myelin basic protein) has been observed in SCI patients [90]. Indeed, the levels of antibodies against 24 different antigens have been reported to be increased in SCI patients [62, 68, 91–96]. Both IgM and IgG isotypes have been identified in the auto-antibodies, suggesting T-B lymphocyte cooperation in SCI patients. These findings support that these patients have a functional activation of autoreactive T and B lymphocytes. Relevant for biomaterial implantation, autoreactive T lymphocytes might migrate to the injury site to recruit and activate additional immune cells responsible for mediating local inflammatory nervous system damage.

2.3.3 Etiopathogenesis of the Immune System Dysfunction Induced by SCI

The immune dysfunction associated with SCI is a pathogenic mechanism to consider for the tolerance and viability of central nervous implanted biomaterials in these patients. Besides, the etiopathogenesis underlying SCI associated immune system dysfunction appears to be complex and exhibits a great individual variability. Several mechanisms may be involved in the induction of the systemic inflammation, immunodeficiency, and autoimmune reactions found in these patients.

The crosstalk between the nervous, the endocrine, and the immune systems has been well established. SCI significantly alters this crosstalk (Fig. 2.4a). Cells of the immune system are under regulation by the nervous system via the direct innervation of primary and secondary lymphoid tissues by autonomic nerve fibers of the sympathetic nervous system [97, 98]. SCIs at levels above T6 may course with an exaggerated and maintained sympathetic response and, thus, an excessive release of norepinephrine in the lymphoid organs [66, 99], which, in turn, may modulate the function of immune cells expressing norepinephrine receptors [100, 101]. Furthermore, SCI increases blood serum glucocorticoids through the activation of the hypothalamic–pituitary–adrenal axis. In addition, mental health diseases such as anxiety and depression are associated with a systemic alteration of the immune system [78]. These disorders are common among SCI patients and could play a role in the progressive worsening of the chronic immune response.

Other mechanisms potentially involved in the dysfunction of the immune system of SCI patients are directly related to the spinal cord lesion. Damage-associated molecular patterns, originating from the host tissue upon injury, may be recognized by pattern recognition receptors, expressed on innate immune and antigen presenting cells [64]. It is also possible to involve the release of sterile particulates from the damaged nervous system that may act as autoantigens recognized by T and B lymphocytes and induce their activation [96].

Additional pathogenic mechanisms of the immune system dysfunction found in SCI patients are related to the secondary neurogenic bowel that leads to infections and bacterial translocation events (Fig. 2.4b). It has been demonstrated that SCI patients suffer a switch in gut microbiome to a proinflammatory profile, damage of the gut barrier, and a leaking intestine. Intestinal dysbiosis and increased intestinal permeability may play a role in the observed increase in bacterial translocation and,

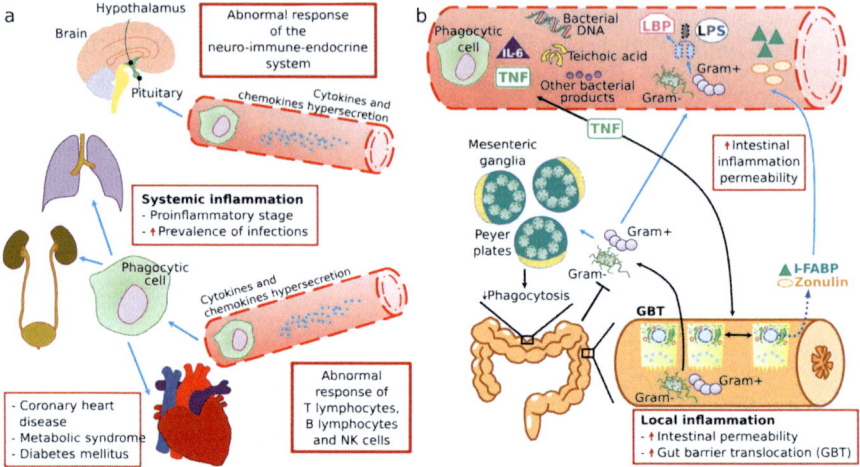

Fig. 2.4 Effects of systemic inflammation during chronic SCI. (**a**) Chronic elevation of cytokines and immune mediators may alter the normal functioning of a variety of organs, including the urinary and cardiovascular systems. (**b**) Gut barrier damage in chronic SCI patients as described by Diaz et al. [84]. Inflammation induces an increased intestinal permeability accompanied by bacterial translocation. This may contribute to the systemic proinflammatory state in chronic SCI patients, including impaired circulating monocyte function and enhanced serum TNF-α and IL-6 levels. GBT: gut barrier translocation; I-FABP: intestinal fatty acid-binding protein; LBP: LPS-binding protein; LPS: lipopolysaccharide

through it, in the disturbances of the innate and adaptive immune responses found in these patients [84].

2.4 Future Perspectives

The use of biomaterials for SCI repair is a blooming field that currently produces a vast amount of literature. However, almost all the data still comes from animal models and the field, conceived to jump into human application as soon as possible, still needs to know what biomaterials will find in the human injured spine to successfully move forward to promote neural regeneration. As shown in this chapter, the lesion type and the stage of the injury at a given time largely determine the cellular and molecular environment inside the cord and in the peripheral blood, conditioning the response of the living tissue to the material and vice versa. In future studies, more detailed information on the variety of cells and cellular states inside the cord, their surroundings and peripheral blood, the exact composition of the ECM, and the cocktail of diffusible substances forming the extracellular milieu along the time course of the injury is still warranted. Nevertheless, the current knowledge already poses useful features for biomaterial researchers, such as the existence of two separate compartments (astrocytosis vs schwannosis) in many chronically injured spinal cords, the sustained activation of the immune system in chronic stages both locally (macrophages) and peripherally (circulating white cells), and the importance of ECM composition at different stages after injury. Much progress is expected from the biomaterials field in the coming years in the path toward more efficient therapeutic approaches for SCI patients.

Acknowledgement We thank Dr. Elisa López-Dolado and Dr. Raquel Madroñero Mariscal for their help with Fig. 2.4 composition.

References

1. Alizadeh A, Dyck SM, Karimi-Abdolrezaee S (2019) Traumatic spinal cord injury: an overview of pathophysiology, models and acute injury mechanisms. Front Neurol 10:282
2. Tran AP, Warren PM, Silver J (2018) The biology of regeneration failure and success after spinal cord injury. Physiol Rev 98:881–917
3. O'Shea TM, Burda JE, Sofroniew MV (2017) Cell biology of spinal cord injury and repair. J Clin Invest 127:3259–3270
4. Brennan FH, Popovich PG (2018) Emerging targets for reprograming the immune response to promote repair and recovery of function after spinal cord injury. Curr Opin Neurol 31:334–344
5. Namimatsu S, Ghazizadeh M, Sugisaki Y (2005) Reversing the effects of formalin fixation with citraconic anhydride and heat: a universal antigen retrieval method. J Histochem Cytochem 53:3–11
6. Fleming JC, Norenberg MD, Ramsay DA et al. (2006) The cellular inflammatory response in human spinal cords after injury. Brain 129:3249–3269
7. Chang A, Nishiyama A, Peterson J et al. (2000) NG2-positive oligodendrocyte progenitor cells in adult human brain and multiple sclerosis lesions. J Neurosci 20:6404–6412
8. Paniagua-Torija B, Norenberg M, Arevalo-Martin A et al. (2018) Cells in the adult human spinal cord ependymal region do not proliferate after injury. J Pathol 246:415–421
9. Bunge RP, Puckett WR, Becerra JL et al. (1993) Observations on the pathology of human spinal cord injury. A review and classification of 22 new cases with details from a case of chronic cord compression with extensive focal demyelination. Adv Neurol 59:75–89
10. Tator CH (1995) Update on the pathophysiology and pathology of acute spinal cord injury. Brain Pathol 5:407–413
11. Kakulas BA (1999) A review of the neuropathology of human spinal cord injury with emphasis on special features. J Spinal Cord Med 22:119–24
12. Norenberg MD, Smith J, Marcillo A (2004) The pathology of human spinal cord injury: defining the problems. J Neurotrauma 21:429–440
13. Choo AM, Liu J, Liu Z et al. (2009) Modeling spinal cord contusion, dislocation, and distraction: characterization of vertebral clamps, injury severities, and node of Ranvier deformations. J Neurosci Methods 181:6–17
14. Steward O, Willenberg R (2017) Rodent spinal cord injury models for studies of axon regeneration. Exp Neurol 287:374–383
15. Domínguez-Bajo A, González-Mayorga A, López-Dolado E et al. (2020) Graphene oxide microfibers promote regenerative responses after chronic implantation in the cervical injured spinal cord. ACS Biomater Sci Eng 6:2401–2414
16. Chang HT (2007) Subacute human spinal cord contusion: few lymphocytes and many macrophages. Spinal Cord 45:174–182

17. Betz R, Biering-Sørensen F, Burns SP et al. (2019) The 2019 revision of the international standards for neurological classification of spinal cord injury (ISNCSCI)—what's new? Spinal Cord 57:815–817

18. Dimitrijevic MR, Faganel J, Lehmkuhl D, Sherwood A (1983) Motor control in man after partial or complete spinal cord injury. Adv Neurol 39:915–926

19. Kakulas BA (1999) The applied neuropathology of human spinal cord injury. Spinal Cord 37:79–88

20. Quencer RM, Bunge RP (1996) The injured spinal cord: Imaging, histopathologic, clinical correlates, and basic science approaches to enhancing neural function after spinal cord injury. Spine (Phila Pa 1976) 21:2064–2066

21. Metz GA, Curt A, van de Meent H et al. (2000) Validation of the weight-drop contusion model in rats: A comparative study of human spinal cord injury. J Neurotrauma 17:1–17

22. Freund P, Seif M, Weiskopf N et al. (2019) MRI in traumatic spinal cord injury: from clinical assessment to neuroimaging biomarkers. Lancet Neurol 18:1123–1135

23. Simard JM, Woo SK, Norenberg MD et al. (2010) Brief suppression of Abcc8 prevents autodestruction of spinal cord after trauma. Sci Transl Med 2:28ra29

24. Hayes KC, Kakulas BA (1997) Neuropathology of human spinal cord injury sustained in sports-related activities. J Neurotrauma 14:235–248

25. Kakulas BA (2004) Neuropathology: the foundation for new treatments in spinal cord injury. Spinal Cord 42:549–563

26. Tator CH, Koyanagi I (1997) Vascular mechanisms in the pathophysiology of human spinal cord injury. J Neurosurg 86:483–492

27. Zrzavy T, Schwaiger C, Wimmer I et al. (2021) Acute and non-resolving inflammation associate with oxidative injury after human spinal cord injury. Brain 144:144–161

28. Buss A, Pech K, Kakulas BA et al. (2007) Growth-modulating molecules are associated with invading Schwann cells and not astrocytes in human traumatic spinal cord injury. Brain 130:940–953

29. Buss A, Pech K, Kakulas BA et al. (2009) NG2 and phosphacan are present in the astroglial scar after human traumatic spinal cord injury. BMC Neurol 9:32

30. Yu WR, Fehlings MG (2011) Fas/FasL - mediated apoptosis and inflammation are key features of acute human spinal cord injury: implications for translational, clinical application. Acta Neuropathol 122:747–761

31. Wolman L (1967) Post-traumatic regeneration of nerve fibres in the human spinal cord and its relation to intramedullary neuroma. J Pathol Bacteriol 94:123–129

32. Ito T, Oyanagi K, Wakabayashi K, Ikuta F (1996) Traumatic spinal cord injury: A neuropathological study on the longitudinal spreading of the lesions. Acta Neuropathol 93:13–18

33. Hashizume Y, Iljima S, Kishimoto H, Hirano A (1983) Pencil-shaped softening of the spinal cord - Pathologic study in 12 autopsy cases. Acta Neuropathol 61:219–224

34. Quencer RM, Bunge RP, Egnor M et al. (1992) Acute traumatic central cord syndrome: MRI-pathological correlations. Neuroradiology 34:85–94

35. Goldstein B, Hammond MC, Stiens SA, Little JW (1998) Posttraumatic syringomyelia: Profound neuronal loss, yet preserved function. Arch Phys Med Rehabil 79:107–112

36. Brodbelt AR, Stoodley MA (2003) Post-traumatic syringomyelia: a review. J Clin Neurosci 10:401–408

37. Klekamp J (2012) Treatment of posttraumatic syringomyelia. J Neurosurg Spine 17:199–211

38. Klekamp J (2002) The pathophysiology of syringomyelia - historical overview and current concept. Acta Neurochir (Wien) 144:649–664

39. Emery E, Aldana P, Bunge MB et al. (1998) Apoptosis after traumatic human spinal cord injury. J Neurosurg 89:911–920

40. Buss A, Brook GA, Kakulas B et al. (2004) Gradual loss of myelin and formation of an astrocytic scar during Wallerian degeneration in the human spinal cord. Brain 127:34–44

41. Becerra JL, Puckett WR, Hiester ED et al. (1995) MR-pathologic comparisons of Wallerian degeneration in spinal cord injury. AJNR Am J Neuroradiol 16:125–133

42. Ackery AD, Norenberg MD, Krassioukov A (2007) Calcitonin gene-related peptide immunoreactivity in chronic human spinal cord injury. Spinal Cord 45:678–686
43. Yang L, Blumbergs PC, Jones NR et al. (2004) Early expression and cellular localization of proinflammatory cytokines interleukin-1β, interleukin-6, and tumor necrosis factor-α in human traumatic spinal cord injury. Spine (Phila Pa 1976) 29:966–971
44. Schmitt AB, Buss A, Breuer S et al. (2000) Major histocompatibility complex class II expression by activated microglia caudal to lesions of descending tracts in the human spinal cord is not associated with a T cell response. Acta Neuropathol 100:528–536
45. Kwon BK, Stammers AMT, Belanger LM et al. (2010) Cerebrospinal fluid inflammatory cytokines and biomarkers of injury severity in acute human spinal cord injury. J Neurotrauma 27:669–682
46. Kwon BK, Streijger F, Fallah N et al. (2017) Cerebrospinal fluid biomarkers to stratify injury severity and predict outcome in human traumatic spinal cord injury. J Neurotrauma 34:567–580
47. Buss A, Pech K, Kakulas B et al. (2008) TGF-β1 and TGF-β2 expression after traumatic human spinal cord injury. Spinal Cord 46:364–371
48. Buss A, Pech K, Kakulas BA et al. (2007) Matrix metalloproteinases and their inhibitors in human traumatic spinal cord injury. BMC Neurol 7:17
49. Scholtes F, Adriaensens P, Storme L et al. (2006) Correlation of postmortem 9.4 tesla magnetic resonance imaging and immunohistopathology of the human thoracic spinal cord 7 months after traumatic cervical spine injury. Neurosurgery 59:671–678
50. Puckett WR, Hiester ED, Norenberg MD et al. (1997) The astroglial response to Wallerian degeneration after spinal cord injury in humans. Exp Neurol 148:424–432
51. Bruce JH, Norenberg MD, Kraydieh S et al. (2000) Schwannosis: role of gliosis and proteoglycan in human spinal cord injury. J Neurotrauma 17:781–788
52. Guest JD, Hiester ED, Bunge RP (2005) Demyelination and Schwann cell responses adjacent to injury epicenter cavities following chronic human spinal cord injury. Exp Neurol 192:384–393
53. González P, González-Fernández C, Campos-Martín Y et al. (2020) Spatio-temporal and cellular expression patterns of PTK7 in the healthy and traumatically injured rat and human spinal cord. Cell Mol Neurobiol 40:1087–1103
54. González P, González-Fernández C, Campos-Martín Y et al. (2020) Frizzled 1 and Wnt1 as new potential therapeutic targets in the traumatically injured spinal cord. Cell Mol Life Sci 77:4631–4662
55. Buss A, Sellhaus B, Wolmsley A et al. (2005) Expression pattern of NOGO-A protein in the human nervous system. Acta Neuropathol 110:113–119
56. Buss A, Pech K, Merkler D et al. (2005) Sequential loss of myelin proteins during Wallerian degeneration in the human spinal cord. Brain 128:356–364
57. Hughes JT, Brownell B (1963) Aberrant nerve fibres within the spinal cord. J Neurol Neurosurg Psychiatry 26:528–534
58. Wang ZH, Walter GF, Gerhard L (1996) The expression of nerve growth factor receptor on Schwann cells and the effect of these cells on the regeneration of axons in traumatically injured human spinal cord. Acta Neuropathol 91:180–184
59. Cawsey T, Duflou J, Weickert CS, Gorrie CA (2015) Nestin-positive ependymal cells are increased in the human spinal cord after traumatic central nervous system injury. J Neurotrauma 32:1393–1402
60. Schwab JM, Zhang Y, Kopp MA et al. (2014) The paradox of chronic neuroinflammation, systemic immune suppression, autoimmunity after traumatic chronic spinal cord injury. Exp Neurol 258:121–129
61. Kopp MA, Druschel C, Meisel C et al. (2013) The SCIentinel study—prospective multicenter study to define the spinal cord injury-induced immune depression syndrome (SCI-IDS)—study protocol and interim feasibility data. BMC Neurol 13:168
62. Davies AL, Hayes KC, Dekaban GA (2007) Clinical correlates of elevated serum concentrations of cytokines and autoantibodies in patients with spinal cord injury. Arch Phys Med Rehabil 88:1384–1393

63. Beck KD, Nguyen HX, Galvan MD et al. (2010) Quantitative analysis of cellular inflammation after traumatic spinal cord injury: evidence for a multiphasic inflammatory response in the acute to chronic environment. Brain 133:433–447
64. Ankeny DP, Lucin KM, Sanders VM et al. (2006) Spinal cord injury triggers systemic autoimmunity: Evidence for chronic B lymphocyte activation and lupus-like autoantibody synthesis. J Neurochem 99:1073–1087
65. Bracchi-Ricard V, Zha J, Smith A et al. (2016) Chronic spinal cord injury attenuates influenza virus-specific antiviral immunity. J Neuroinflammation 13:125
66. Brommer B, Engel O, Kopp MA et al. (2016) Spinal cord injury-induced immune deficiency syndrome enhances infection susceptibility dependent on lesion level. Brain 139:692–707
67. Kopp MA, Watzlawick R, Martus P et al. (2017) Long-term functional outcome in patients with acquired infections after acute spinal cord injury. Neurology 88:892–900
68. Hayes KC, Hull TCL, Delaney GA et al. (2002) Elevated serum titers of proinflammatory cytokines and CNS autoantibodies in patients with chronic spinal cord injury. J Neurotrauma 17:753–761
69. Frost F, Roach MJ, Kushner I, Schreiber P (2005) Inflammatory C-reactive protein and cytokine levels in asymptomatic people with chronic spinal cord injury. Arch Phys Med Rehabil 86:312–317
70. Bank M, Stein A, Sison C et al. (2015) Elevated circulating levels of the pro-inflammatory cytokine macrophage migration inhibitory factor in individuals with acute spinal cord injury. Arch Phys Med Rehabil 96:633–644
71. Segal JL, Gonzales E, Yousefi S et al. (1997) Circulating levels of IL-2R, ICAM-1, and IL-6 in spinal cord injuries. Arch Phys Med Rehabil 78:44–47
72. Sambrano GR, Steinberg D (1995) Recognition of oxidatively damaged and apoptotic cells by an oxidized low density lipoprotein receptor on mouse peritoneal macrophages: role of membrane phosphatidylserine. Proc Natl Acad Sci USA 92:1396–1400
73. Khallou-Laschet J, Varthaman A, Fornasaet G et al. (2010) Macrophage plasticity in experimental atherosclerosis. PLoS One 5:e8852
74. Pello OM, Silvestre C, De Pizzol M, Andrés V (2011) A glimpse on the phenomenon of macrophage polarization during atherosclerosis. Immunobiology 216:1172–1176
75. Franceschi C, Garagnani P, Vitale G et al. (2017) Inflammaging and 'Garb-aging'. Trends Endocrinol Metab 28:199–212
76. Rawji KS, Mishra MK, Michaels NJ et al. (2016) Immunosenescence of microglia and macrophages: impact on the ageing central nervous system. Brain 139:653–661
77. Jackaman C, Tomay F, Duong L et al. (2017) Aging and cancer: the role of macrophages and neutrophils. Ageing Res Rev 36:105–116
78. Alvarez-Mon MA, Gómez AM, Orozco A et al. (2017) Abnormal distribution and function of circulating monocytes and enhanced bacterial translocation in major depressive disorder. Front Psychiatry 10:812
79. Stein A, Panjwani A, Sison C et al. (2013) Pilot study: elevated circulating levels of the proinflammatory cytokine macrophage migration inhibitory factor in patients with chronic spinal cord injury. Arch Phys Med Rehabil 94:1498–1507
80. Farkas GJ, Gorgey AS, Dolbow DR, Berg AS, Gater DR (2018) The influence of level of spinal cord injury on adipose tissue and its relationship to inflammatory adipokines and cardiometabolic profiles. J Spinal Cord Med 41:407–415
81. Dumitriu IE, Araguás ET, Baboonian C, Kaski JC (2009) CD4$^+$CD28^{null} T cells in coronary artery disease: when helpers become killers. Cardiovasc Res 81:11–19
82. Kigerl KA, Gensel JC, Ankeny DP et al. (2009) Identification of two distinct macrophage subsets with divergent effects causing either neurotoxicity or regeneration in the injured mouse spinal cord. J Neurosci 29:13435–13444
83. Murray PJ (2018) Immune regulation by monocytes. Semin Immunol 35:12–18
84. Diaz D, Lopez-Dolado E, Haro S et al. (2021) Systemic inflammation and the breakdown of intestinal homeostasis are key events in chronic spinal cord injury patients. Int J Mol Sci 22:744

85. Campagnolo DI, Bartlett JA, Keller SE, Sanchez W, Oza R (1997) Impaired phagocytosis of Staphylococcus aureus in complete tetraplegics. Am J Phys Med Rehabil 76:276–280
86. Cruse JM, Lewis RE, Bishop GR et al. (1993) Decreased immune reactivity and neuroendocrine alterations related to chronic stress in spinal cord injury and stroke patients. Pathobiology 61:183–192
87. Cruse JM, Lewis RE, Bishop GR et al. (1992) Neuroendocrine-immune interactions associated with loss and restoration of immune system function in spinal cord injury and stroke patients. Immunol Res 11:104–116
88. Zha J, Smith A, Andreansky S et al. (2014) Chronic thoracic spinal cord injury impairs CD8$^+$ T-cell function by up-regulating programmed cell death-1 expression. J Neuroinflammation 11:65
89. Monahan R, Stein A, Gibbs K, Bank M, Bloom O (2015) Circulating T cell subsets are altered in individuals with chronic spinal cord injury. Immunol Res 63:3–10
90. Kil K, Zang YC, Yang D, Markowski J et al. (1999) T cell responses to myelin basic protein in patients with spinal cord injury and multiple sclerosis. J Neuroimmunol 98:201–207
91. Mizrachi Y, Ohry A, Aviel A et al. (1983) Systemic humoral factors participating in the course of spinal cord injury. Spinal Cord 21:287–293
92. Taranova NP, Makarov AI, Amelina OA et al. (1992) The production of autoantibodies to nerve tissue glycolipid antigens in patients with traumatic spinal cord injuries. Zh Vopr Neirokhir Im N N Burdenko 4–5:21–24
93. Palmers I, Ydens E, Put E et al. (2016) Antibody profiling identifies novel antigenic targets in spinal cord injury patients. J Neuroinflammation 13:243
94. Hergenroeder GW, Moore AN, Schmitt KM et al. (2016) Identification of autoantibodies to glial fibrillary acidic protein in spinal cord injury patients. Neuroreport 27:90–93
95. Zajarías-Fainsod D, Carrillo-Ruiz J, Mestre H et al. (2012) Autoreactivity against myelin basic protein in patients with chronic paraplegia. Eur Spine J 21:964–970
96. Arevalo-Martin A, Grassner L, Garcia-Ovejero D et al. (2018) Elevated autoantibodies in subacute human spinal cord injury are naturally occurring antibodies. Front Immunol 9:2365
97. Nance DM, Sanders VM (2007) Autonomic innervation and regulation of the immune system (1987–2007). Brain Behav Immun 21:736–745
98. Pavlov VA, Tracey KJ (2017) Neural regulation of immunity: molecular mechanisms and clinical translation. Nat Neurosci 20:156–166
99. Campagnolo DI, Bartlett JA, Keller SE (2000) Influence of neurological level on immune function following spinal cord injury: a review. J Spinal Cord Med 23:121–128
100. Ueno M, Ueno-Nakamura Y, Niehaus J et al. (2016) Silencing spinal interneurons inhibits immune suppressive autonomic reflexes caused by spinal cord injury. Nat Neurosci 19:784–787
101. Zhang Y, Guan Z, Reader B et al. (2013) Autonomic dysreflexia causes chronic immune suppression after spinal cord injury. J Neurosci 33:12970–12981
102. Pukos N, Goodus MT, Sahinkaya FR et al. (2019) Myelin status and oligodendrocyte lineage cells over time after spinal cord injury: what do we know and what still needs to be unwrapped? Glia 67:2178–2202

Chapter 3
Tailoring 3D Biomaterials for Spinal Cord Injury Repair

André F. Girão, Joana Sousa, Mónica Cicuéndez, María Concepción Serrano, María Teresa Portolés, and Paula A. A. P. Marques

Abstract Spinal cord injury (SCI), with either traumatic or non-traumatic aetiology, brings lifetime health, economic and social consequences to thousands of people worldwide. Tragically, there are no available therapies capable of reversing the condition of SCI patients, who experience their daily routines becoming nearly impossible tasks due to the abrupt decrease in their mobility and independence. During the last decades, biomaterials have continuously been tested as central players for a wide range of SCI regenerative strategies, particularly the development of highly biocompatible 3D tissue-engineered scaffolds proficient to bridge the lesion site. Importantly, the clinical success of such constructs deeply relies on the generation of functional neural circuits that resemble the spinal cord network. In this chapter, we overview the most promising methodologies for tailoring biomaterials towards the recreation of biochemical and biomechanical gradients capable of boosting neural cell responses in vitro and in vivo. Relevant research topics regarding scaffolding approaches such as microfabrication techniques and some functionalization strategies are presented and critically discussed. Furthermore,

A. F. Girão
TEMA, Department of Mechanical Engineering, University of Aveiro, Aveiro, Portugal

Instituto de Ciencia de Materiales de Madrid (ICMM), Consejo Superior de Investigaciones Científicas (CSIC), Madrid, Spain
e-mail: andrefgirao@ua.pt

J. Sousa · P. A. A. P. Marques (✉)
TEMA, Department of Mechanical Engineering, University of Aveiro, Aveiro, Portugal
e-mail: joanapmsousa@ua.pt; paulam@ua.pt

M. Cicuéndez · M. T. Portolés
Departamento de Bioquímica y Biología Molecular, Facultad de Ciencias Químicas, Universidad Complutense de Madrid, Instituto de Investigación Sanitaria del Hospital Clínico San Carlos and CIBER de Bioingeniería, Biomateriales y Nanomedicina, CIBER-BBN, Madrid, Spain
e-mail: mcicuendez@ucm.es; portoles@quim.ucm.es

M. C. Serrano
Instituto de Ciencia de Materiales de Madrid (ICMM), Consejo Superior de Investigaciones Científicas (CSIC), Madrid, Spain
e-mail: mc.terradas@csic.es

57

decisive parameters commonly used to assess the biocompatibility of biomaterials for SCI repair are also reviewed.

Keywords Biocompatibility · Biomaterials · Fibres · Hydrogels · Microfabrication · Neural tissue engineering

3.1 Biomaterials Criteria Applied to SCI

The complexity of the central nervous system (CNS), in general, and the spinal cord, in particular, obligates to devote an important attention to biomaterials design in order to maximize the chances to success. Advances during the last decades have proved that successful biomaterial-based approaches to repair damaged tissues and organs imply an optimal matching of their physicochemical and biological properties [1]. The spinal cord tissue is not an exemption for this rule, with particularly sophisticated features that are both unique and pivotal for the initiation of reparative mechanisms in such a detrimental scenario when injured. Contrarily to other tissues and organs in the body (e.g. skin and liver), which are characterized by a large intrinsic capacity for regeneration, nervous tissues show a minimal ability for self-repair, particularly poor in the CNS. Then, it turns crucial to incorporate features in the therapeutic biomaterials to both block inhibitory elements at the injury site and boost/initiate these limited self-reparative routes.

Based on the past and present progress in the field, main criteria that biomaterials envisioned for spinal cord tissue engineering (TE) should fulfil are listed below. For further details on the general requirements for CNS implantable biomaterials, readers are referred to Chap. 1 in this book.

3.1.1 Biocompatibility

As in any biomaterial-based therapeutics no matter the targeted tissue/organ, novel neural devices should be biocompatible, so no toxic effects are induced while eliciting desirable tissue responses for regeneration. Importantly, this requirement does not reduce importance by the presence of the highly selective barrier defined by the meninges. These membranes do not represent an impermeable frontier between the CNS and the rest of the body. As they are commonly disrupted because of the traumatic injury, elements (e.g. inflammatory components, extracellular matrix (ECM) constituents, cells, biomaterial fragments) can more freely move across it accessing an eventual systemic distribution. Moreover, the spinal cord is a highly vascularized tissue, so submicron biomaterial elements can travel through the blood stream and reach other organs. In the spinal cord scenario, biocompatibility not only means absence of direct damage to all neurons and glial, vascular and immune cells provoked by harmful interactions with the bulk biomaterial and/or its surface, but also indirect toxicity related to degradation sub-products and leachables. Based on this, biomaterials produced from natural components seem, a priori, advantageous to fulfil this biocompatibility requirement. In this sense, natural polymers such as collagen [2], chitosan [3], hyaluronic acid [4] and alginate [5] have been extensively investigated for SCI, all eliciting biocompatible responses. Nonetheless, a wide range of synthetic polymers are also proving comparable biocompatible behaviour once implanted at the injured spinal cord. Polycaprolactone (PCL) [6], poly(lactic-co-glycolide) (PLGA) [7] and methylmethacrylate [8] can be cited among the most explored. In any case, biocompatibility of biomaterials can be further improved by diverse strategies such as coating with highly biocompatible materials and functionalization with adhesive molecules, as discussed further in this chapter.

3.1.2 Mechanical Compliance

Neural tissues have been classically considered soft tissues in the body. Recent technological advances, such as those provided by the development of instruments capable of measuring mechanical properties at the nanoscale (e.g. atomic force microscopy (AFM)), have indeed proved it. For instance, the Young's modulus of pyramidal neurons is in the range of 480–970 Pa, while astrocytes display values of 300–520 Pa [9]. Fresh spinal cord slices have Young's modulus values of 97 ± 69 Pa at white matter regions and 275 ± 99 Pa at the grey matter [10]. Interestingly, damaged CNS tissues significantly soften after injury, both at the neocortex and the spinal cord [11]. Biomaterials should then match this range of tissue softness in order not to induce friction forces and mechanical stresses, which are largely responsible for augmenting scar tissue formation and biomaterial encapsulation.

3.1.3 Mimicking the 3D Architecture

It has been extensively proved that cells behave differently in 2D versus 3D systems [12, 13]. As the body is composed of 3D constituents, biomaterials should also mimic the 3D architecture of native spinal tissues to prompt optimized biological responses. In this sense, advances provided by the use of hydrogels [14] and 3D-printing techniques [15] are offering new avenues for improvement in the design of SCI biomaterials.

3.1.4 Implantability

As in any other clinical application, SCI biomaterials should be easily handled by clinicians and implantable through a relatively simple procedure in which their stability and integrity is not compromised. They should also comply with the size and shape dimensions of the injury site or, alternatively, provide an adaptable architecture in situ, so the formation of cavities is prevented. For this kind of biomaterials, suturing is not needed and their placement inside the injury site typically conveys adequate integration with the neural tissue stumps if displaying compatible physicochemical and biological properties.

3.1.5 Degradability

An optional but attractive requirement that is being frequently pursued in biomaterials for SCI is their capacity to be progressively degraded in accordance with the time frame of neural regeneration. This implies the disassembly and dissociation of the biomaterial in fragments of micron and submicron size that can be either degraded by phagocytic cells around, then entering metabolic routes, or transported to the blood stream for posterior elimination through detoxification routes at the liver or, preferentially, discarded in the urine by the kidneys if water soluble. Importantly, this requirement is closely related to the fulfilment of biocompatibility as such by-products must also satisfy being biocompatible until their final removal.

3.1.6 Electrical Conductivity

Optionally, biomaterials can incorporate electrical properties to either stimulate neural electrical signalling (for instance, by the use of an external electrical source) or boost electrical transmission between neurons within the injury site. Biomaterials under consideration to this regard include carbon-based materials (e.g. carbon nanotubes [16] and graphene [10]) and conductive polymers (e.g. poly(3,4-ethylenedioxythiophene) [17], polypyrrole (PPy) [18]), as the most extensively explored. However, despite the interest and potential utility of this type of properties, research on the use of electrically active biomaterials for SCI treatment in vivo is rare and the mechanisms behind most of the improved biological features found still poorly understood. For further details on this particular topic, readers are referred to Chap. 5 in this book.

3.2 3D biomaterials Design for Neural Repair After SCI

Biomaterials together with biochemical signals, cells and nano/microfabrication techniques are the most significant cornerstones for designing and fabricating 3D biomimetic scaffolds for TE applications. Particularly, the recreation of cellular microenvironments with similar anatomical and functional organization relatively to the spinal cord is a titanic challenge. In this way, a promising solution relies on developing new scaffolds with smart designs able to support the regeneration of neural cells and, consequently, the reconstruction of complex 3D neuronal circuits. The concept of an ideal scaffold for SCI is possible yet limited because the site, geometry and extension of the injury are highly variable, leading to outcomes that are difficult to determine/evaluate considering the high complexity of SCI physiopathology. Herein, examples of the most prominent techniques to prepare 3D biomaterials for SCI repair are described and discussed.

3.2.1 Fibrous Scaffolds

Since fibrous proteins, such as collagen, are abundant components of neural tissues, the development of biomimetic fibre-based constructs is widely used to recreate biochemical and biophysical gradients analogous to each specific cellular microenvironment present in the nervous system, including the spinal cord. Among the available methods to fabricate fibrous scaffolds (e.g. self-assembly [19], phase separation [20] and decellularization of fibrous animal tissues [21]), electrospinning is currently the most popular due not only to the morphological resemblance between electrospun fibres and the native ECM, but also to the cost-efficiency, reproducibility and versatility in controlling physical parameters and functionalization with electrochemical and biological cues.

In this way, numerous physical characteristics of the electrospun scaffolds can be tailored with the purpose of modulating cell behaviour and fate. Indeed, by tuning the solution (e.g. concentration), process (e.g. voltage and collector geometry) and environmental (e.g. temperature and humidity) parameters, it is possible to adjust the topography, diameter (from nano to micro range) and orientation of the fibres at the microscopic level and the porosity, pattern, mechanical properties and dimensionality of the final fibrous system at the macroscopic level. Complementarily, there is a wide range of natural (e.g. collagen and chitosan) and synthetic (e.g. PCL and PLGA) biomaterials that can be successfully electrospun, although desirable mechanical strength and biodegradability are usually achieved combining both [22]. These properties can be further improved by welding the fibres at their cross-points through post-processing crosslinking and thermal treatment, or by the incorporation of micro- and nanoparticles. Moreover, the introduction of particles into the polymer matrix of the electrospun fibres can confer additional functionalities such as electrical conductivity and the possibility to customize their orientation according to a magnetic field [23]. For example, by adding spherical PPy nanoparticles to the polymer matrix of polylactic acid (PLA) fibres, Shu et al. [24] were able to raise the scaffold surface conductivity by twelve orders of magnitude (Fig. 3.1) and, consequently, increase the latency and amplitude of motor evoked-potentials in rats after SCI. Typically, conventional electrospinning is limited to the fabrication of 2D membranes that are not suitable for filling the defects of the injured spinal cord. Therefore, strategies have been developed to produce 3D scaffolds and better

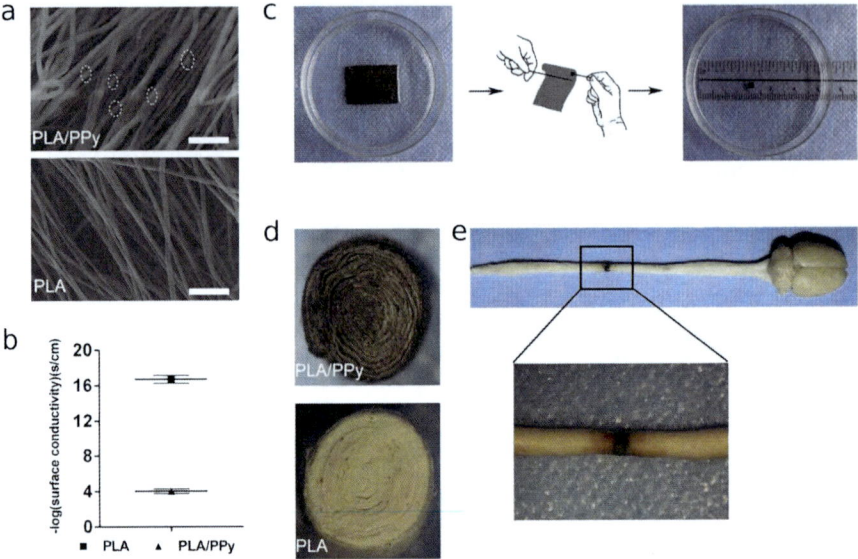

Fig. 3.1 PLA/PPy scaffold fabrication. (**a**) Micrographs of oriented PLA/PPy fibres by scanning electron microscopy. PPy particles coated on aligned PLA fibres (shown in white circles). Scale bars: 5 μm. (**b**) Surface conductivity of PLA/PPy and PLA nanofibre scaffolds. (**c**) Schematic of conduit fabrication. (**d**) Cross-sectional morphology of PLA/PPy and PLA scaffolds. (**e**) Picture of a spinal cord specimen carrying the scaffold implant. Reprinted from [24], Copyright (2018), with permission from Elsevier

mimic the ECM native environment. Those strategies comprise setup modifications (e.g. liquid reservoirs as collectors [19] and air-gap electrospinning [25]) and/or assembly by post-processing (e.g. rewinding the fibrous mat [26], rolling the mat into a conduit (Fig. 3.1) [24, 27]) and encapsulating fibres into a hydrogel [23]). In some of the above approaches, it is also possible to produce scaffolds with nanofibres extended along the vertical direction, mimicking the longitudinally anisotropic structure of the spinal cord neural tracts and, consequently, provide directional growth cues through topographic guidance. Omidinia-Anarkoli et al. [28] showed that neurite extension in hydrogels with magnetically aligned fibres was 21% and 55% higher compared to hydrogels with randomly oriented fibres and hydrogels alone, respectively. Additionally, neural cells exhibited spontaneous electrical activity along the anisotropy axis of the aligned hydrogels while in random hydrogels the propagation was multidirectional. It is also important to notice that scaffolds with aligned fibres promoted a faster and more robust cellular infiltration and vascularization comparatively to random orientated fibres, leading to an enhanced restoration of neuronal networks together with an improved locomotor recovery [19–25].

Importantly, the high surface-area-to-volume ratio and encapsulation capability of electrospun fibres make them ideal for the functionalization with biomolecules and drugs [29]. Indeed, such biochemical cues can be integrated onto the surface (e.g. via physical adsorption, electrostatic interaction and covalent grafting) or into the polymer matrix (e.g. emulsion and coaxial electrospinning) of the electrospun fibres with the purpose of being, respectively, immediately available for cell receptors or released as the fibres degrade. Likewise, growth factors, signal inhibitors and even cells can also be loaded as cues. In particular, Colello et al. [25] have successfully encapsulated nerve growth factor and chondroitinase ABC into electrosprayed alginate microspheres, before adding them to the polydioxanone electrospinning solution. The resulting electrospun scaffold reduced cell death of neurons with injured axons and mitigated the inhibitory effect of proteoglycans on neurite outgrowth in vivo.

Although electrospinning is currently boosting a wide range of strategies for neural TE applications, including the regeneration of the injured spinal cord, some issues of this microfabrication technique have yet to be addressed in the next couple of years. For example, biological materials added to the working solution can be deactivated during the electrospinning process due to the high density of charges, high voltage applied and use of organic solvents [30].

3.2.2   Hydrogels

Hydrogels are 3D polymeric networks capable of using their intrinsic water-retaining capacities to efficiently boost the regeneration of injured tissues. The popularity of this class of biomedical platforms is intimately related to their tunable characteristics, as the selection of particular synthesis parameters (e.g. polymer origin and crosslinking degree) directly predicts important features of the final hydrogel (e.g. biocompatibility and biodegradability rate) [31, 32]. Relatively to therapeutic strategies aiming SCI repair, hydrogels can be designed to either moderate the effects of the damage and inflammatory cascade initiated immediately after trauma—neuroprotective approach—or support the restoration of the damaged neural circuits during the later stages of the disease—neuroregenerative approach [33].

Considering that traumatic events are the major cause of SCI, injectable hydrogels are gaining interest as therapeutic agents mainly due to their less-invasive character and superior ability to match the irregular geometry of the lesion. By encouraging the concept of in situ gelation, the synthesis of these hydrogels often involves the physical crosslinking of polymer chains according to specific conditions of the injured microenvironment (e.g. ionic interactions and temperature). For example, it was hypothesized that, during the secondary injury stage, the neurotoxic effects associated with the abrupt increase of extracellular Ca^{2+} ions could be blocked, as the cations were used to trigger the gelation of a biocompatible alginate-based composite also responsible for enhancing the astrocyte response [34]. Alternatively, a highly functional injectable hydrogel was not only capable of protecting patient-derived Schwann cells during transplantation into the lesion site, but also ensuring their attachment and spreading within the target area after in situ thermo-responsive crosslinking [35]. Following this trend, hydrogels with temperature-dependent sol-gel transition properties are progressively becoming attractive candidates to integrate SCI therapies. Indeed, engineering polymer systems proficient to transit from a solution to a gel state at body temperature could successfully bridge cystic cavities, boosting a synergetic remodelling of the ECM and, consequently, assisting a significant functional recovery [36]. The sol-gel behaviour of thermosensitive hydrogels could be also used to precisely control the release of biomolecules responsible for minimizing the neurotoxic impact of the inflammatory cascade triggered by SCI [37].

Yet, it is noteworthy to mention that the promising results of injectable hydrogels are also commonly associated with some drawbacks, including limited control on both morphological and mechanical properties of the implanted constructs. Therefore, some studies have suggested that nanofibres could be used to counterbalance the lack of structural elements of hydrogels. For instance, hydrophobic poly(D,L-lactide) nanofibres wrapped around an injectable hydrogel were able to ensure a restrict delivery of nanoparticles to the surfaces contacting the bone, enhancing spinal fusion [38]. In a different approach, shape-memory poly(D,L-lactic acid-co-trimethyl carbonate) nanofibres incorporated within a gelatin composite hydrogel were capable of providing biomimetic topographic features to encourage and guide neurite outgrowth [39]. Alternatively, functionalized PCL nanofibres were successfully combined with hyaluronic acid molecules to generate a biomimetic nanofibre-containing hydrogel proficient to enhance macrophage response, angiogenesis, neurogenesis and axon growth (Fig. 3.2) [40].

Diverging from injectable hydrogels, an accurate design and gelation of 3D systems before surgical implantation could guarantee the addition of topographical cues and/or suitable mechanical reinforcement. For example, a 3D alginate hydrogel presenting a capillary structure shaped from a mould ensured effective cell transplantation and delivery of neurotrophic factors within the lesion area and mediated a highly orientated axonal regeneration [41]. Alternatively, by controlling the photo-crosslinking degree of a 3D gelatin methacrylate hydrogel, it was possible to customize its mechanical properties and, consequently, to influence the differentiation process of the encapsulated neural stem cells into either neuronal (soft hydrogels) or glial (stiff hydrogels) cell lineages towards spinal cord regeneration [42].

Overall, hydrogels present indisputable design versatility and a tremendous capability to mimic important biochemical and biophysical gradients of the natural spinal cord microenvironment. Therefore, these TE scaffolds are currently frontrunners to unlock viable therapeutic strategies for reversing the tragic sequence of pathophysiological events triggered by SCI.

3.2.3 Porous Scaffolds

The porous architecture and consequent mechanical properties of bioengineered scaffolds designed for SCI are two crucial factors for the guidance of regenerating

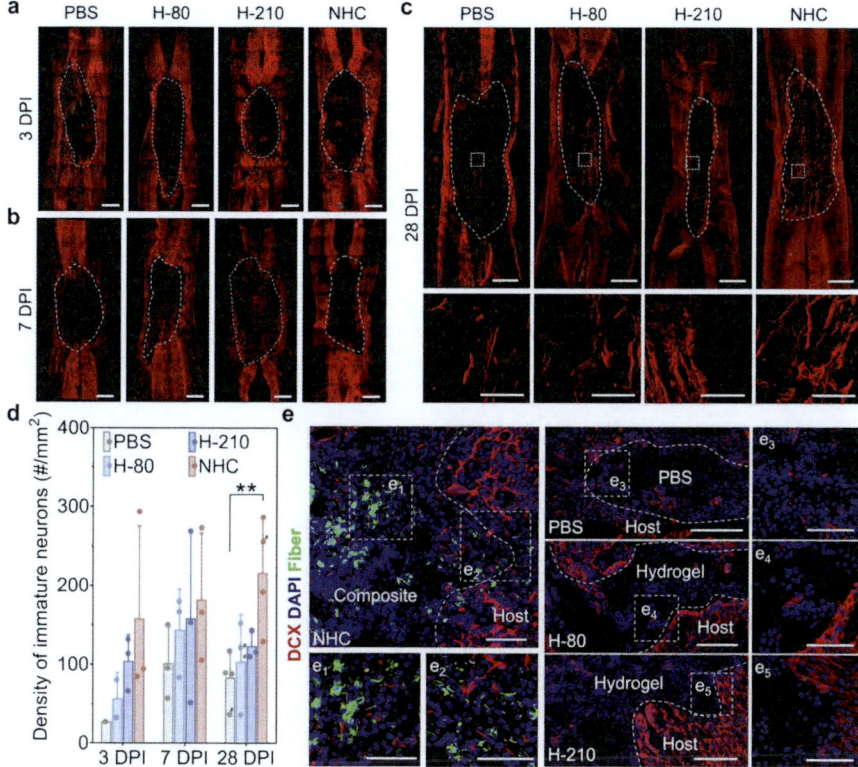

Fig. 3.2 An injectable nanofibre-hydrogel composite (NHC) stimulated neurogenesis in the injury site. Microphotographs showing immature neurons stained for βIII-tubulin at the injury site and around it in representative horizontal sections from each treatment group at 3 (**a**), 7 (**b**) and 28 (**c**) days post-injury (dpi). Scale bar: 200 μm. Representative area for each group at 28 dpi is shown in high magnification in c. Scale bar: 100 μm. (**d**) Bar graph showing the average density of immature neurons per mm^2 in the injury for each experimental group at 3, 7 and 28 dpi ($n = 3$; $p < 0.01^{**}$). Bars represent standard deviation. (**e**) Microphotographs showing neural precursor cells stained for doublecortin (DCX, red) in the injury and surroundings in horizontal sections from each treatment group at 28 dpi. PCL fibres display green fluoresce due to F8BT. Sections were counterstained with DAPI (blue). 4 experimental groups were tested: NHC (G' = 210 Pa; $n = 25$), hyaluronic acid hydrogel-210 (HA-210; G' = 210 Pa; $n = 25$), hyaluronic acid hydrogel-80 (HA-80; G' = 80 Pa; $n = 25$) and phosphate-buffered saline (PBS, 0.1 M, pH 7.4; $n = 25$). For each group, inserts are depicted in high magnification images. Scale bar: 100 μm. Reprinted from [40], Copyright (2020), with permission from Elsevier

axons through the injury site. It has been largely evidenced that, besides chemical cues, physical ones play an important role in promoting axon outgrowth and elongation [43]. Therefore, the design of 3D scaffolds with porous interconnected architectures should replicate some of the most important functions of the native ECM and vasculature, allowing an enhanced dispersion of cells as well as an efficient exchange of nutrients and waste products. However, this is a challenge since there is a sensible dependence between the presence of pores (e.g. holes and channels) and the structural/mechanical integrity required to guarantee the functionality of the scaffold. Indeed, the creation of hierarchical porous features from the nanometre to millimetre scales will determine how well the scaffold satisfies mechanical compliance and mass-transport needs. Material chemistry, together with processing techniques, defines the maximum functional properties that a scaffold can achieve, as well as how cells can interact with it [44].

Ice-templating techniques are upfront methodologies to prepare 3D porous scaffolds based on the freezing of a water-based solution/suspension followed by lyophilization. The final structure, formed during the freezing stage, can be either isotropic or anisotropic. Therefore, the control of ice nucleation and growth is of utmost importance to tailor the scaffold structure. Complete reviews concerning the freezing step before lyophilization can be found in the literature [45, 46], where the physics of ice solidification and the numerous variables which can be reworked during the process are explained in detail. For example, the construction of aligned porous architectures through directional freezing is an interesting option for neural TE applications. For this, freezing temperature gradients are applied in one certain direction, leading to the generation of regular ice columns within the samples [47–49]. Many water-soluble natural and synthetic polymers may be used to prepare aligned porous scaffolds using these techniques [50–53].

The gas foaming-based fabrication process is another approach to fabricate a suitable accommodating cellular environment for axon extension by creating a bridging architecture with channels and porosity [54, 55]. In fact, gas foaming methods minimize the use of organic solvents, and the process can be carried out at low temperatures [56]. The use of supercritical CO_2 foaming is a recent valuable and widespread choice to design and fabricate 3D porous bioactive scaffolds [57–59].

Alternatively, sacrificial templating approaches aim to overcome the technical challenge to maximally retain in the scaffolds the channel structures present in native tissues [43, 60–62]. Microarchitecture templates are used as support sacrificial polymers, usually extruded as fibres of combined diameters [63, 64]. Combinatorial technical approaches such as tubes filled with porous hydrogels are able to provide uniaxial tissue growth, a feature that is essential for neural regeneration following SCI [65].

3.2.4 3D Printed Scaffolds

3D bioprinting is considered a top technology for the production of biologically complex microstructures providing control over shape and microarchitecture across conventional manufacturing techniques [66]. This bio-fabrication strategy allows to precisely position biological elements, including living cells and ECM components, combined and used synergistically, in a specified 3D hierarchical organization and to create artificial multi-cellular tissues/organs through computer-aided design/computer-aided manufacturing (CAD/CAM) [67]. There are different 3D bioprinting strategies depending on their fundamental working principles for building functional scaffolds, namely inkjet, laser-assisted, pressure-assisted (extrusion), acoustic bioprinting, stereolithography, fused deposition modelling and magnetic bioprinting [67]. These approaches can be used alone or in combination to achieve the desired additive manufacturing objective. Additionally, 4D printing is proving to be a very promising microfabrication methodology by enabling the production of dynamic 3D structures able to change their shape/colour, produce an electrical current, become bioactive and perform an intended function in response to an external stimulus [68–70]. In this way, besides having the advantages of 3D printing, such as the production of scaffolds with well-defined internal organization, 4D printing benefits from the capability of smart materials to closely imitate the dynamic behaviour of natural tissues against natural stimuli [71].

Still, strategies to engineer structures of the CNS are tricky due to the difficulty to replicate such a heterogeneous tissue with multiple cell types and complex architectures [15, 72]. Among the major challenges, the bioink choice is of extreme importance. Specifically, it should show biocompatibility, cell adhesiveness and printability and impart appropriate mechanical properties to the printed scaffolds [73]. Having this into account, hydrogels are considered the primary candidates for bioinks since they can create 3D hydrophilic polymer networks recapitulating several features of the natural cellular microenvironment, allowing an efficient and homogeneous cell seeding in a highly hydrated mechanically supportive 3D environment [74, 75].

A remarkable study concerning 3D printing was recently reported by Koffler et al. [15], who developed a microscale continuous projection printing method (μCPP) to fabricate 3D biomimetic hydrogel scaffolds with the dimensions of the rodent spinal cord and scalable to human spinal cord sizes and lesion geometries. These μCPP 3D-printed polyethylene glycol diacrylate–gelatin methacrylate scaffolds were loaded with neural progenitor cells (NPCs) before implantation in a complete spinal cord transection model. Notably, results showed that the injured host axons were capable of regenerating through the scaffolds and stablish synapses on the seeded NPCs. Complementarily, NPCs-derived axons were able to extend out of the scaffold, outgrowing towards the host spinal cord below the injury to restore synaptic transmission and significantly improve functional outcomes. Another pioneer strategy targeting SCI was recently developed by Kaplan et al. [76], who were able to successfully use an anatomical Magnetic Resonance Imaging (MRI) scan to precisely reconstruct the lesion site and, posteriorly, personalize the 3D printing of a PLA/PLGA microchannel scaffold according to the specific geometry and size of the injury. Results showed that, after 4 weeks in vivo, these scaffolds induced the formation of new tissue, including the growth and guidance of viable axons into the aligned microchannels (Fig. 3.3).

Taking all this into account, biomimetic scaffolds fabricated by 3D printing are promising entities to boost CNS regeneration in the context of personalized medicine.

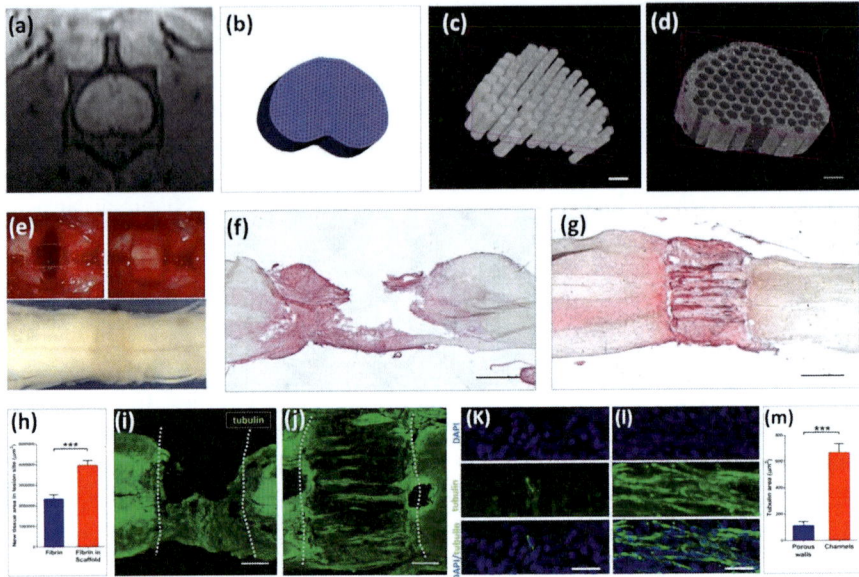

Fig. 3.3 Exemplary scaffold integration in vivo. (**a**) Coronal MRI section of a healthy segment of the spinal cord caudal to the lesion site used to perform a 3D anatomical reconstruction of the cavity (**b**). (**c**) μCT scan of the printed linear array pattern designed according to the MRI scan. (**d**) Resulting anatomical PLA/PLGA scaffolds constructed based on the MRI scan. (**e**) Scaffold implantation: Complete spinal cord transection (upper left), scaffold implantation in the lesion site (upper right) and ventral aspect of the explant two weeks post implantation (below). (**f, g**) Haematoxylin and eosin stained images of the injured spinal cord 4 weeks after implantation of a fibrin gel (**f**) and a fibrin-loaded scaffold (**g**). (**h**) Quantification of new tissue area in the lesion between the transected stumps of the spinal cord (***$p < 0.001$). (**i, j**) Immuno-labelling of regenerating axons labelled with anti-βIII tubulin staining. Images show injured spinal cords injected with fibrin (**i**) or implanted with scaffolds (**j**). Dashed lines indicate the borders of the injury. (**k, l**) High magnification images of βIII tubulin staining and DAPI inside either porous walls of the scaffold (**k**) or guidance channels (**l**). (**m**) Quantification of βIII tubulin area in porous walls and channels (***$p < 0.001$). Scale bars: 500 μm (**c, d, i,** and **j**), 1 mm (**f** and **g**) and 30 μm (**k** and **l**). Reprinted from [76], Copyright (2020), with permission from Elsevier

3.3 Biocompatibility Assessment of Biomaterials for SCI Repair

The use of 3D biomaterials for repairing injured neural structures and restoring neurological functions after SCI requires the previous evaluation of their biocompatibility. As D. F. Williams redefined in 2008 [77],

> *"biocompatibility refers to the ability of a biomaterial to perform its desired function with respect to a medical therapy, without eliciting any undesirable local or systemic effects in the recipient or beneficiary of that therapy, but generating the most appropriate beneficial cellular or tissue response in that specific situation, and optimizing the clinically relevant performance of that therapy."*

Thus, the biocompatibility assessment of biomaterials for SCI repair makes it necessary to consider two essential factors: I) the type of cells and tissues with which the biomaterial will come into contact in this scenario and II) the processes to be evaluated to know the biomaterial effects on these cells/tissues that allow to demonstrate the absence of cytotoxicity and the induction of appropriate responses.

Regarding cell/tissue types with which the implanted biomaterial will come in contact after SCI, fibroblasts, neuroglia and vascular cells such as pericytes can rapidly contribute to the formation of the fibroglial scar, inhibiting axon regeneration and acting as a mechanical and chemical barrier, as extensively described in previous chapters. In addition, damaged cells, nerves and blood vessels release biomolecules that can also induce secondary injuries in the spinal cord tissue by damaging healthy neurons and oligodendrocytes [78]. Importantly, macrophages are involved in the immune response after SCI and can play both beneficial and detrimental roles depending on their functional plasticity. These cells present a spectrum of phenotypes characterized by specific cell surface markers, genes, cytokines, chemokines and enzymes related to their functions. Figure 3.4 shows the macrophage phenotypes involved in either injury or repair responses after SCI. Thus, M1 macrophages promote secondary injury and moderate axon growth, but M2 macrophages are not neurotoxic and can promote repair by long-distance axon growth [79, 80]. Kokaia et al. have evidenced that the cross-talk between neural stem cells and immune cells is necessary for a reparative response because neural stem

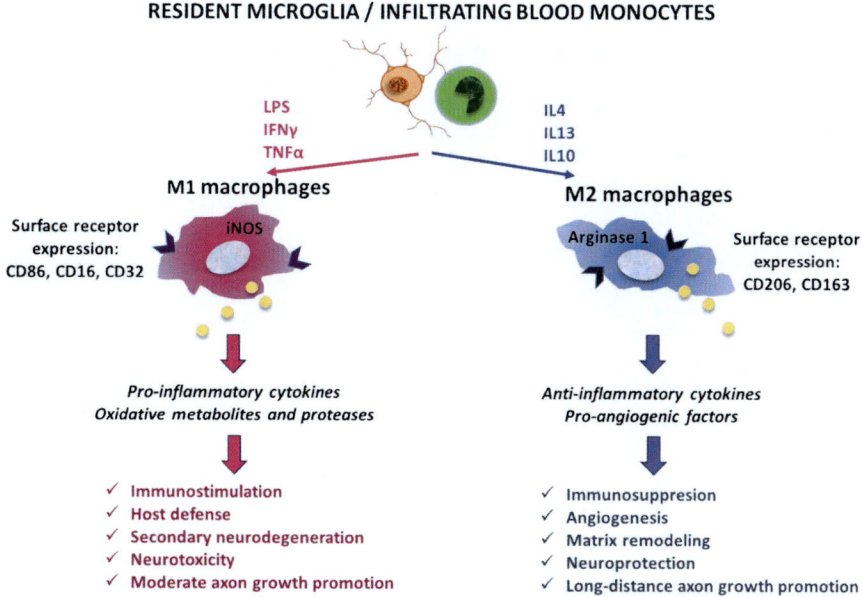

Fig. 3.4 Macrophage phenotypes involved in either injury or repair responses after SCI

cells are involved in cell replacement but also in immunomodulation and trophic support [81].

The cellular activities to be evaluated in order to know the biomaterial effects on all these cell types after its implantation for SCI repair can be classified in: (a) cellular processes which are common to all cell types such as proliferation, apoptosis/necrosis and oxidative stress and (b) more specific cellular responses such as neural differentiation and macrophage polarization.

The main parameters analysed for the biocompatibility assessment of different biomaterials designed to regenerate the neural tissue after SCI are described below. As discussed in detail previously in this chapter and elsewhere throughout the book, it is worth to note that the physicochemical properties of biomaterials largely influence cellular responses. In general terms, SCI treatment mainly requires inhibiting apoptosis and necrosis, minimizing local inflammatory reactions and reducing glial scar formation. In order to enhance the regenerative response after contact with biomaterials designed for SCI repair, different aspects are currently being optimized regarding mechanical properties, cell adhesiveness, biodegradability, electrical activity and topography of different kinds of scaffolds (Fig. 3.5) [78].

As discussed in Chap. 5, the application of conductive biomaterials is a promising approach to increase neurite extension and axonal regeneration. In this sense, fibrous scaffolds with conductive properties have been recently designed to transmit physiological electrical signals and to aid SCI repair. For example, in work reported by Shu et al., the biocompatibility parameters evaluated, after nanofibrous scaffold implantation, were apoptosis and autophagy in neural cells, activation of astrocytes and axonal regeneration through the use of immunofluorescent markers in the injury [24]. Apoptotic cells in the injured spinal cord were identified under fluorescence

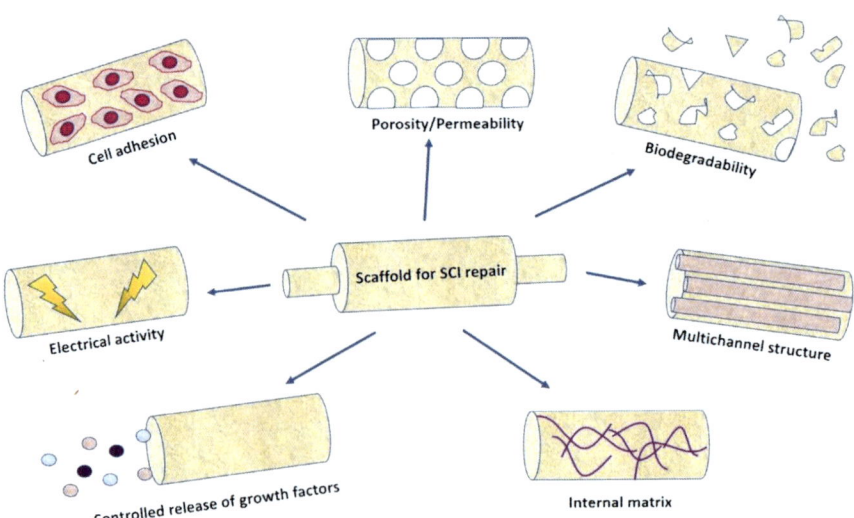

Fig. 3.5 Key aspects in the design of biocompatible scaffolds for SCI repair

microscopy after terminal deoxynucleotidyl transferase dUTP nick end labelling (TUNEL) and the expression of caspase 3 by western blotting. To know the scaffold effects on the autophagy of neural cells after implantation, LC3B$^+$ cells were detected by immunofluorescent staining and western blot of LC3 and Beclin1 proteins expression. The immunofluorescent staining of glial fibrillary acidic protein (GFAP) and neurofilament 200 (NF-200) was performed to detect astrocyte activation and axonal regeneration, respectively. Autophagy plays important roles in neuronal cell survival and function. In this cellular degradation pathway, cytosolic components are sequestered in autophagosomes and degraded upon their fusion with lysosomal components. Several findings evidence that the imbalance of autophagic flux leads to neurodegeneration, so the abnormal accumulation of autophagosomes in neurons has been related to neurodegenerative diseases [82]. In studies by Shu et al. [24], nanofibrous PLA/PPy scaffolds significantly increased axonal regeneration in the injury and alleviated secondary tissue damage by reducing apoptosis and autophagy in neural cells and decreasing the activation of astrocytes.

Alternatively, drug loading in scaffolds designed for the treatment of SCI is a strategy for achieving more effective therapeutic effects. Recently, Zhang et al. carried out the encapsulation of the glucocorticoid methylprednisolone (MP) in PCL/polysialic acid (PSA) nanofibre scaffolds for SCI treatment [83]. The biocompatibility of this PCL/PSA/MP nanofibre scaffold was first evaluated in vitro, followed by its in vivo implantation in a rat spinal cord transection model to study nerve regeneration and functional recovery. Primary astrocytes and the SH-SY5Y human neuroblastoma cell line were used as in vitro models of neuronal function and differentiation. The anti-inflammatory and anti-apoptotic effects induced by these nanofibre scaffolds were analysed in vivo through the secretion of pro-inflammatory cytokines, as tumour necrosis factor (TNF-α) and interleukin 6 (IL-6), and the expression of caspase-3, GFAP and NF-200, respectively.

In a different approach, PLLA nanofibrous scaffolds in combination with graphene oxide (GO) nanosheets have been designed for nerve regeneration by combining GO physicochemical properties with the topological features of these scaffolds, able to direct cell migration, proliferation and differentiation and to induce neurite outgrowth [84]. Regarding the biocompatibility assessment of these aligned scaffolds coated with GO, rat Schwann cells and PC-12 cells were cultured in the presence of nerve growth factor (NGF) on these biomaterials to evaluate their cytocompatibility and their capacity to promote cell differentiation and neurite outgrowth.

Other kind of structures such as 3D flexible and porous scaffolds composed of partially reduced GO have been implanted in the injured rat spinal cord showing the absence of local and systemic toxic responses at the subacute state and after their chronic implantation at the lesion site [85, 86]. A very complete study of the biocompatibility of these scaffolds was carried out including immunofluorescence staining of specific markers to identify the nature of both the cells infiltrating the scaffold and those forming the interface tissue in comparison to injured rats without implants and non-lesion spinal areas. Thus, MAP-2 and tau were used as markers for neurons (MAP-2 for somas and dendrites in neurons, tau for axons);

vimentin for glial cells and connective tissue cells; Iba1 for microglia; ED1 for macrophages; platelet-derived growth factor receptor β (PDGFRβ) for pericytes, precursors of oligodendrocytes and connective tissue cells such as fibroblasts and smooth muscle cells; and GFAP for astrocytes [85]. In order to detect different macrophage phenotypes, CD80 and CD86 were used as markers of M1 pro-inflammatory macrophages and CD163 and arginase I for M2 reparative ones. For immunofluorescence detection of neuronal axons, SMI-311 (cocktail for neurofila-ments) and βIII tubulin (playing a critical role in axon guidance and maintenance) were used. Mature blood vessels were evidenced with specific antibodies (SMI-71 and RECA-1) for endothelial cells in blood–brain, blood–spinal cord and blood–nerve barriers [86].

Regarding other biomaterials designed to direct the re-organization of cells into the injured spinal cord tissue, hydrogels have mechanical properties similar to those of soft tissues and can be prepared in situ to conform to a defect. In this sense, aligned hydrogel tubes have been recently prepared to guide regeneration following SCI in mice [65]. In this case, a comprehensive biocompatibility study with a wide range of markers was carried out. Thus, GFAP and CD45 were used as markers to detect astrocytes and leukocytes infiltrating the injury site, respectively. Further evaluation of the leukocyte phenotypes included CD11c$^+$ dendritic cells, F4/80$^+$ macrophages, F4/80$^+$arginase$^+$ M2 macrophages, Lyg6$^+$ neutrophils and CD4$^+$ T-cells. Additionally, glial scar thickness was quantified 2 weeks after injury through GFAP$^+$ astrocyte staining and axons were identified by NF-200 staining. Moreover, to characterize the extent of myelinated axons which ensure signal propagation, the myelin basic protein (MBP) was used to discern total myelinated axons, while myelin protein zero (MPZ) was used to identify myelin from Schwann cells of the peripheral nervous system. Myelinated axons from both oligodendrocytes (MBP$^+$) and Schwann cells (MBP$^+$MPZ$^+$) were observed in each condition.

Table 3.1 summarizes biocompatibility markers that can be tested to characterize cell responses to biomaterials designed for promoting neural repair after SCI, as described above in exemplary studies.

3.4   Conclusion and Future Perspectives

The modulation of biomaterials towards 3D constructs with designs resembling the spinal cord ECM is assisting the understanding of SCI and, consequently, providing important guiding lights for improved clinical treatments. Biomaterials are directly responsible for modulating the SCI microenvironment since their characteristics deeply affect their biocompatibility and biodegradability profiles. Furthermore, the mechanical and swelling performances of the implantable constructs are also defined by the selected biomaterial(s), influencing the regeneration and ingrowth of viable neural tissues and blood vessels across the injured area. Thus, a major challenge for SCI repair is to develop an optimal material composition, going from the combination of natural and synthetic polymers, to the incorporation of

Table 3.1 Selection of biocompatibility markers used to characterize cell responses to biomaterials designed to mediate neural repair after SCI

Biomaterial	Fabrication method	Biocompatibility parameters in vitro and in vivo	Cell type	Ref.
PPy-PLA nanofibrous scaffolds	Electrospinning	Apoptosis Autophagy Axonal regeneration	Neural cells and astrocytes	[24]
PCL/PSA hybrid nanofibre scaffolds	Electrospinning	Secretion of TNF-α and IL-6 Expression of caspase-3 protein, GFAP and NF-200 Basso/Beattie/Bresnahan test	Primary astrocytes and SH-SY5Y cells	[83]
Aligned PLA nanofibrous scaffolds / GO nanosheets	Electrospinning	Cell migration, proliferation and differentiation Neurite outgrowth	Rat Schwann cells and PC-12 cells	[84]
Reduced GO scaffolds	Ice segregation induced self-assembly (ISISA)	Immunofluorescence studies of specific markers: MAP-2, tau, vimentin, iba1, ED1, PDGFRβ, GFAP	Neurons, glial cells, connective tissue cells, macrophages and pericytes	[85]
Reduced GO scaffolds	Freeze-casting	Immunofluorescence studies of specific markers: CD80, CD86, CD163, arginase I, βIII tubulin, SMI-311, SMI-71, RECA-1	Macrophages, neurons and endothelial cells	[86]

nanoparticles and bioactive molecules. In this sense, considering the high panoply of biomaterials being currently tested both in vitro and in vivo, it is crucial to uniformize cell culture protocols, functionalization strategies and injury models for guaranteeing comparable data towards a sustainable clinical translation.

It is noteworthy to mention that the experimental success in animal models could not be immediately reflected in effective medical treatments, considering the divergence of such models and the complexity of neural plasticity, immune responses and specific molecular/cellular pathways of the human CNS. In parallel, as described in this chapter, each microfabrication technique used to tailor the shape of biomaterials presents specific design and scaling opportunities, offering, subsequently, particular advantages concerning biomimetics (e.g. electrospinning and ice-templating), implantability (e.g. injectable hydrogels) and reproducibility (e.g. 3D bioprinting). Although these methodologies have already proved to be suitable for constructing highly promising scaffolds for SCI, it is possible that some decisive advances could only be reached by combined strategies involving more than one microfabrication approach. Indeed, the standardization of coherent relationships between techniques and scaffolding functionalities could enrich computational models and boost the creation of multi-step sequences able to adapt the design, processing and customization of the final scaffold according to particular SCI clinical requirements. From a commercial perspective, together with an indisputable clinical efficiency, it is mandatory that scaffolds aiming the repair of the injured spinal cord could also guarantee key product development requirements such as scalability and cost/time-effectiveness.

Taking all these pivotal aspects into consideration, there is a high probability that 3D-shaped biomaterials could achieve clinical translation in a near future by improving or restoring the functionality of the injured spinal cord, leading to a robust support of neuroregenerative therapies as well as the optimization of some clinical management approaches for SCI.

Acknowledgments A.F.G. thanks the *Fundação para a Ciência e a Tecnologia* (FCT) for the Ph.D. grant SFRH/BD/130287/2017, which is carried out in collaboration between TEMA-UA and ICMM-CSIC. J.S. thanks FCT for the Ph.D. grant SFRH/BD/144579/2019. P.A.A.P.M., M.T.P. and M.C. acknowledge the European Union's Horizon 2020 Research and Innovation Programme for funding the project NeuroStimSpinal (grant agreement No. 829060). The projects UIDB/00481/2020 and UIDP/00481/2020 (FCT) and CENTRO-01-0145-FEDER-022083 (Centro Portugal Regional Operational Programme, Centro2020), under the PORTUGAL 2020 Partnership Agreement, through the European Regional Development Fund, are also acknowledged for supporting TEMA Research Unit.

References

1. Langer R, Tirrell DA (2004) Designing materials for biology and medicine. Nature 428(6982):487–492
2. Fan C, Li X, Xiao Z et al. (2017) A modified collagen scaffold facilitates endogenous neurogenesis for acute spinal cord injury repair. Acta Biomater 51:304–316
3. Oudega M, Hao P, Shang J et al. (2019) Validation study of neurotrophin-3-releasing chitosan facilitation of neural tissue generation in the severely injured adult rat spinal cord. Exp Neurol 312:51–62
4. Wen Y, Yu S, Wu Y et al. (2016) Spinal cord injury repair by implantation of structured hyaluronic acid scaffold with PLGA microspheres in the rat. Cell Tissue Res 364:17–28
5. Grulova I, Slovinska L, Blaško J et al. (2015) Delivery of alginate scaffold releasing two trophic factors for spinal cord injury repair. Sci Rep 5:13702
6. Shahriari D, Koffler JY, Tuszynski MH et al. (2017) Hierarchically ordered porous and high-volume polycaprolactone microchannel scaffolds enhanced axon growth in transected spinal cords. Tissue Eng Part A 23:415–425
7. Reis KP, Sperling LE, Teixeira C et al. (2018) Application of PLGA/FGF-2 coaxial microfibers in spinal cord tissue engineering: an *in vitro* and *in vivo* investigation. Regen Med 13:785–801
8. Kubinová S, Horák D, Hejčl A et al. (2011) Highly superporous cholesterol-modified poly(2-hydroxyethyl methacrylate) scaffolds for spinal cord injury repair. J Biomed Mater Res A 99:618–629
9. Lu YB, Franze K, Seifert G et al. (2006) Viscoelastic properties of individual glial cells and neurons in the CNS. Proc Natl Acad Sci USA 103:17759–17764
10. Domínguez-Bajo A, González-Mayorga A, Guerrero CR et al. (2019) Myelinated axons and functional blood vessels populate mechanically compliant rGO foams in chronic cervical hemisected rats. Biomaterials 192:461–474
11. Moeendarbary E, Weber IP, Sheridan GK et al. (2017) The soft mechanical signature of glial scars in the central nervous system. Nat Commun 8:14787
12. Yamada KM, Sixt M (2019) Mechanisms of 3D cell migration. Nat Rev Mol Cell Biol 20:738–752

13. Duval K, Grover H, Han LH et al. (2017) Modeling physiological events in 2D vs. 3D cell culture. Physiology (Bethesda) 32:266–277
14. Drury JL, Mooney DJ (2003) Hydrogels for tissue engineering: scaffold design variables and applications. Biomaterials 24:4337–4351
15. Koffler J, Zhu W, Qu X et al. (2019) Biomimetic 3D-printed scaffolds for spinal cord injury repair. Nat Med 25:263–269
16. Usmani S, Aurand ER, Medelin M et al. (2016) 3D meshes of carbon nanotubes guide functional reconnection of segregated spinal explants. Sci Adv 2:e1600087
17. Alves-Sampaio A, García-Rama C, Collazos-Castro JE (2016) Biofunctionalized PEDOT-coated microfibers for the treatment of spinal cord injury. Biomaterials 89:98–113
18. Raynald, Shu B, Liu XB et al. (2019) Polypyrrole/polylactic acid nanofibrous scaffold cotransplanted with bone marrow stromal cells promotes the functional recovery of spinal cord injury in rats. CNS Neurosci Ther 25:951–964
19. Yao S, Yu S, Cao Z et al. (2018) Hierarchically aligned fibrin nanofiber hydrogel accelerated axonal regrowth and locomotor function recovery in rat spinal cord injury. Int J Nanomedicine 13:2883–2895
20. Liu S, Sun X, Wang T et al. (2018) Nano-fibrous and ladder-like multi-channel nerve conduits: Degradation and modification by gelatin. Mater Sci Eng C 83:130–142
21. Liu D, Li X, Xiao Z et al. (2019) Different functional bio-scaffolds share similar neurological mechanism to promote locomotor recovery of canines with complete spinal cord injury. Biomaterials 214:119230
22. Nemati S, Kim S-j, Shin YM, Shin H (2019) Current progress in application of polymeric nanofibers to tissue engineering. Nano Converg 6:36
23. Johnson CDL, Ganguly D, Zuidema JMet al. (2019) Injectable, magnetically orienting electrospun fiber conduits for neuron guidance. ACS Appl Mater Inter 11:356–372
24. Shu B, Sun X, Liu R et al. (2019) Restoring electrical connection using a conductive biomaterial provides a new therapeutic strategy for rats with spinal cord injury. Neurosci Lett 692:33–40
25. Colello RJ, Chow WN, Bigbee JW et al. (2016) The incorporation of growth factor and chondroitinase ABC into an electrospun scaffold to promote axon regrowth following spinal cord injury. J Tissue Eng Regen Med 10:656–668
26. Chen C, Tang J, Gu Y et al. (2019) Bioinspired hydrogel electrospun fibers for spinal cord regeneration. Adv Funct Mater 29:1806899
27. Hurtado A, Cregg JM, Wang HB et al. (2011) Robust CNS regeneration after complete spinal cord transection using aligned poly-l-lactic acid microfibers. Biomaterials 32:6068–6079
28. Omidinia-Anarkoli A, Boesveld S, Tuvshindorj U et al. (2017) An injectable hybrid hydrogel with oriented short fibers induces unidirectional growth of functional nerve cells. Small 13:1702207
29. Yang G, Li X, He Y et al. (2018) From nano to micro to macro: electrospun hierarchically structured polymeric fibers for biomedical applications. Prog Polym Sci 81:80–113
30. Liu W, Thomopoulos S, Xia Y (2012) Electrospun nanofibers for regenerative medicine. Adv Healthc Mater 1:10–25
31. Rose JC, De Laporte L (2018) Hierarchical design of tissue regenerative constructs. Adv Healthc Mater 7:1701067
32. Annabi N, Tamayol A, Uquillas JA et al. (2014) 25th Anniversary article: rational design and applications of hydrogels in regenerative medicine. Adv Mater 26:85–124
33. Higuchi A, Suresh Kumar S, Benelli G et al. (2019) Biomaterials used in stem cell therapy for spinal cord injury. Prog Mater Sci 103:374–424
34. McKay CA, Pomrenke RD, McLane JS et al. (2014) An injectable, calcium responsive composite hydrogel for the treatment of acute spinal cord injury. ACS Appl Mater Inter 6:1424–1438
35. Marquardt LM, Doulames VM, Wang AT et al. (2020) Designer, injectable gels to prevent transplanted Schwann cell loss during spinal cord injury therapy. Sci Adv 6:eaaz1039
36. Hong LTA, Kim YM, Park HH et al. (2017) An injectable hydrogel enhances tissue repair after spinal cord injury by promoting extracellular matrix remodeling. Nat Commun 8:533

37. Wang Q, He Y, Zhao Y et al. (2017) A thermosensitive heparin-poloxamer hydrogel bridges aFGF to treat spinal cord injury. ACS Appl Mater Inter 9:6725–6745

38. Qu Y, Wang B, Chu B et al. (2018) Injectable and thermosensitive hydrogel and PDLLA electrospun nanofiber membrane composites for guided spinal fusion. ACS Appl Mater Inter 10:4462–4470

39. Wang C, Yue H, Feng Q et al. (2018) Injectable nanoreinforced shape-memory hydrogel system for regenerating spinal cord tissue from traumatic injury. ACS Appl Mater Inter 10:29299–29307

40. Li X, Zhang C, Haggerty AE et al. (2020) The effect of a nanofiber-hydrogel composite on neural tissue repair and regeneration in the contused spinal cord. Biomaterials 245:119978

41. Günther MI, Weidner N, Müller R, Blesch A (2015) Cell-seeded alginate hydrogel scaffolds promote directed linear axonal regeneration in the injured rat spinal cord. Acta Biomater 27:140–150

42. Fan L, Liu C, Chen X et al. (2018) Directing induced pluripotent stem cell derived neural stem cell fate with a three-dimensional biomimetic hydrogel for spinal cord injury repair. ACS Appl Mater Inter 10:17742–17755

43. Scott JB, Afshari M, Kotek R, Saul JM (2011) The promotion of axon extension *in vitro* using polymer-templated fibrin scaffolds. Biomaterials 32:4830–4839

44. Hollister SJ (2005) Porous scaffold design for tissue engineering. Nat Mater 4:518–524

45. Pawelec KM, Husmann A, Best SM, Cameron RE (2014) Ice-templated structures for biomedical tissue repair: From physics to final scaffolds. Appl Phys Rev 1:021301

46. Kasper JC, Friess W (2011) The freezing step in lyophilization: Physico-chemical fundamentals, freezing methods and consequences on process performance and quality attributes of biopharmaceuticals. Eur J Pharm Biopharm 78:248–263

47. Lin W, Lan W, Wu Y et al. (2019) Aligned 3D porous polyurethane scaffolds for biological anisotropic tissue regeneration. Regen Biomater 7:19–27

48. Francis NL, Hunger PM, Donius AE et al. (2013) An ice-templated, linearly aligned chitosan-alginate scaffold for neural tissue engineering. J Biomed Mater Res A 101:3493–3503

49. Yuan NY, Lin YA, Ho MH et al. (2009) Effects of the cooling mode on the structure and strength of porous scaffolds made of chitosan, alginate, and carboxymethyl cellulose by the freeze-gelation method. Carbohyd Polym 78:349–356

50. Zhang Q, Shi B, Ding J et al. (2019) Polymer scaffolds facilitate spinal cord injury repair. Acta Biomater 88:57–77

51. Zhu J, Marchant RE (2011) Design properties of hydrogel tissue-engineering scaffolds. Expert Rev Med Devic 8:607–626

52. Li X, Liu D, Xiao Z (2019) Scaffold-facilitated locomotor improvement post complete spinal cord injury: Motor axon regeneration versus endogenous neuronal relay formation. Biomaterials 197:20–31

53. Stokols S, Tuszynski MH (2006) Freeze-dried agarose scaffolds with uniaxial channels stimulate and guide linear axonal growth following spinal cord injury. Biomaterials 27:443–451

54. Thomas AM, Kubilius MB, Holland SJ et al. (2013) Channel density and porosity of degradable bridging scaffolds on axon growth after spinal injury. Biomaterials 34:2213–2220

55. Sakiyama-Elbert S, Johnson PJ, Hodgetts SI, Plant GW, Harvey AR (2012) Chapter 36—scaffolds to promote spinal cord regeneration. In: Verhaagen J, McDonald JW (eds) Handbook of clinical neurology, vol 109. Elsevier, Amsterdam, pp 575–594

56. Prieto EM, Guelcher SA (2014) Chapter 5—tailoring properties of polymeric biomedical foams. In: Netti PA (ed) Biomedical foams for tissue engineering applications. Woodhead Publishing, pp 129–162

57. Salerno A, Leonardi AB, Pedram P et al. (2020) Tuning the three-dimensional architecture of supercritical CO_2 foamed PCL scaffolds by a novel mould patterning approach. Mater Sci Eng C 109:110518

58. Duarte RM, Correia-Pinto J, Reis RL, Duarte ARC (2018) Subcritical carbon dioxide foaming of polycaprolactone for bone tissue regeneration. J Supercrit Fluid 140:1–10

59. Kuang T, Chen F, Chang L (2017) Facile preparation of open-cellular porous poly(l-lactic acid) scaffold by supercritical carbon dioxide foaming for potential tissue engineering applications. Chem Eng J 307:1017–1025

60. Zhu M, Li W, Dong X (2019) *In vivo* engineered extracellular matrix scaffolds with instructive niches for oriented tissue regeneration. Nat Commun 10:4620

61. Nikolova MP, Chavali MS (2019) Recent advances in biomaterials for 3D scaffolds: A review. Bioact Mater 4:271–292

62. Samadian H, Maleki H, Fathollahi A et al. (2020) Naturally occurring biological macromolecules-based hydrogels: Potential biomaterials for peripheral nerve regeneration. Int J Biol Macromol 154:795–817

63. Stokols S, Tuszynski MH (2004) The fabrication and characterization of linearly oriented nerve guidance scaffolds for spinal cord injury. Biomaterials 25:5839–5846

64. Koffler J, Samara RF, Rosenzweig ES (2014) Using templated agarose scaffolds to promote axon regeneration through sites of spinal cord injury. In: Murray AJ (ed) Axon growth and regeneration: methods and protocols. Springer, New York, pp 157–165

65. Dumont CM, Carlson MA, Munsell MK et al. (2019) Aligned hydrogel tubes guide regeneration following spinal cord injury. Acta Biomater 86:312–322

66. Bedir T, Ulag S, Ustundag CB, Gunduz O (2020) 3D bioprinting applications in neural tissue engineering for spinal cord injury repair. Mater Sci Eng C 110:110741

67. Matai I, Kaur G, Seyedsalehi A et al. (2020) Progress in 3D bioprinting technology for tissue/organ regenerative engineering. Biomaterials 226:119536

68. Ashammakhi N, Ahadian S, Zengjie F et al. (2018) Advances and future perspectives in 4D bioprinting. Biotechnol J 13:e1800148

69. Wan Z, Zhang P, Liu Y et al. (2020) Four-dimensional bioprinting: current developments and applications in bone tissue engineering. Acta Biomater 101:26–42

70. Miao S, Cui H, Esworthy T et al. (2020) 4D Self-morphing culture substrate for modulating cell differentiation. Adv Sci 7:1902403

71. Tamay DG, Usal TD, Alagoz AS et al. (2019) 3D and 4D printing of polymers for tissue engineering applications. Front Bioeng Biotechnol 7:164

72. Joung D, Truong V, Neitzke CC et al. (2018) 3D Printed stem-cell derived neural progenitors generate spinal cord scaffolds. Adv Funct Mater 28:1801850

73. Knowlton S, Anand S, Shah T, Tasoglu S (2018) Bioprinting for neural tissue engineering. Trends Neurosci 41:31–46

74. Li J, Wu C, Chu PK, Gelinsky M (2020) 3D printing of hydrogels: rational design strategies and emerging biomedical applications. Mat Sci Eng R 140:100543

75. Jungst T, Smolan W, Schacht K et al. (2016) Strategies and molecular design criteria for 3D printable hydrogels. Chem Rev 116:1496–1539

76. Kaplan B, Merdler U, Szklanny AA et al. (2020) Rapid prototyping fabrication of soft and oriented polyester scaffolds for axonal guidance. Biomaterials 251:120062

77. Williams DF (2008) On the mechanisms of biocompatibility. Biomaterials 29:2941–2953

78. Straley KS, Foo CW, Heilshorn SC (2010) Biomaterial design strategies for the treatment of spinal cord injuries. J Neurotrauma 27:1–19

79. Kigerl KA, Gensel JC, Ankeny DP et al. (2009) Identification of two distinct macrophage subsets with divergent effects causing either neurotoxicity or regeneration in the injured mouse spinal cord. J Neurosci 29:13435–13444

80. Brown BN, Ratner BD, Goodman SB et al. (2012) Macrophage polarization: an opportunity for improved outcomes in biomaterials and regenerative medicine. Biomaterials 33:3792–3802

81. Kokaia Z, Martino G, Schwartz M, Lindvall O (2012) Cross-talk between neural stem cells and immune cells: the key to better brain repair? Nat Neurosci 15:1078–1087

82. Lee JA (2012) Neuronal autophagy: a housekeeper or a fighter in neuronal cell survival? Exp Neurobiol 21:1–8

83. Zhang S, Wang XJ, Li WS et al. (2018) Polycaprolactone/polysialic acid hybrid, multifunctional nanofiber scaffolds for treatment of spinal cord injury. Acta Biomater 77:15–27

84. Zhang K, Zheng H, Liang S, Gao C (2016) Aligned PLLA nanofibrous scaffolds coated with graphene oxide for promoting neural cell growth. Acta Biomater 37:131–142

85. López-Dolado E, González-Mayorga A, Portolés MT et al. (2015) Subacute tissue response to 3D graphene oxide scaffolds implanted in the injured rat spinal cord. Adv Healthc Mater 4:1861–1868

86. López-Dolado E, González-Mayorga A, Gutiérrez MC, Serrano MC (2016) Immunomodulatory and angiogenic responses induced by graphene oxide scaffolds in chronic spinal hemisected rats. Biomaterials 99:72–81

Chapter 4
Nanostructures to Potentiate Axon Navigation and Regrowth in the Damaged Central Nervous Tissue

Sadaf Usmani and Laura Ballerini

Abstract Neural interfaces as prosthetic devices are engineered in order to achieve neural recording and stimulation, to promote neural regeneration and to assist therapeutic delivery of bioactive molecules. By tailoring the interface architecture with nanoscale geometries, it is possible to mimic topographical cues able to adapt neuronal growth and redirect neurite navigation to a functional recovery. In this chapter, we present an overview of nano-dimensional strategies focusing on the new generation of artificial implantable scaffolds that can provide potential opportunities in brain and spinal cord healing. We strive to discuss how miniaturization of tools and prostheses at the nanoscale will help exploring the central nervous system (CNS) at subcellular scales to exploit artificial devices adaptation to neuronal biology and functions. Finally, some of the key advancements and hurdles currently emerging in the use of such artificial nanodevices in vitro and in vivo are discussed.

Keywords Carbon nanotubes · Graphene · Nanofiber · Nanomaterial · Nanostructure · Neural growth · Neural regeneration · Silicon-based materials

4.1 Nanoengineering and Neuroscience

Engineering materials at the nanoscale have provided a bridge from traditional material science to biology with the development of miniaturized devices for biomedical applications. In neurology and neuroscience, great deal of attention was directed to the discovery and design of nanostructures for the treatment and diagnosis of neurological diseases. Nanoengineered materials, under what is nowadays named the neuronanotechnology field, offer the potential to interact with biological systems at the subcellular scale, with exceptional levels of control over physiological processes. In vivo, cells are surrounded by the extracellular matrix

S. Usmani · L. Ballerini (✉)
International School for Advanced Studies (SISSA/ISAS), Trieste, Italy
e-mail: susmani@sissa.it; ballerin@sissa.it

© The Author(s), under exclusive license to Springer Nature Switzerland AG 2022
E. López-Dolado, M. Concepción Serrano (eds.), *Engineering Biomaterials for Neural Applications*, https://doi.org/10.1007/978-3-030-81400-7_4

(ECM), which provides biological, mechanical, and structural support to cells and tissues. ECM can be described as a spatio-dynamic bioscaffold generated and maintained by the surrounding cells, which in return influences cellular signaling, architecture, and functionality [1]. In the central nervous system (CNS), the core players in the neuronal machinery involved in axon navigation and regeneration, such as adhesion and signal proteins, ionic channels, filopodia and other elements of the cytoskeleton, are all within the nanoscale dimension. It is therefore hypothesized that engineering extracellular spaces at the nano-dimension will allow interfacing these cellular machineries and their operations [2]. Regenerative neural interfaces, once miniaturized and enriched by ad hoc designed nanoscale features, will better probe biological systems by mimicking bio-environments most acceptable for CNS cells, facilitating biocompatibility and axon regrowth and regeneration [3, 4].

4.2 Recent Advancements on the Application of Nanomaterials and Nanotools in Neuroengineering

The foundation of the "nanoworld" was established in 1980s with the invention of the scanning tunneling microscope and techniques such as atomic force microscopy. Thus, for the first time, scientists were able to see the tiny brick of matter, i.e., the atom in the 3D real space [5]. In neuroscience, these advanced technologies led to a better understanding of the nervous system and promoted the progression from recognizing the organization of the nervous system in the centimeter range (keeping into account various lobes of the brain and spinal segments and their physiological functions) to a deeper understanding at the sub-micrometer scale. Going further toward the nanometer range, distinct sub-neuronal compartments and their interactions with surroundings at the molecular levels started to be investigated (Fig. 4.1).

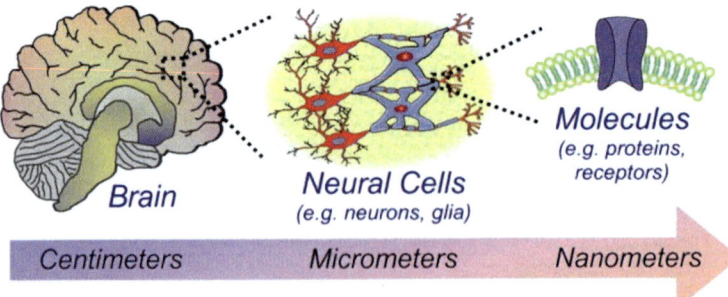

Fig. 4.1 Anatomical organization of the CNS at different size scales. Reproduced by permission from [4]. Copyright© 2016, Springer Nature

We should not forget, however, that CNS components are complex systems. To achieve functional recovery, tissue growth and repair are not enough. The correct reestablishment of axonal connections within neuronal circuits is necessary. The arena of nanomaterials, nanotools, and nanodevices more recently introduced has led to significant developments in the field of bioengineering, drug delivery, diagnostics, optogenetics, and regenerative medicine [3, 6]. The miniaturization trend by the discovery and manufacturing of controlled nanostructures brought to the table the most promising nanomaterials crucial in interfacing neural circuits [7]. Significant improvements in sensing, stimulating, recording, and monitoring neural functions by the use of nanoscale array devices made possible, for example, increasing signal resolution by implementing the number of electrodes with a much reduced device size [8].

The discovery of nanomaterials, such as carbon nanotubes (CNTs), nanofibers, and nanowires in particular, allowed the development of scaffolds characterized by electrical conductivity while enabling axonal interfacing and growth. CNTs interfaced to neural tissues have not only shown potential to drive neurite growth but also improved synaptic performance and reconnection in segregated neural networks [9–12]. Graphene oxide (GO), another carbon-based nanomaterial that has emerged with promising potential in neuroscience when fabricated into grid patterns, was shown to control effective stem cell differentiation. Such grid-pattern arrays have been suggested to mimic interconnected/elongated neuronal networks [13]. Similarly, hydrogels (self-assembled into a nanofibrous morphology encapsulated with neural stem cells) demonstrated enhanced reconstruction of injured brain tissue [14]. On their turn, magnetic manipulation by doping supermagnetic iron oxide nanoparticles on microgels allowed alignment of nerve cells by external magnetic fields [1].

Alternatively, silicon-based materials offer wide clinical application for their use as implants, electrodes, and tissue scaffolds [8, 15]. However, their dimensions and mechanical properties do not perfectly match those of the CNS tissue resulting in poor tissue integration. To overcome this limitation, the use of nano/microfabrication techniques (e.g., introduction of pores, patterns, grids, or channels) to implement subcellular features is currently pursued [16, 17]. The introduction of such nanopores and channels into silica-based materials has shown to further enhance properties like cellular adhesion and tissue integration [17–19]. Similarly, nanostructures like gold nanowires, CNTs, and nanofibers promoted cell adhesion, tissue polarity, and control over nerve growth factors deposition, thus mimicking ECM functions [20].

4.3 Nanomaterial Design Strategies and Their Impact on Neural Growth and Regeneration

Interfacing materials with neural tissues possess multiple challenges that strongly delayed translational applications of nanomaterial-based implants, such as the high level of control and precision demanded by delivering nanovectors through neuroprotective barriers [2]. The complex inhibitory post-traumatic CNS milieu intricates designing, source and fate of the nanoengineered materials.

To develop a successful biomimetic approach, there are four major requirements to be accounted: (1) a permissive growth environment, (2) a neuro-stimulatory ECM, (3) provision of neurotrophic factors, and (4) glial and/or other supportive cells. Biocompatibility of the material interface to the cellular machinery greatly governs the success of the biomaterial [21]. In this sense, the performance of a material and its biocompatibility is directly dependent on how cells interact with its surface. Tailoring of nanomaterials to be accommodated within the CNS depends on material composition, biomechanical properties, dimensions, and related biofunctionalities, among other major parameters. Table 4.1 summarizes major aspects of nanomaterials design for biomedical applications targeting CNS tissues.

It has been extensively proved that topographical modifications of a substrate largely influence cell adhesion, infiltration, and signaling. For instance, engineered micro- and nanopatterns on the substrate surface induce changes in the directional persistence of cells (Fig. 4.2). By doing so, cell motility, migration, and directionality could be manipulated [39].

Cell adhesion plays a key role in determining the success of the material implant. Specifically, biomaterials that do not support cell adhesion show poor integration and thus fail to achieve regenerative targets. Improving cell affinity and attachment, by the use of laminin coatings, ECM-derived proteins functionalization, neural adhesion molecules, and active peptide sequences, to mention a few, enhanced biomaterial neural affinity and repair [32].

Additionally, modifications in the architecture of the material by the inclusion of topographical features, in the form of pores, channels, ridges, and pillars, for instance, increase bioactivity and biocompatibility via nano/micro topographical cues that provide landscapes for neural networking, biological cues, and communication. The presence of such topographical cues serves as artificial ECM scaffolds and contributes in promoting neural regeneration. A wide range of implants derived from collagen, matrigel, agarose, CNTs, silicon, and graphene, among many others, fabricated as nanofibers, nanopillars, foams, and sponges, have demonstrated enhanced biomimetic and regenerative properties. For instance, collagen hydrogel blocks carrying microchannels permitted growth and penetration of motor axons emerging from spinal explants in vitro, an effect absent when a non-porous block was used [26]. For spinal cord injury regeneration, scaffolds designed with channels and fibers with longitudinal orientation showed tremendous potential in redirecting neural outgrowth [33]. Patterning of naturally derived and synthetic materials by

Table 4.1 Summary of major nanomaterial design strategies for CNS tissues

Factors	Description	Application/remarks	Limitations
Nanomaterial source and fate	Degradable: – Slowly degrades into the biological milieu – Usually composed of natural components (e.g., chitosan, collagen, poly(α-hydroxyacids), poly(β-hydroxybutyrate) [22, 23]	Integrate better with the biological milieu due to their natural derivation [23] – Chitosan: Exhibits capacities to tune properties like degradation rate, strength, and cell adhesivity [22, 24] – Collagen: Enhances neural growth and support fiber penetration [25, 26]	– Uniformity between batches and controlled fabrication pose problems – Obtaining an acceptable level of purity is an issue leading to enhanced activation of immune responses [23]
Nanomaterial source and fate	Non-degradable: – Usually synthetic – Require less complex designs – Offer more uniform and controlled synthesis techniques – Not inherently cell adhesive, modification of surface properties by means of coating, nanotechnology, and functionalization are commonly employed (e.g., silicone, polyacrylonitrile/polyvinylchloride (PAN/PVC), poly(tetrafluoroethylene) (PTFE), poly(2-hydroxyethyl methacrylate) (PHEMA), carbon-based substrates) [27]	– Tunable mechanical properties – Easily shaped with controlled dimensions and morphology – Facile incorporation of bioactive compounds [28]	– Higher risk of inflammation due to permanent implantation – Despite inherent limitations, a simplified design or modification can result in overcoming such limitations [27]

(continued)

Table 4.1 (continued)

Factors	Description	Application/remarks	Limitations
Physical properties	– Morphological tailoring to resemble CNS milieu achieved by controlling bulk physical characteristics (e.g., porosity, geometry, dimension, mechanical compliance) and surface properties (e.g., chemical and topographical cues for cell organization) [16]	– Governs substrate ability to promote cell adhesion, guide cell behavior (including migration, proliferation, differentiation, maintenance of phenotype, and apoptosis) – Act as a mechanical support for tissue regeneration – Improve adhesion, cell colonization, and signaling [23]	– Careful and complex engineering and design is of utmost importance – Ease of physical manipulation is most common in artificial non-degradable materials, and fate of such implants is an issue that remains to be critically addressed
Mechanical cues	Material mechanical properties (e.g., elasticity, stiffness, fatigue life) must specifically match those of the neural tissues and guarantee integrity [29, 30]	– Allow cell infiltration and vascularization – Approaches like crosslinking, tuning material composition, and coil reinforcement can control mechanical properties of materials [31]	Failure to achieve accurate mechanical mismatch can result in enhanced activation of immune responses, thus leading to rejection and further damage to the injury site
Cell adhesion	– Material interactions with cells determine success of any scaffold – Material properties such as roughness, hydrophilicity, static forces, and electrostatic interactions largely affect cell adhesion to nanostructures [24]	– Enable contact-mediated guidance for regeneration – Determine biocompatibility – Can alter synaptic activity and performance – Functionalization with laminin, ECM-derived proteins, neural adhesion molecules, and active peptide sequences can enhance cell affinity and the potential of materials as regenerative implants [32, 33]	

Material architecture	Incorporation of specific topographies via nano/micromodifications (e.g., pores, channels, ridges, channels, fibers, pillars)	– Provide increased surface area for neuronal attachment and infiltration – Offer landscapes for cell supplementation and functionalization – Impart topographical closeness to CNS milieu – Enhance bioactivity and biocompatibility [26, 34]	
3D configuration	3D scaffolds compared with conventional 2D substrates demonstrate more similarity with native CNS topography [35]	– Contribute in cell migration, adhesion, and proliferation – Permit penetration of neuronal processes – Behave as more interactive surfaces – Ensure ECM deposition and tissue formation – Provide larger surfaces for functionalization and capturing of neuroprotective agents [1, 17, 36]	3D bio-fabrication demands smart and accurate construction strategies and a high degree of control in fine tuning of geometry and mechanical properties
Electrical conductance	Electrically charged materials generate electric fields in ECM able to promote and control growth, remodeling and protein adsorption (e.g., PTFE, polyvinylidene fluoride (PVDF), polypyrrole (PPy), CNTs, nanowires) [23, 35, 37]	– Interfacing neural cells/tissues show neurite extension, controlled and enhanced outgrowth and improved synaptic performance [9–11] – Application in brain–machine interfaces as stimulation and/or recording electrodes	– Fate and toxicity of most electrically active materials still remain under discussion – Accurate conductance compatible with physiological conditions and engineering is of utmost importance [2, 38]

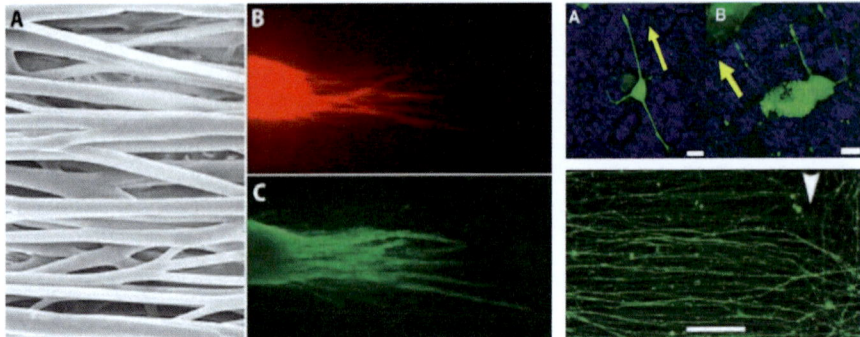

Fig. 4.2 Left: (**a**) SEM image of aligned poly acrylonitrile-co-methylacrylate (PAN-MA) nanofibers. (**b**) NF-160 staining of neurite outgrowth from a dorsal root ganglion (DRG) cultured on an aligned nanofiber film. (**c**) S-100 staining of Schwann cell migration from the same DRG [2]. Right: (Top) (**a, b**) Examples of two different neurites (green) extending from neuron cell bodies aligned in the direction of CNT orientation (yellow arrow). Scale bar: 20 μm. (Bottom) Alignment of axons of adult mouse sympathetic and sensory ganglia on horizontally imprinted patterns made by nanoscale contact printing. The arrow indicates the border of the pattern. The more random growth of axons outside the pattern is clearly visible. Scale bar: 50 μm. Adapted from [2], with permission from John Wiley & Sons

electrospinning (i.e., electric charge used to form fine aligned fibers from polymers) and microcontact printing (i.e., production of patterns with controlled printing) are alternative techniques in the design of polymer fibers from materials such as collagen and polycaprolactone [6, 40].

4.4 Advantages of 3D Versus 2D Nanomaterial-Based Scaffold Configurations

In 1911, Tello and colleagues first demonstrated that recovery and regrowth of CNS neurons could take place if provided with a permissive environment [41]. By doing so, functional recovery after CNS injuries could be facilitated, guided, or promoted [42, 43]. In this sense, a three-dimensional (3D) environment is more advantageous than two-dimensional (2D) interfaces in many different aspects. Specifically, 3D topologies positively contribute to cell adhesion, migration, and proliferation. Also, the penetration of neuronal processes into the third dimension is expected to provide more interactive surfaces and realistic structures for ECM deposition and tissue growth. In addition, physical properties including mechanical features offered by 3D architectures promote more physiological cell–cell interactions and tissue-like formation. Therefore, all this is hypothesized to generate microenvironments that better mimic in vivo situations, enhance cell adhesion, and facilitate functions. For instance, 3D configurations such as hydrogels [36, 44], self-assembling peptide nanofiber scaffolds [14, 45, 46], carbon, silicon, and collagen-based scaffolds

[8, 17, 26] are just few examples of 3D-shaped materials with large potential for their application in neural tissue engineering, regenerative medicine, and biomedicine in general.

4.5 Electrically Active Materials

An emerging strategy extensively adopted in excitable tissues is the use of electrically active materials in CNS regeneration approaches [20]. It is believed that electrically charged materials generate electric fields in the ECM that act as signals to promote and control growth, remodeling, and protein adsorption [47, 48]. Control and enhancement of neurite extension and outgrowth have been shown in vitro when cells and tissues were interfaced to electrically active materials [9–11, 49]. Moreover, eliciting electrical stimulation in vivo has been proved to promote recovery of motor and CNS nerves [50, 51].

Some examples of widely investigated electrically active materials include poly(tetrafluoroethylene) (PTFE), polyvinylidene fluoride (PVDF), and polypyrrole (PPy) [23, 52]. While these electrically active polymers are characterized by low, if any, cell adhesion [53], next generation of conductive artificial materials such as graphene- and CNTs-derived substrates might better enhance cell and neurite adhesion, as tested in vitro with cells and tissues [2]. Use of conductive materials like CNTs, microfabricated silicon, and nanowires hold promise for their potential application in brain–machine interface as stimulation and/or recording electrodes that, in turn, can aid the treatment of motor and sensory disabilities caused by neurological disorders [54].

4.6 Next Generation of Nanomaterial-Based Scaffolds for Neural Regeneration

The new generation of manufactured nanomaterials is gaining more and more acceptance due to their enhanced properties and huge range of applications and engineering. Such materials include carbon-based substrates, silicon nanowires, hydrogels, and nanofibers, among others, which have been extensively used as tools, injectable devices, electrodes, and, eventually, components for regenerative interfaces [9, 37, 55, 56]. Nanowires and nanoneedles, for instance, are excellent devices for recording and eliciting signals to brain and spinal cord circuits in vitro and in vivo [57]. Hydrogel blocks, with or without functionalization, are used for drug delivery, replacement therapy, and regenerative scaffolds [6].

Within this area of applications, much of the current approaches in CNS repair are shifted to the development of electrically active artificial materials and their modifications to provide a permissive and stimulating environment for neural regeneration and circuit repair [58]. Reportedly, in engineering artificial ECM components, aspects such as the use of nanoscale tools, the introduction of three dimensionality in scaffolds, and a correct choice of composition play essential roles [20].

CNTs- and graphene-derived substrates are among the most popular candidates for the new generation of artificial materials within neural biomedical applications. With their outstanding ability to adsorb molecules, they serve large surface areas as attractants to adhesive moieties, cell nourishing components, and carriers of therapeutic agents with potential in neural guidance and regeneration.

4.6.1   Carbon-Based Nanostructured Scaffolds

CNTs, made of graphene sheets rolled up to form hollow tubes with diameters typically within the nanometer range, are considered (among carbon-based nano-materials characterized by high electrical conductivity and strength) an example of outstanding candidates for neural engineering. This is motivated by their inherent mechanical and thermal properties, electrical conductivity, and dimensional and chemical compatibility with biomolecules [59]. Due to the unique properties of CNTs, various designs have been implemented in producing probes, conductive composites, electrodes, and microflexible devices that can adapt to the physiological movements of the brain and spinal cord [60, 61]. Indeed, one of the key properties of CNTs is their ability to make intimate contacts with biological membranes. Tight interactions between cell membranes and CNTs have been identified by ultramicroscopy [11, 62].

This particular feature has been described in isolated cultured neurons [9, 63], as well as in more complex neural systems like organotypic spinal tissue cultures. In this last case, CNTs have been reported to instruct axonal growth leading to desirable organization and reconstruction of neural networks in vitro (Fig. 4.3) [11, 12]. In cultured spinal cord explants grown on multiwalled CNTs (MWCNTs), a significant increase in the number of outgrowing fibers was reported. Moreover, emerging axons from these spinal cord explants traveled longer distances in radial fashion. However, increased growth is not the only governing factor for a successful regenerative approach. Functional rewiring of neural circuitry and synaptic contacts with appropriate targets are necessary for functional regeneration. By exploiting variable patterns of CNTs (e.g., vertically aligned patterns), these were shown to guide neural growth with remarkably improved functional performance [46, 64, 65]. This improved neuronal signaling, leading to an electrically favorable environment, might play a role in sustaining neuronal network formation and synapse maturation.

These findings obtained in vitro also suggest that mechanically derived microen-vironments can affect biochemical signaling, ultimately playing a significant role in CNS reconstruction. More specifically, neurites that were in contact with CNTs displayed different elasto-mechanical properties [11]. In a different approach, Gui et al. molded CNTs into a porous 3D sponge with a very high porosity while retaining

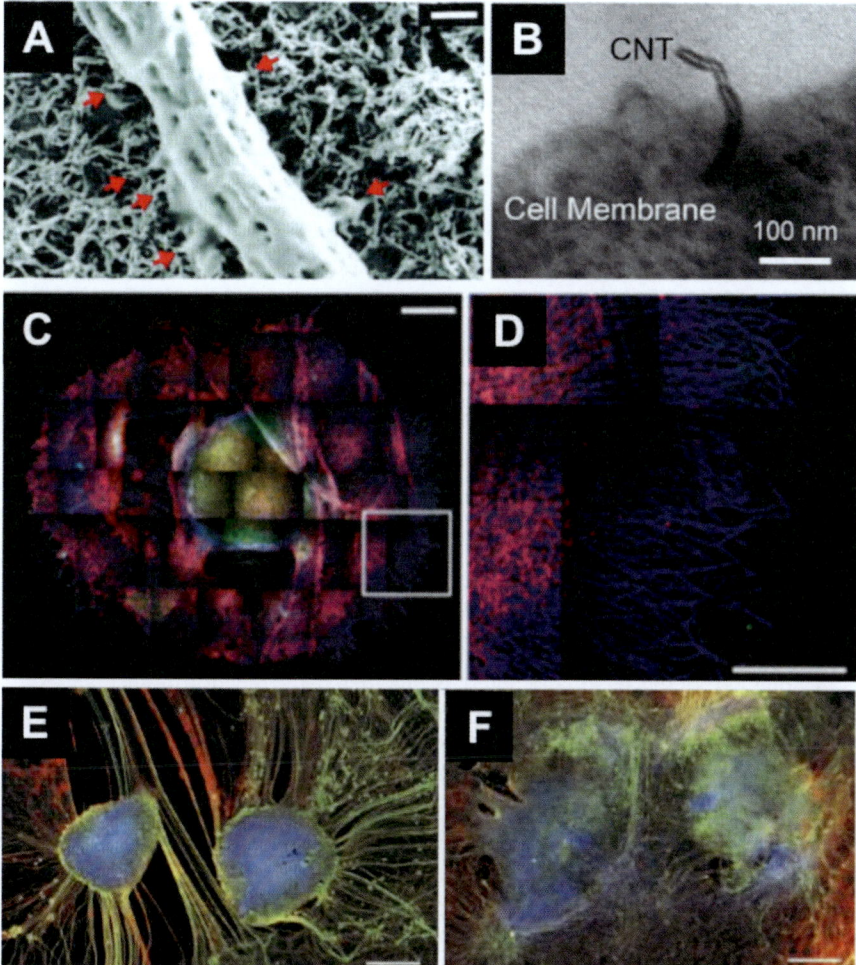

Fig. 4.3 (a) SEM image shows a peripheral neuronal fiber emerging from a spinal cord explant tightly anchored to a MWCNTs-based substrate. Red arrows demonstrate intimate contacts. Scale bar: 500 nm. (b) TEM image of HeLa cells exposed to ammonium-functionalized MWCNTs. CNTs successfully traverse cell membranes, producing gene transfection. (c, d) Impact of MWCNTs interfaces on neurite outgrowth from organotypic spinal cultures. Confocal image reconstructions of spinal slice cultures at 8 days in vitro on a 2D CNTs carpet. Immuno-labeling of specific cytoskeletal components: F-actin, β-tubulin III, and GFAP. β-tubulin III positive neuronal processes radially exiting the growth area in a MWCNTs interfaced explant are larger in number and travel longer distances. Red: F-actin, blue: β-tubulin III, and green: GFAP. Scale bars: 1 mm (c) and 500 µm (d). (e, f) Free standing MWCNTs scaffolds redirect neurite outgrowth between spinal organotypic slices in control (e) and 3D carbon nanofibers scaffolds (f) after 14 days of growth. Immunofluorescence is for neuron-specific microtubules (β-tubulin III; red), neurofilament H (SMI-32; green), and nuclei (4',6-diamidino-2-phenylindole, DAPI; blue). Scale bar: 500 µm. (a, c, d) Reprinted with permission from [11]. Copyright (2012) American Chemical Society. (b) Courtesy of Prof. M. Prato, University of Trieste, Trieste, Italy. Adapted from [62], with permission from John Wiley & Sons. (e, f) from [12]. © The Authors, some rights reserved; exclusive licensee American Association for the Advancement of Science

desired mechanical properties [66]. The sponge structure obtained was very stable and allowed excellent compressibility and volume recovery by free expansion. These 3D scaffolds maintained good percolation and polymer infiltration capability.

While such 2D substrates show potential in guiding axon navigation, a pure 3D mesh of the same material improved synaptic reconnection among cultured spinal explants, acting as a supportive grid and exploiting the elaborated surface area of the scaffold architecture [12]. Neuronal fibers displaying tight contacts with the 3D mesh of MWCNTs-based nanofibers formed a dense hybrid network presenting a favorable environment for neural growth and the formation of functional contacts via multi-synaptic pathways. In this approach, CNTs seem to be the key component, possibly acting as an electrical attractant and providing the necessary nano-surface features for cell colonization of the implant [9, 67–69], also sustaining neurite migration into the implanted scaffolds at distances away from the source [12, 70].

Neural abnormalities derived from the interaction with 3D CNTs scaffolds can be attributed to physical, chemical, and biological alterations leading to axonal regrowth, synapse formation, and/or ECM deposition [71]. Importantly, following CNS injury in vivo, the ability of these neural interfaces to reduce the inhibitory environment and favor neural circuit formation is the determinant factor for the success of an implant. Thus, behavioral and functional investigation in in vivo models of injured CNS is pivotal to carefully elucidate the usefulness of any future neural substrate.

Among carbon-derived nanomaterials, another rising star for biomedical applications is graphene. With its excellent physical properties including mechanical flexibility, graphene has been proved as a potential neural interface with the capacity to affect biological processes of the neural machinery by physical interactions. As a bioactive interface, graphene has impacted growth and activity of neural cells in vitro [73–75]. In a different approach, a hybrid polyacrylamide hydrogel embedded with graphene supported the growth and development of synaptic activity, thus enhancing neuronal adhesion by means of the graphene coating and then improving polyacrylamide biocompatibility [76]. In its reduced form, the so-called reduced graphene oxide (rGO), a more hydrophilic form of graphene demonstrated regenerative features such as axonal and vascular growth in chronically injured spinal cord tissue in vivo in the shape of 3D porous scaffolds and microfibers [30, 72] (Fig. 4.4). In combination with electrical stimulation, a 3D foam of graphene oxide used as electrically conductive scaffold allowed directional growth of neural fibers and differentiation of human neural stem cells into neurons [73].

4.6.2 Silicon-Derived Nanostructures

Silicon-based scaffolds have been widely used in clinical applications as neural electrode devices and implants. One such derivative is a highly flexible and tunable polymer called as polydimethylsiloxane (PDMS). Due to the inherent flexible capacity of PDMS, it can be molded into different shapes and space filling agents

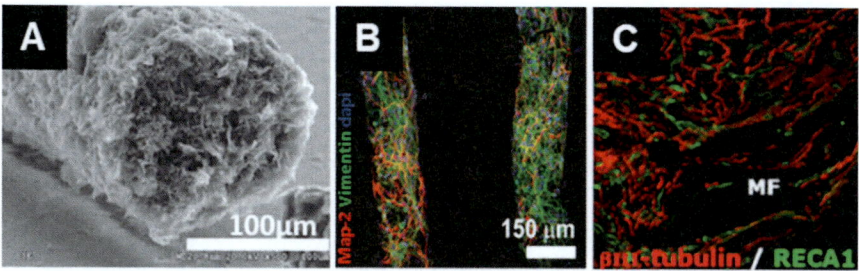

Fig. 4.4 Examination of reduced graphene oxide microfibers (rGO-MFs) for neural tissue engineering applications. (**a**) Representative SEM image of a dense and compact microfiber composed of rGO nanosheets. (**b**) Confocal fluorescence images of highly interconnected cultures composed of neurons (map-2; green) and non-neuronal cells (vimentin; red) maintained up to 21 days in vitro. Cell nuclei were labeled with DAPI (blue). (**c**) Vascular and neuronal invasion of rGO-MFs scaffolds implanted into a chronic spinal cord injury model demonstrating integration with the neural tissue. Reprinted with permission from [72]. Copyright (2020) American Chemical Society

making it more ideal for implantation into complex CNS injuries in the spinal cord and/or the brain. Another important property of PDMS is the ease of fabrication into soft 3D structures that can mimic in vivo ECM environment. Therefore, PDMS can readily be tailored into any size or shape. It is soft, flexible, penetrable, and elastic and, most importantly, can provide landscapes for functionalization/customization options [77, 78].

3D structures of PDMS are easily customized into desirable structures of micro- or nano-dimension with high porosities. This is due to the mechanical characteristics of this polymer, which can serve to design porous polymer network platforms for the purpose of neural tissue interfaces. PDMS has been successfully used as an artificial dura matter with microchannels to deliver drugs and electrically conductive gold coatings for eliciting electrical stimulation [15, 78]. In this same work, it has also been tested as a substitute for the skull and *dura mater* to introduce a cranial window allowing ease of insertion of microelectrodes and micropipettes into the cortical tissue for electrophysiological recording and chemical injection [15, 78]. While in vivo studies have shown limited astroglial reaction to PDMS when implanted into the rat cortex [18, 79], the surface properties of this material, however, often failed to properly integrate with neural tissues.

Cellular adhesion and tissue integration were further enhanced in PDMS substrates by customization techniques, functionalization with bioadhesive materials, and/or various nanofabrication techniques. For instance, the use of laminin peptides or CNTs improved the properties of silicon-based materials for therapeutic applications [18, 19]. Since the use of PDMS as a neural substrate ranges from cell seeding to neural electrical devices, customization of PDMS with electrically active materials such as CNTs aims at improving surface interaction with biological components as well as providing properties attributed to electrically active materials. Specifically, the incorporation of CNTs into the polymer matrix served to enhance its physical properties as well as offer a favorable electrical environment both in vitro and in vivo [70].

4.7 General Pathophysiological and Biocompatibility Aspects of Nanostructures Interfacing the Injured CNS

The ultimate potential of any implantable device resides in its integration with biological milieu in vivo. As extensively exposed in previous chapters, following a CNS trauma, a series of pathophysiological events occur that not only pose non-permissive conditions for regenerative axon regrowth but also result in tissue damage beyond the site of trauma [80]. Such a cascade of deleterious events like edema, ischemia, excitotoxicity, and inflammation adversely affect the integrity of spared neurons and result in continued dysfunction and prolonged degeneration [81]. Specifically, therapeutic strategies can fail due to inflammatory responses initiated by the trauma lesion, leading to release of proinflammatory cytokines, migration of microglia, and formation of fibroglial scars. Acutely, proinflammatory cytokines pose harmful effects like blood–brain barrier (BBB) and blood–spinal cord barrier (BSCB) dysfunction, leukocyte infiltration, edema development, and, ultimately, promotion of neuronal death [81]. Microglia, which are the first cells to respond (minutes to hours), proliferate, activate, migrate to the area of injury, and, along with astrocytes, form a tight interpenetrating network known as reactive glial scar. Initially, the glial response helps to buffer excitotoxicity and cytotoxicity, repair BBB and/or BSCB, and isolate the injury site, but it produces inhibitory factors that limit axonal regeneration and abort sprouting chronically. This dynamic interplay between events promoting repair and regeneration and those leading to damage and inhibition calls for a greater need to achieve effective approaches toward injured CNS repair.

Facilitated and increased access to cellular and nuclear compartments by nano-materials may be responsible for toxic responses posing issue in their application. Various factors such as (i) composition of the material, (ii) surface-to-volume ratio, (iii) purification method, (iv) metabolism mechanism, and (v) administration method play an important role in determining potential toxic effects. For instance, hydrophobic materials tend to aggregate in body fluids. In such cases, it is critical to ensure accurate and rapid targeting, functionalization for sustained effects, and/or controlled degradation into non-toxic by-products [2, 38, 82]. While toxicity of nanomaterials remains an open debate [83–85], outstanding effects like enhance-ment of neurite branching, length and density, number of growth cones, synaptic activity, and neuronal firing make them widely accepted as neural scaffolds.

In the particular case of CNTs, for instance, it has been reported that their biocompatibility is strongly influenced by geometry and surface chemistry [86]. Therefore, by understanding better the immune system responses and defining exposure standards, one can engineer nanostructure geometry into cytocompatible architectures. Functionalization of CNTs has also been performed by different approaches to modify their solubility and reduce toxic effects in adaptation for biomedical applications [86].

4.8 Future Perspectives

Nanomaterials, due to their outstanding physical and chemical properties, offer versatile potential in drug delivery, gene transfer, neural rehabilitation, basic science, and clinical applications. Based on current interventions, the future of nanomedicine with respect to therapeutics, drug delivery, toxicity, axon guidance, and regenerative medicine relies on addressing challenges faced in application of such smart materials.

It is clear that multidisciplinary approaches are essential in next therapeutic strategies to CNS healing. Advances in development of biohybrid systems, merging artificial scaffolds and nerve tissue, combined with cell replacement therapy, pharmacological delivery, and electrical stimulations pave future directions toward CNS repair. Such combinatorial approaches include the development of novel nanomaterials to exploit biomimetic interfaces or scaffolds.

Very little is known about the mechanisms behind the outstanding potential of nanostructures at sub-molecular and subcellular levels in complex systems such as our nervous milieu. A major observation revealed that there are contrasting regenerative potentials in the CNS, certainly a target to be explored to manipulate CNS environment by nanostructures [87].

Yet, limited studies have translated the use of nanomaterials-based interfaces in primates and humans. An intensive research on the biomimetic potential of new generation of nanomaterials has demonstrated exceptional outcomes in neurobiology and requires further studies for translation in clinics.

Acknowledgement This work was funded by the European Union's Horizon 2020 research and innovation programme under grant agreement No. 737116 to L. B.

References

1. Rose JC, Cámara-Torres M, Rahimi K et al. (2017) Nerve cells decide to orient inside an injectable hydrogel with minimal structural guidance. Nano Lett 17:3782–3791
2. Kotov NA, Winter JO, Clements IP et al. (2009) Nanomaterials for neural interfaces. Adv Mater 21:3970–4004
3. Alivisatos AP, Andrews AM, Boyden ES et al. (2013) Nanotools for neuroscience and brain activity mapping. ACS Nano 7:1850–1866
4. Shah S (2016) The nanomaterial toolkit for neuroengineering. Nano Convergence 3:25
5. Binnig G, Rohrer H, Gerber C et al. (1982) Surface studies by scanning tunneling microscopy. Phys Rev Lett 49:57

6. Pampaloni NP, Giugliano M, Scaini D et al. (2019) Advances in nano neuroscience: from nanomaterials to nanotools. Front Neurosci 12:953

7. Shah S, Solanki A, Lee KB (2016) Nanotechnology-based approaches for guiding neural regeneration. Acc Chem Res 49:17–26

8. Lu C, Park S, Richner TJ et al. (2017) Flexible and stretchable nanowire-coated fibers for optoelectronic probing of spinal cord circuits. Sci Adv 3:e1600955

9. Cellot G, Cilia E, Cipollone S et al. (2009) Carbon nanotubes might improve neuronal performance by favouring electrical shortcuts. Nat Nanotechnol 4:126–133

10. Cellot G, Toma FM, Varley ZK et al. (2011) Carbon nanotube scaffolds tune synaptic strength in cultured neural circuits: novel frontiers in nanomaterial-tissue interactions. J Neurosci 31:12945–12953

11. Fabbro A, Villari A, Laishram J et al. (2012) Spinal cord explants use carbon nanotube interfaces to enhance neurite outgrowth and to fortify synaptic inputs. ACS Nano 6:2041–2055

12. Usmani S, Aurand ER, Medelin M et al. (2016) 3D meshes of carbon nanotubes guide functional reconnection of segregated spinal explants. Sci Adv 2:e1600087

13. Kim TH, Shah S, Yang L et al. (2015) Controlling differentiation of adipose-derived stem cells using combinatorial graphene hybrid-pattern arrays. ACS Nano 9:3780–3790

14. Cheng TY, Chen MH, Chang WH et al. (2013) Neural stem cells encapsulated in a functionalized self-assembling peptide hydrogel for brain tissue engineering. Biomaterials. 34:2005–2016

15. Minev IR, Musienko P, Hirsch A et al. (2015) Electronic dura mater for long-term multimodal neural interfaces. Science 347:159–163

16. Xie C, Liu J, Fu TM et al. (2015) Three-dimensional macroporous nanoelectronic networks as minimally invasive brain probes. Nat Mater 14:1286–1892

17. Bosi S, Rauti R, Laishram J et al. (2015) From 2D to 3D: novel nanostructured scaffolds to investigate signalling in reconstructed neuronal networks. Sci Rep 5:9562

18. Pennisi CP, Zachar V, Gurevich L et al. (2010) The influence of surface properties of plasma-etched polydimethylsiloxane (PDMS) on cell growth and morphology. Conf Proc IEEE Eng Med Biol Soc 2010:3804–3807

19. Choi HK, Im SH, Park OO (2009) Shape and feature size control of colloidal crystal-based patterns using stretched polydimethylsiloxane replica molds. Langmuir 25:12011–12014

20. Dvir T, Timko BP, Kohane DS et al. (2011) Nanotechnological strategies for engineering complex tissues. Nat Nanotechnol 6:13–22

21. Bellamkonda RV (2006) Peripheral nerve regeneration: an opinion on channels, scaffolds and anisotropy. Biomaterials 27:3515–3518

22. Rodríguez-Vázquez M, Vega-Ruiz B, Ramos-Zúñiga R et al. (2015) Chitosan and its potential use as a scaffold for tissue engineering in regenerative medicine. Biomed Res Int 2015:821279

23. Straley KS, Foo CW, Heilshorn SC (2010) Biomaterial design strategies for the treatment of spinal cord injuries. J Neurotrauma 27:1–19

24. Freier T, Koh HS, Kazazian K et al. (2005) Controlling cell adhesion and degradation of chitosan films by N-acetylation. Biomaterials 26:5872–5878

25. Ucar B, Humpel C (2018) Collagen for brain repair: therapeutic perspectives. Neural Regen Res 13:595–598

26. Gerardo-Nava J, Hodde D, Katona I et al. (2014) Spinal cord organotypic slice cultures for the study of regenerating motor axon interactions with 3D scaffolds. Biomaterials 35:4288–4296

27. Belkas JS, Munro CA, Shoichet MS et al. (2005) Long-term *in vivo* biomechanical properties and biocompatibility of poly(2-hydroxyethyl methacrylate-co-methyl methacrylate) nerve conduits. Biomaterials 26:1741–1749

28. Dalton PD, Flynn L, Shoichet MS (2002) Manufacture of poly(2-hydroxyethyl methacrylate-co-methyl methacrylate) hydrogel tubes for use as nerve guidance channels. Biomaterials 23:3843–3851

29. O'Brien FJ (2011) Biomaterials & scaffolds for tissue engineering. Materials today 14:88–95

30. Domínguez-Bajo A, González-Mayorga A, Guerrero CR et al. (2019) Myelinated axons and functional blood vessels populate mechanically compliant rGO foams in chronic cervical hemisected rats. Biomaterials 192:461–474
31. Itoh S, Takakuda K, Kawabata S et al. (2002) Evaluation of cross-linking procedures of collagen tubes used in peripheral nerve repair. Biomaterials 23:4475–4481
32. Suzuki M, Itoh S, Yamaguchi I et al. (2003) Tendon chitosan tubes covalently coupled with synthesized laminin peptides facilitate nerve regeneration *in vivo*. J Neurosci Res 72:646–659
33. Verdú E, Labrador RO, Rodríguez FJ et al. (2002) Alignment of collagen and laminin-containing gels improve nerve regeneration within silicone tubes. Restor Neurol Neurosci 20:169–179
34. Nomura H, Zahir T, Kim H et al. (2008) Extramedullary chitosan channels promote survival of transplanted neural stem and progenitor cells and create a tissue bridge after complete spinal cord transection. Tissue Eng Part A 14:649–665
35. Scaini D, Ballerini L (2018) Nanomaterials at the neural interface. Curr Opin Neurobiol 50:50–55
36. Madhusudanan P, Raju G, Shankarappa S (2020) Hydrogel systems and their role in neural tissue engineering. J R Soc Interface 17:20190505
37. Shur M, Fallegger F, Pirondini E et al. (2020) Soft printable electrode coating for neural interfaces. ACS App Bio Mater 3:4388–4397
38. Mohanta D, Patnaik S, Sood S et al. (2019) Carbon nanotubes: evaluation of toxicity at biointerfaces. J Pharm Anal 9:293–300
39. Tang QY, Tong WY, Shi J et al. (2014) Influence of engineered surface on cell directionality and motility. Biofabrication 6:015011
40. Cooper DR, Nadeau JL (2009) Nanotechnology for *in vitro* neuroscience. Nanoscale 1:183–200
41. Fawcett JW (2018) The paper that restarted modern central nervous system axon regeneration research. Trends Neurosci 41:239–242
42. David S, Aguayo AJ (1981) Axonal elongation into peripheral nervous system "bridges" after central nervous system injury in adult rats. Science 214:931–933
43. Langer R, Vacanti JP (1993) Tissue engineering. Science 260:920–926
44. George J, Hsu CC, Nguyen LTB et al. (2019) Neural tissue engineering with structured hydrogels in CNS models and therapies. Biotechnol Adv. 2020:107370
45. Guo J, Leung KK, Su H et al. (2009) Self-assembling peptide nanofiber scaffold promotes the reconstruction of acutely injured brain. Nanomedicine 5:345–351
46. Zhang S, Gelain F, Zhao X (2005) Designer self-assembling peptide nanofiber scaffolds for 3D tissue cell cultures. Semin Cancer Biol 15:413–420
47. Fine EG, Valentini RF, Bellamkonda R et al. (1991) Improved nerve regeneration through piezoelectric vinylidene fluoride-trifluoroethylene copolymer guidance channels. Biomaterials 12:775–780
48. Kotwal A, Schmidt CE (2001) Electrical stimulation alters protein adsorption and nerve cell interactions with electrically conducting biomaterials. Biomaterials 22:1055–1064
49. Patel N, Poo MM (1982) Orientation of neurite growth by extracellular electric fields. J Neurosci 2:483–496
50. Borgens RB, Roederer E, Cohen MJ (1981) Enhanced spinal cord regeneration in lamprey by applied electric fields. Science 213:611–617
51. Nix WA, Hopf HC (1983) Electrical stimulation of regenerating nerve and its effect on motor recovery. Brain Res 272:21–25
52. Wan Y, Yu A, Wu H et al. (2005) Porous-conductive chitosan scaffolds for tissue engineering II. *In vitro* and *in vivo* degradation. J Mater Sci Mater Med 16:1017–1028
53. Rivers T, Hudson T, Schmidt C (2002) Synthesis of a novel, biodegradable electrically conducting polymer for biomedical applications. Adv Funct Mater 12:33–37
54. Patil AC, Thakor NV (2016) Implantable neurotechnologies: a review of micro- and nanoelectrodes for neural recording. Med Biol Eng Comput 54:23–44

55. Schuhmann TG Jr, Zhou T, Hong G et al. (2018) Syringe-injectable mesh electronics for stable chronic rodent electrophysiology. J Vis Exp 137:58003
56. Pampaloni NP, Lottner M, Giugliano M et al. (2018) Single-layer graphene modulates neuronal communication and augments membrane ion currents. Nat Nanotechnol 13:755–764
57. Fang Y, Jiang Y, Ledesma HA et al. (2018) Texturing silicon nanowires for highly localized optical modulation of cellular dynamics. Nano Lett 18:4487–4492
58. Fattahi P, Yang G, Kim G et al. (2014) A review of organic and inorganic biomaterials for neural interfaces. Adv Mater 26:1846–1885
59. Raphey VR, Henna TK, Nivitha KP et al. (2019) Advanced biomedical applications of carbon nanotube. Mater Sci Eng C Mater Biol Appl 100:616–630
60. Vitale F, Summerson SR, Aazhang B et al. (2015) Neural stimulation and recording with bidirectional, soft carbon nanotube fiber microelectrodes. ACS Nano 9:4465–4474
61. Vitale F, Vercosa DG, Rodriguez AV et al. (2018) Fluidic microactuation of flexible electrodes for neural recording. Nano Lett 18:326–335
62. Pantarotto D, Singh R, McCarthy D et al. (2004) Functionalized carbon nanotubes for plasmid DNA gene delivery. Angew Chem Int Ed Engl 43:5242–5246
63. Mazzatenta A, Giugliano M, Campidelli S et al. (2007) Interfacing neurons with carbon nanotubes: electrical signal transfer and synaptic stimulation in cultured brain circuits. J Neurosci 27:6931–6936
64. Nguyen-Vu TD, Chen H, Cassell AM et al. (2006) Vertically aligned carbon nanofiber arrays: an advance toward electrical-neural interfaces. Small 2:89–94
65. Fabbro A, Prato M, Ballerini L (2013) Carbon nanotubes in neuroregeneration and repair. Adv Drug Deliv Rev 65:2034–2044
66. Gui X, Wei J, Wang K et al. (2010) Carbon nanotube sponges. Adv Mater 22:617–621
67. Mattson MP, Haddon RC, Rao AM (2000) Molecular functionalization of carbon nanotubes and use as substrates for neuronal growth. J Mol Neurosci 14:175–182
68. Shein M, Greenbaum A, Gabay T et al. (2009) Engineered neuronal circuits shaped and interfaced with carbon nanotube microelectrode arrays. Biomed Microdevices 11:495–501
69. Huang YJ, Wu HC, Tai NH et al. (2012) Carbon nanotube rope with electrical stimulation promotes the differentiation and maturity of neural stem cells. Small 8:2869–2877
70. Aurand E R, Usmani S, Medelin M et al. (2018) Nanostructures to engineer 3D neural-interfaces: Directing axonal navigation toward successful bridging of spinal segments. Adv Func Mater 28:1700550
71. Mwenifumbo S, Shaffer MS, Stevens MM (2007) Exploring cellular behaviour with multi-walled carbon nanotube constructs. J Mater Chem 19:1894–1902
72. Domínguez-Bajo A, González-Mayorga A, López-Dolado E et al. (2020) Graphene oxide microfibers promote regenerative responses after chronic implantation in the cervical injured spinal cord. ACS Biomat Sci Eng 6:2401–2414
73. Akhavan O, Ghaderi E, Shirazian SA et al. (2016) Rolled graphene oxide foams as three-dimensional scaffolds for growth of neural fibers using electrical stimulation of stem cells. Carbon 97:71–77
74. Li N, Zhang X, Song Q et al. (2011) The promotion of neurite sprouting and outgrowth of mouse hippocampal cells in culture by graphene substrates. Biomaterials 32:9374–9382
75. He Z, Zhang S, Song Q et al. (2016) The structural development of primary cultured hippocampal neurons on a graphene substrate. Colloids Surf B Biointerfaces 146:442–451
76. Martín C, Merino S, González-Domínguez JM et al. (2017) Graphene improves the biocompatibility of polyacrylamide hydrogels: 3d polymeric scaffolds for neuronal growth. Sci Rep 7:10942
77. Liang Guo, Guvanasen GS, Xi Liu et al. (2013) A PDMS-based integrated stretchable microelectrode array (isMEA) for neural and muscular surface interfacing. IEEE Trans Biomed Circuits Syst 7:1–10
78. Heo C, Park H, Kim YT et al. (2016) A soft, transparent, freely accessible cranial window for chronic imaging and electrophysiology. Sci Rep 6:27818

79. Lu Y, Wang D, Li T et al. (2009) Poly(vinyl alcohol)/poly(acrylic acid) hydrogel coatings for improving electrode-neural tissue interface. Biomaterials 30:4143–4151
80. Brennan FH, Anderson AJ, Taylor SM et al. (2012) Complement activation in the injured central nervous system: another dual-edged sword? J Neuroinflammation 9:137
81. Shoichet MS, Tate CC, Baumann MD et al. (2008) Strategies for regeneration and repair in the injured central nervous system. In: Reichert WM (ed) Indwelling neural implants: strategies for contending with the in vivo environment. CRC Press/Taylor & Francis, Boca Raton. Chapter 8. https://www.ncbi.nlm.nih.gov/books/NBK3941/
82. Gilmore JL, Yi X, Quan L et al. (2008) Novel nanomaterials for clinical neuroscience. J Neuroimmune Pharmacol 3:83–94
83. Teleanu DM, Chircov C, Grumezescu AM et al. (2019) Neurotoxicity of nanomaterials: an up-to-date overview. Nanomaterials (Basel) 9:96
84. Baldrighi M, Trusel M, Tonini R et al. (2016) Carbon nanomaterials interfacing with neurons: An *in vivo* perspective. Front Neurosci 10:250
85. Yang Z, Liu ZW, Allaker RP et al. (2010) A review of nanoparticle functionality and toxicity on the central nervous system. J R Soc Interface 7:S411–S422
86. Bianco A, Kostarelos K, Prato M (2011) Making carbon nanotubes biocompatible and biodegradable. Chem Commun (Camb) 47:10182–10188
87. Usmani S, Franceschi Biagioni A et al. (2020) Functional rewiring across spinal injuries via biomimetic nanofiber scaffolds. PNAS 117:25212–25218

Chapter 5
Nanostructured Electroactive Materials with Large Charge Capacity: Direct Field Electrostimulation Through Connected and Non-connected Electrodes

Ann M. Rajnicek, Cristina Suñol, and Nieves Casañ-Pastor

Abstract Electric field stimulation protocols depend on the electrode material used, but the material characteristics are often not considered or sufficiently described for optimization. Furthermore, charge capacity is considered only in capacitor-like systems, without taking into account that intercalation materials offer an internal faradaic charge delivery advantage, with substantially less risk for biological systems. This chapter describes new materials with high charge capacities, appropriate electric field protocols for using them, and examples of neural cultures that can be used to elucidate the biological effects of fields. Mammalian neurons, neuron–astrocyte co-cultures, and amphibian spinal neurons are used in vitro, often as scratch wound models, to assess their potential for stimulating tissue repair. Importantly, remote control of dipoles induced in conducting implanted materials is shown to be a new promising approach and a breakthrough.

Keywords Carbon-based materials · Charge capacity · Electroactive materials · Electrode · Electrostimulation · Iridium oxide · Nanostructured hybrid materials · PEDOT · PPy

A. M. Rajnicek
School of Medicine, Medical Sciences and Nutrition, University of Aberdeen, Aberdeen, Scotland, UK
e-mail: a.m.rajnicek@abdn.ac.uk

C. Suñol
Institut de Investigacions Biomèdiques de Barcelona (IIBB), Consejo Superior de Investigaciones Científicas (CSIC), Barcelona, Spain

Institut d'Investigacions Biomèdiques August Pi i Sunyer (IDIBAPS), Barcelona, Spain
e-mail: cristina.sunyol@iibb.csic.es

N. Casañ-Pastor (✉)
Institut de Ciència de Materials de Barcelona (ICMAB), Consejo Superior de Investigaciones Científicas (CSIC), Barcelona, Spain
e-mail: nieves@icmab.es

E. López-Dolado, M. Concepción Serrano (eds.), *Engineering Biomaterials for Neural Applications*, https://doi.org/10.1007/978-3-030-81400-7_5

5.1 Introduction

5.1.1 Electric Fields and Electrode Materials in Electrostimulation and Repair

The electric characteristics of biological systems observed two centuries ago [1] inspired substantial discoveries in physics and are now the basis of modern technologies dependent on electrochemical interactions at the material biological entities interface. Such matter–biosystems interfaces allow sensing, electrical stimulation, and design of systems for bionic actuation in a large variety of medical applications with increased complexity. In particular, electric eel actuation modes offer new insights. Although empirically observed during the dawn of animal electricity and electromagnetism theories, only recently they have been studied further [2]. Pulsed fields of the same polarity are emitted from stored energy electrocells in the eel's body. These paralyze the prey from a distance but also create involuntary movements that betray its location and facilitate its detection [2]. The long-range effects can only be carried out through ionic gradients and correlated capacitor effects. The actual effect on involuntary muscle contractions correlates with a disturbance of action potentials and transmission of electric signals in the prey and suggests that pulsed fields allow the creation of sustained ionic gradients. The same fundamental behavior is the basis for electrostimulation procedures now used in the medical world. Additional underlying effects are also being discovered [3–5].

Although the study of interactions between biosystems and external fields or electrodes is a rather empirical world, electrostimulation protocols may be established that modify cell behavior and action potentials. They are used with the objective of minimizing functional symptoms (as in Parkinson's disease tremor through deep brain stimulation, DBS) or inducing normal function (as in targeted muscle stimulation, for example, bladder sphincter control), thus stopping tremor or unconscious movements [5].

Nonetheless, the process of interacting with living systems through applied fields has clear restrictions. Aqueous-based electrolytes that are present in all living systems elicit oxygen radical formation if charge is delivered through implanted electrodes. This is derived from water oxidation or dissolved O_2 reduction, so protocols are defined to minimize charge delivery and to achieve net zero charge in short time frames. Generally, little attention is paid however to the actual electrode material even when neurosurgeons inform of specific inflammatory and encapsulation problems.

Electrostimulation protocols for functional therapies in the nervous system rely mainly on capacitive and relatively high-impedance electrodes including platinum (Pt) and its alloys, steel, and titanium nitride. They minimize net charge delivery by using alternating positive and negative pulses (AC fields) with large frequencies [3]. On the other hand, in living systems, directional ion transport across tightly sealed epithelia generates electric fields within tissues with a constant or mostly sustained field direction (DC electric field). These endogenous DC fields influence cell behaviors, including control of neural growth and differentiation; therefore, the electric field applications that aim to repair damaged tissues often mimic this by using DC fields to achieve a sustained ionic gradient. At present, the type of ionic gradient that favors neural growth is not clear, but evidence suggests a significant intracellular role for Ca^{2+} ions [6] during DC field effects on neural cell development. K^+ and Na^+ ions are also known to have growth effects [7].

Empirically, electrical stimulation is also gaining prominence, in addition to functional therapies, as a therapy for conditions as diverse as non-healing wounds and traumatic injury [8, 9]. Generally, these procedures use DC stimulation, either pulsed or steady, through direct connection to implanted metal electrodes. If an external field is to be applied, the DC field may be continuous or it may use repetitive pulses of the same polarity to prevent heating effects and capacitor draining effects in the setup. Pulses of opposite polarity could also be used as long as a big portion of one polarity is sustained. In any case, application of an electric field involves capacitive charging at the electrodes (with very small charge capacity) and a faradaic electronic–ionic transfer between the electrode and the ionic electrolyte medium of variable charge depending on electrode materials and potentials used. In all cases, AC pulses and the DC continuous and pulsed fields, a real charge delivery within the time frame of possible chemical reactions (micro to milliseconds) is occurring. Reported AC field protocols [3] calculated to be net zero charge in the long term still have charge transfer in short time scales.

Such electron–ion transfer at the electrode surfaces involves redox chemical processes and therefore radical formation related to O_2 reduction and H_2O oxidation, eventually leading to tissue damage and enhanced encapsulation and inflammation. Purely capacitive electrodes, like Pt, have, as expected, very low charge capacity, basically the formation of a double layer on the surface of the electrode, yielding immediately faradaic reactions from the aqueous biological media at the interface. The possibility of conferring an extra 3D structure or nanostructure would be desirable, enhancing the exchange area without affecting the geometric area. Including internal faradaic processes in the material through redox intercalation behavior may also enhance the final capacity.

New 3D structures currently being developed to increase their exposed surface, vascularization potential, and charge capacity have in fact demonstrated some success when the materials are carbon-based [10, 11], fiber electrodes based on carbon [12], and carbon materials coated with conducting polymers. In neither of those cases are charge capacities reported. It is worth remarking here that charges used for electrodeposition of those polymers are not the final delivery capacities [13, 14], which are much lower.

A more ambitious alternative to purely capacitive materials in electrodes (e.g., Pt or carbon) is the use of electrode materials that offer an extra charge capacity during electron–ion exchange through an internal redox intercalation process, therefore minimizing the electrochemical transfer delivered to the ionic cell media. Indeed, the material should have a large ion-intercalation capacity and permit a large degree of reversibility in the process to have a significant advantage as electrode material. Among the possible existing materials, some also allow electrodeposition

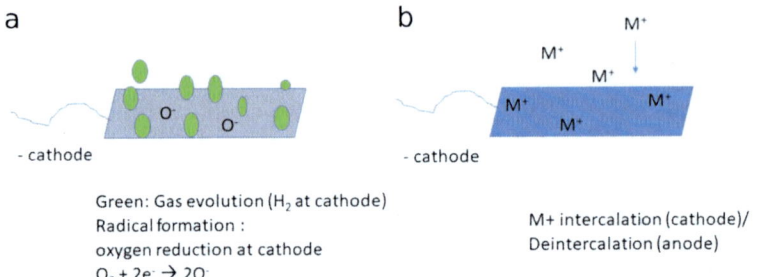

Green: Gas evolution (H₂ at cathode)
Radical formation :
oxygen reduction at cathode
$O_2 + 2e^- \rightarrow 2O^-$

M+ intercalation (cathode)/
Deintercalation (anode)

Fig. 5.1 Simplified scheme of electrodes: (**a**) Capacitive electrodes connected as cathode where hydrogen gas and oxygen radical formation (shown as O^-) occur at the interface. (**b**) Connected intercalation mixed valence electrodes as cathode where cations from the solution intercalate in the material (e.g., Na^+), instead of gas and radical formation. Horizontal configuration drawn is arbitrary

on previously formed substrates, offering a wide variety of possibilities (like iridium oxide being deposited on Deep Brain Stimulation (DBS) electrodes). See Fig. 5.1, comparing electrochemical processes at capacitive electrodes and intercalation electrode surfaces.

Electrodeposited iridium oxide, named IrOx here, is an amorphous open structure with oxohydroxide character derived from a hydrolysis process of iridium salts combined with anodic deposition. Its chemical mixed valence structure renders a mixed conducting material that allows oxidation state changes along with intercalation and deintercalation processes of H^+, OH^-, Na^+ and other ions, as recently proven in [15]. Its charge capacity reaches 5 mC/cm^2 (vs. 0.1 mC/cm^2 charge capacity for Pt) [16], when deposited through constant current anodic processes, and 20 to 30 mC/cm^2 when prepared through dynamic anodic potential sweeps [15]. Other materials are also being tested including conducting polymers based on poly(3,4-ethylene dioxythiophene) (PEDOT) and polypyrrole (PPy), with a variable number of oxidation states, and nanocarbons like graphene and carbon nanotubes, all with unknown or similar charge capacities to those of IrOx [12].

Through the mixed valence states possible in those materials and the corresponding intercalation processes induced by the applied electric field, an alternative charge delivery process is possible at the interface between the conducting material and the electrolyte, canceling most of the radical formation effects. An added advantage usually seen in those materials is a lower impedance and larger tissue tolerance, very possibly related to the same "shock absorber" effect of the intercalation processes. Along with this alternative mechanism for charge transfer, each material offers a peculiar chemistry and microstructure that may result in different prevailing intercalation processes and a variety of possible conductivity changes for the same applied fields.

Still, IrOx and conducting polymers, like PPy or polythiophenes with various counterions, fall short of the charge capacities desired for longer period stimulations, mostly because of strong limitations on coating thickness and substantial changes in conductivity during preparation. In our scientific approach, we could foresee a large scope for significant improvement in conductivity, charge capacity, and kinetics of ion exchange through the formation of hybrid materials by the use of nanostructuring processes that offer larger conducting interfaces. In particular, carbons are known as additives in many electrochemical devices and typically improve the conductivity of oxides by segregating on the surface of the particles and allowing percolation when forming composites [17].

Thus, with the aim of expanding electrode properties to allow longer periods and larger stimulation rates of electric stimulation, a variety of new materials and processes to prepare them is being explored. Composites with bioactive molecules and processes that modify the final structure are also being tested. Here, we will discuss some examples of electrode materials that have been successful under the application of electric fields. When comparing different material types, a range of micro- and macro-scale surface topographies will be present. Although neurons have been reported to be sensitive to the macroscopic topography of insulating materials [18, 19], the observations described in this chapter for conducting materials have shown no effect in that respect, so this aspect will not be discussed.

5.1.2 Cell Response and Repair Under Electrostimulation vs. Electrode Materials

In recent years, new microstructures in coatings of IrOx with improved adhesion [15] and doping [20], conducting polymers containing biological molecules [21, 22], and new hybrid materials of IrOx-nanocarbons have been developed and tested in neural cell cultures from mammalian and amphibian (*Xenopus laevis*) embryos [23–27]. Their charge capacities are two orders of magnitude larger than Pt and 10 times larger than those of the individual components. Importantly, they have allowed short-term electrostimulation with positive neurite outgrowth and have permitted new protocols for electric field application [26, 27].

As mentioned above, electrical stimulation is gaining prominence as a therapy for conditions as diverse as non-healing wounds and nervous system disease and traumatic injury. Generally, these procedures use AC pulses, and few use DC stimulation, either pulsed or steady, through direct connection to implanted metal electrodes. Such therapies are initially extracted from observations of isolated cells and cell cultures, often modeling in vitro regeneration of a nervous system wound by using a scratch wound model with an external electric field applied. Although this model was first used for glial proliferating cells [28], the observations shown suggest that the study of electric field effects based on these scratch models in neuron cultures results a very powerful alternative as a simplified in vitro repair model [26], if compared with sole cell cultures on electrode materials.

Alternatively, cultures of dissociated neurons from the embryonic spinal cord of *Xenopus* on insulating materials had been used to demonstrate a variety of

Fig. 5.2 *Xenopus* neuronal studies under an electric field on both insulating and conducting substrates. Materials were not "wired" directly to the power supply. Arrows indicate the imposed external electric field (solid red arrow) in the culture medium, and the dipole (dotted red arrow) of opposite polarity induced within the conducting materials on which the neurons grew. Reproduced from [27], with permission from John Wiley & Sons. © 2018 WILEY-VCH Verlag GmbH & Co. KGaA, Weinheim

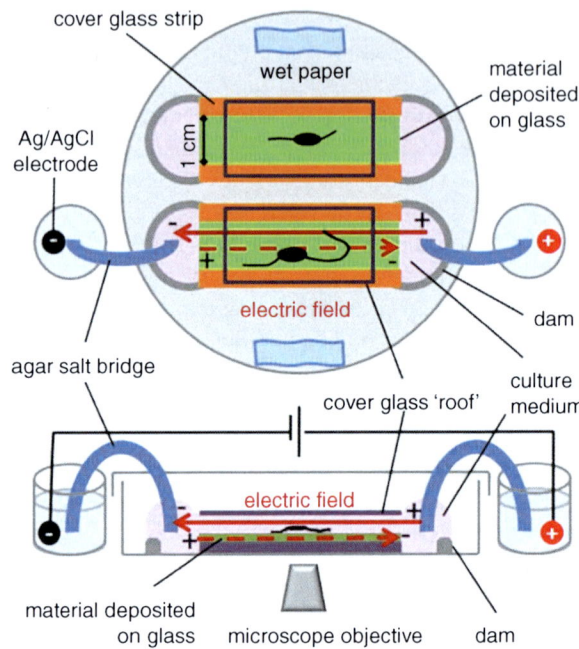

reproducible cell responses during electric field stimulation, including axonal migration parallel to the field vector and faster migration toward the external cathode (see Fig. 5.2). Cells survived well during field exposure (150 mV/mm) because agar salt bridges physically separated the neurons from the electrodes and their potentially harmful electrode reactions [27].

Although results from these fundamental studies are crucial to understand how cells respond to electrical stimulation, the use of salt bridges is not feasible for electrostimulation in biological systems. In terms of neural repair, the evaluation of the effects of delivered charge relative to the maximum charge capacity of each material has been performed using DC pulsed fields with constant polarization. The time scale of pulses has been in the order of minutes and relaxation times sufficient to dissipate heat effects, with total charges delivered above and below charge storage capacity (CSC) values. Specifically, DC fields with preprogrammed charge control protocols have been reported for various materials using cultures of mouse cortical neural cells [26], inducing wound repair in vitro in very short periods (less than an hour) using standard connected electrodes and without salt bridges to separate immersed electrodes. Repair is particularly significant when wounds are created on the IrOx-graphene hybrid material substrates, even if the same charge is delivered as in other materials with lower capacity limits [26]. To evaluate the possible differences in repair at the anode and the cathode, cells from both areas are separated by salt bridges. Surprisingly, cathode secondary reactions (e.g., O_2 reduction to radicals) seem to be more significant and slightly decrease neurite outgrowth.

Again, by using *Xenopus* neurons, this time grown on unwired conducting materials (see Fig. 5.2), an even more interesting observation was made [27]. The induced dipoles at the immersed conductive material created by the application of an external driving voltage are sufficient to stimulate cells through distant external electrodes. But it is not only the magnitude of the dipole that induces neural growth. Cells behaved differently depending on the specific conducting material on which they were grown. Indeed, the induced field either stimulated faster neurite growth (on IrOx and iridium-doped titanium oxide (Ir-Ti)Ox) or biased the direction of growth toward the cathode of the external field, on PEDOT-polystyrene sulfonate (PEDOT-PSS).

On the basis of the dipolar effects observed, direct dipolar interaction between conducting materials and neurons is also expected in the absence of electric fields. Because cells create dynamic fields through action potentials and signal transmission, conducting materials immersed in the cell cultures or implanted could respond to them through dipoles that would also be dynamic. Other previously reported experiments on various materials may be explained now through that dipolar interpretation [23–26, 29], although dipolar interactions had not been offered as the mechanistic interpretation at the time. Furthermore, behavioral changes have been observed in in vivo experiments in rats with electrodes simply implanted, but not connected [30, 31].

Finite-element theoretical studies have been made in an attempt to map the field gradients sensed by neurons in the presence of external fields. This work could aid the specific design of field magnitudes and geometries of implants [32].

Together, experimental and simple theoretical models suggest that the existence of dipolar interactions between materials and cells may be of use for additional types of electrostimulation devices. Both direct dipolar interactions with conducting materials and remote control of the induced dipoles on the materials may facilitate alternative designs in new clinical electrostimulation devices and also offer new interpretations on the neural interfaces established.

5.2 Materials with Large Charge Capacity and Nanostructured Hybrid Materials as Electrodes and Electrode Coatings

5.2.1 Oxides-Oxohydroxides, Coatings, and Electrochemical Methods to Achieve Good-Quality Coatings

Typical biocompatible oxides are based on TiO_2 and derivatives [33]. However, the lack of conductivity in those cases prevents their use as electrodes. On their turn, highly reduced Ti oxides have electronic conductivity, but they are not biocompatible. Among conducting oxides, Iridium is the only case where cell cultures reflect biocompatibility. While the IrOx coatings and precursor solutions are still being

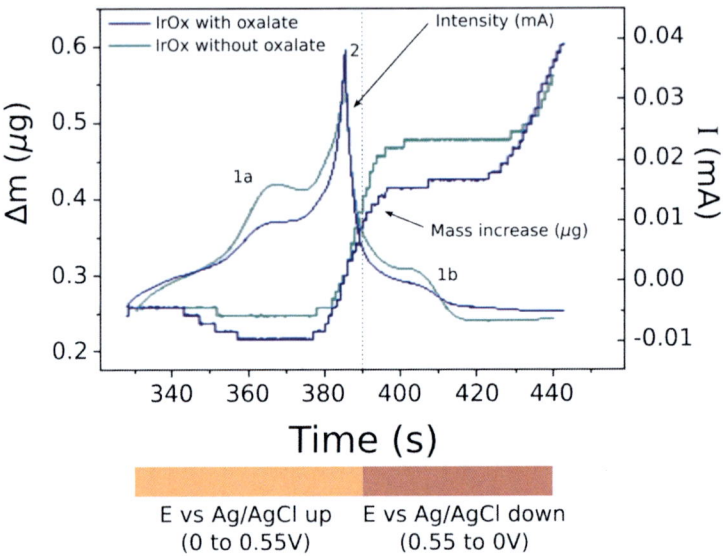

Fig. 5.3 Electrodeposition of iridium oxide coatings through dynamic anodic process (voltammetry response and simultaneous mass deposited through quartz microbalance studies) using aged oxalate and non-oxalate solutions (pH = 11) [34]. Intensity shows voltammetry, and mass change shows quartz microbalance signal simultaneously measured during deposition. The same mechanism is evident for the two solutions. X Time scale corresponds to an increase in voltage followed by a decrease in the original resting potential, as indicated in figure. (Reprinted from [34] under Creative Commons Attribution Licence)

studied, new electrodeposition processes have been found to achieve well-adhered stable IrOx coatings on Pt (or Pt-coated glass, carbons, and steel) that are stable during cell culture studies [15]. Notably, when such coating is obtained, the charge capacity increases five times (from 4 to 20 mC/cm^2), even with a surface roughness, and exposed area, several orders of magnitude lower (Fig. 5.3). Possible differences in the Ir oxidation state may arise, in our opinion, by changing the deposition method, but no detection of such changes has been possible. Clear evidence on the mechanism involved in anodic deposition and redox response is well observed, however, in the quartz microbalance study coupled with cyclic voltammetry in rich K$^+$ media [34]. Figure 5.3 shows the quartz microbalance response (along with the voltammetric response) during the dynamic deposition process from the precursor Ir solutions. In the anodic direction, a mass increase is observed, coupled with mass decrease (peaks 1a and 2), which corresponds to iridium oxide deposition and oxidation changes in it that come along with deintercalation of K. Thus, we observed simultaneously the deposition of the hydrated oxide and the redox intercalation processes. As the potential returns back in the cathodic direction, a new mass increase is observed (peak 1b), corresponding to K$^+$ intercalation in the solid. The dynamic deposition described yields to a conducting hydrated IrOx coating of low impedance that, in turn, can still be oxidized and reduced further (e.g., during electric field application in the presence of cells). K^+ ions in the solid are easily exchanged by H^+ upon immersion in water or sterilization [15]. Figure 5.4 shows transmission electron microscope images of the dried precursor solutions that allow such deposition process, while Fig. 5.5 shows the resulting transparent coatings (SEM profile image of IrOx coating in (a) and macroscopic images corresponding to various oxidation states in (b), (c), (d)) [34].

The precursor Ir solutions that yield IrOx coatings are significant in the development of more complex hybrid materials, as will be described below. Despite the claims of some publications [35], the solution does not contain IrOx nanoparticles. Although 2 nm nanoparticles are always observed at high-intensity electron beams typically used in TEM, when lowering voltage and current, small portions of dried solutions evidence clusters of iridium oxo species (dark in Fig. 5.4a) that evolve within minutes to species with local hollandite-type structures $K_x IrO_n$ [36], and eventually to 2 nm metallic Ir particles, with no presence of oxygen or potassium in EDX [34].

This Ir solution will be later used for the fabrication of hybrid, and IrOx-nanocarbon material and the IrOx deposition have been proven to be the driving process for the deposition of many hybrids. IrOx coatings show the same response than iridium oxo solutions under electron beams, suggesting that the oxo character is also preserved in the anodic coating and the possible existence of polyoxometalates in the precursor solutions (IrOn)x, with variable oxidation states. In all cases, dynamic deposition yields well-adhered and stable coatings during electrochemical cycling and cell culture, with an empirical formula $K_{1.7}(CO_3)_{0.2}IrO_{1.1}(OH)_{2.7} \cdot 0.4H_2O$, which, after soaking in pure water or neural cells culture media prior to any assay, exchanges K^+ for H^+ and yields $IrO_{0.3}(OH)_3 \cdot 0.4H_2O$. The non-stoichiometry of the phase described here results essential. If crystallized, IrO_2, without hydration or mixed valence states, does not offer the same electrode behavior [15].

5.2.2 Conducting Polymers with Various Counterions (X), PPy-X and PEDOT-X Bilayers and Composites

Thanks to the use of the same dynamic electrodeposition scheme developed on IrOx, other coatings have been prepared with good adhesion and stability upon cell culture studies, something that had not been easy before [21]. In particular, conducting polymers such as PPy-X and PEDOT-X, obtained by oxidation of the monomers, pyrrole and EDOT, are the best candidates for bioelectrodes, if the appropriate X counterions are chosen. Given that PEDOT-PSS is rather erratic as a substrate for mammalian neuron cell cultures [21], depending on the coatings used on active materials to induce cell adhesion [21], the authors approached electrodeposition using alternative counterions in the understanding that a surfactant like PSS, commercial and mostly used, could generate heterogeneities in the surface. In particular, using amino acids in aqueous solutions with pH above their isoelectric

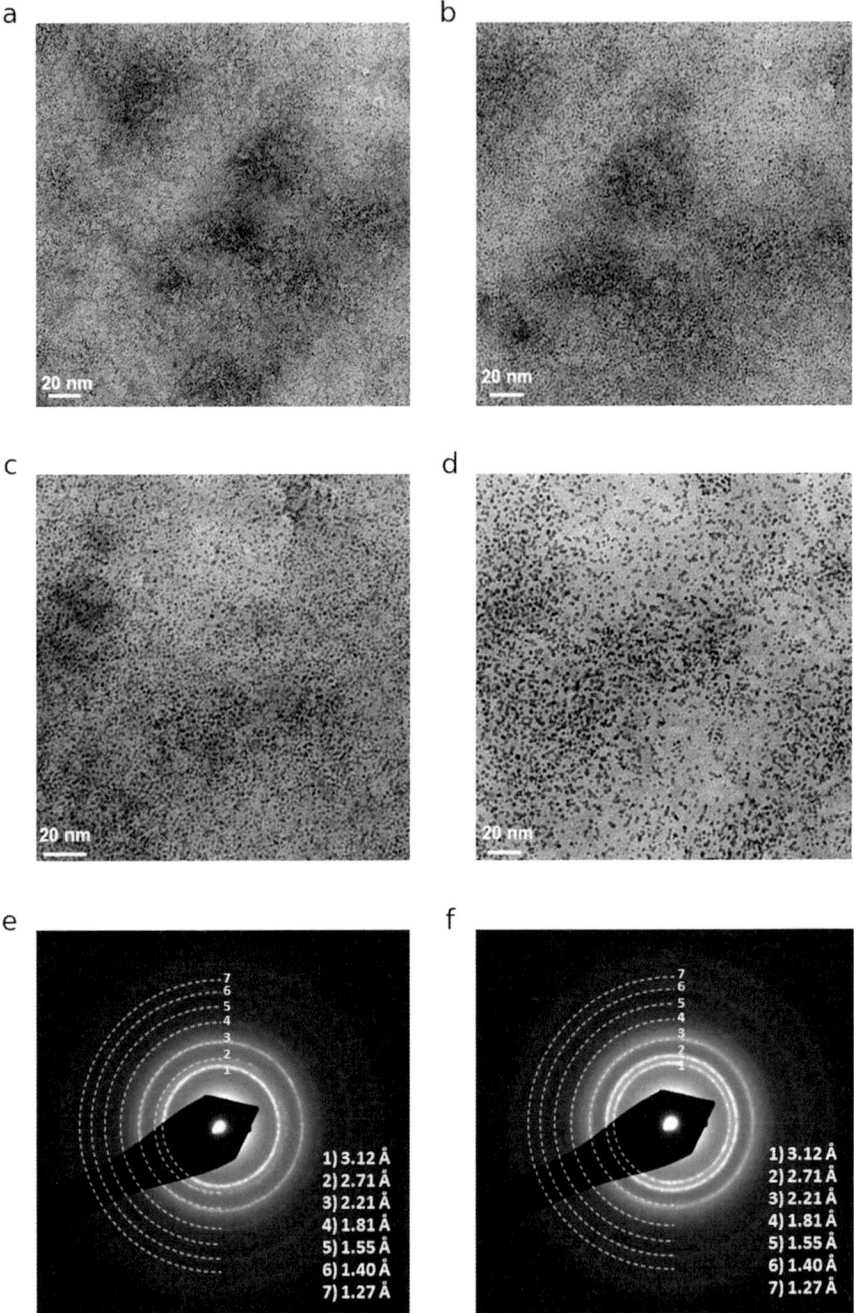

Fig. 5.4 (**a**–**d**) Transmission electron microscope images of dried precursor iridium solutions that allow anodic deposition. Sequential time evolution (5 min) in the same zone from oxoiridium species in solution to metallic iridium 2 nm nanoparticles, by using an 80 kV electron beam in a 120 kV Jeol TEM microscope. (**e**–**f**) Examples of electron diffraction circles observed in the same time evolution scheme [34]. Circles involve several nanoparticles being observed, suggesting hollandite quasiamorphous structures and metallic Ir. Evolution of oxygen and potassium is clear. Acknowledgement to Dr. J. Oro (ICMAB-CSIC, Spain). The final EDX signal (not shown) only contains Ir. (Reprinted from [34] under Creative Commons Attribution Licence)

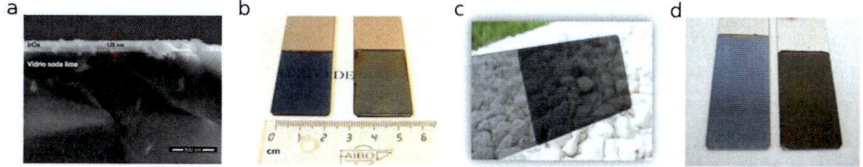

Fig. 5.5 (a) SEM profile image of IrOx coatings using dynamic potential sweeps, (b) macroscopic color changes observed upon reduction for IrOx coatings on Pt-coated glass (involving simultaneous Na^+ intercalation [21]); blue as prepared through anodic sweeps deposition, and green upon reduction in sodium phosphate buffer, (c) PEDOT-PSS coating con Pt-coated glass, and (d) PEDOT-lysine on PPy-lysine bilayers on Pt-coated glass. (a, b) Adapted with permission from [15]. Copyright (2012) American Chemical Society. (c, d) Reprinted from [21], Copyright (2013), with permission from Elsevier

point, highly compatible coatings of conducting polymers have been achieved with the amino acid being a counterion. Although deposition is more difficult to achieve, the final polymer is more biocompatible and supports neural cell growth better than other reported polymers. Furthermore, stabilization of PPy versus atmospheric oxidation is achieved by forming bilayers with a common counterion (e.g., PPy-lysine deposited on PEDOT-lysine), with better response in mammalian neuron growth than any polymer reported [21]. Although having optimal cytocompatibility, charge capacities for all conducting polymers tested remain similar or inferior to that of dynamically deposited IrOx [15, 21].

5.2.3 The Role of Carbons and Nanocarbons in the Formation of Hybrid Materials

With the aim of achieving materials with larger charge capacities, structural stability, and biocompatibility, a number of hybrid materials taking IrOx as a basis have provided successful results. Using the same dynamic electrodeposition processes and variable limiting potentials, precursors of iridium oxosolutions have been mixed with pyrrole, EDOT, graphite, carbon nanotubes, and various graphene varieties, in combination of either two or three elements.

Even though both conducting polymers and IrOx are obtained by anodic depositions, when mixed, electrodeposition does not work. A direct chemical reaction of the monomers and iridium yields a hybrid IrOx-PPy or IrOx-PEDOT hybrid in powder form, where the polymer encapsulates the oxide, evidencing that the polymers form around the nucleating oxidizing species (IrOx) and such particles prevent electrodeposition [37]. In turn, this suggests that IrOx electrodeposition is hindered in this case by the oxidized polymer and that IrOx deposition is the key process to develop hybrid coatings.

Using a carbon component in an IrOx-based material is an attractive twist given the known behavior of carbon in supercapacitors and batteries and the improvement of the conducting properties of known oxides when admixed with small amounts of carbon [17]. In general, it is observed that carbon segregates at the oxide surfaces, offering a percolation road possible for electron conduction. Such resulting microstructure results in better conductivity in the final composite material. By using nanocarbons instead, we expected to find the possibility of nanostructuring new hybrid materials offering significant improvement in their properties.

Thus, biocompatible hybrid materials based on IrOx and carbons have been obtained with graphite, graphene oxide, exfoliated graphene, and carbon nanotubes (CNTs). In all cases, evidence exists [23–25] (Fig. 5.6) that, this time, Ir species adheres to carbon in the precursor solution (a and b, pristine graphene exfoliated electrochemically from graphite), as opposed to the case of polymers [37]. Figure 5.6 shows SEM images indicating unambiguously the nanostructures achieved through the interaction and electrodeposition process driven by IrOx (nanocarbons do not deposit in the absence of IrOx). IrOx is sustained by CNTs in one of the hybrids in a nanostructure that resembles reinforced concrete. In the case of pristine graphene, graphene oxide, and N-doped graphenes, a millfeuille nanostructure is formed. Noteworthy, hybrids of IrOx with nitrogen-doped graphene oxide do not allow neuron growth [38] and will not be described in this chapter. In the presence of carbons, conducting polymers may also be incorporated into the nanostructure forming a trihybrid phase.

In all cases, cyclic voltammetry for all IrOx-nanocarbons hybrids (Fig. 5.7a) shows the expected Ir redox processes (with the exception of the trihybrid containing PEDOT), evidencing that the material has that charge capacity available during stimulation. However, if PEDOT is incorporated, the trihybrid case IrOx-CNT-PEDOT, only PEDOT processes are available electrochemically, suggesting encapsulation of all components by the polymers [39]. Accordingly, IrOx-nanocarbons hybrids, except the trihybrid IrOx-CNT-PEDOT, have charge capacities ten times higher than the dynamic IrOx electrodeposit, reaching 120 mC/cm^2. The cycling stability after 1000 cycles remains about 70%, implying a 10-fold increase in stimulation time possibilities. The IrOx-CNT-PEDOT trihybrid, however, has the charge capacity previously observed for PEDOT-PSS, again confirming encapsulation. It is significant that, if calculated per carbon atom, the capacity of the IrOx-pristine graphene hybrids doubles that of the one obtained with graphene oxide, despite the larger roughness and surface exposed in the last case, implying that the effect of conducting pristine graphene is larger [34]. On the other hand, the same enlargement of charge capacities is reached in fresh IrOx-graphite hybrids but, after a few electrochemical cycles, the capacity fades to 10% of the original, through disintegration of the hybrid. Therefore, the formation of nanostructured hybrids offers new alternatives in charge delivery for IrOx-nanocarbons combinations, where capacity retention is above 70%.

As described below, several models for cell growth evaluation have been chosen, but it is significant to mention here that the different nanostructures described reflect also in cell cultures. In all IrOx-nanocarbon hybrids, with the exception of the

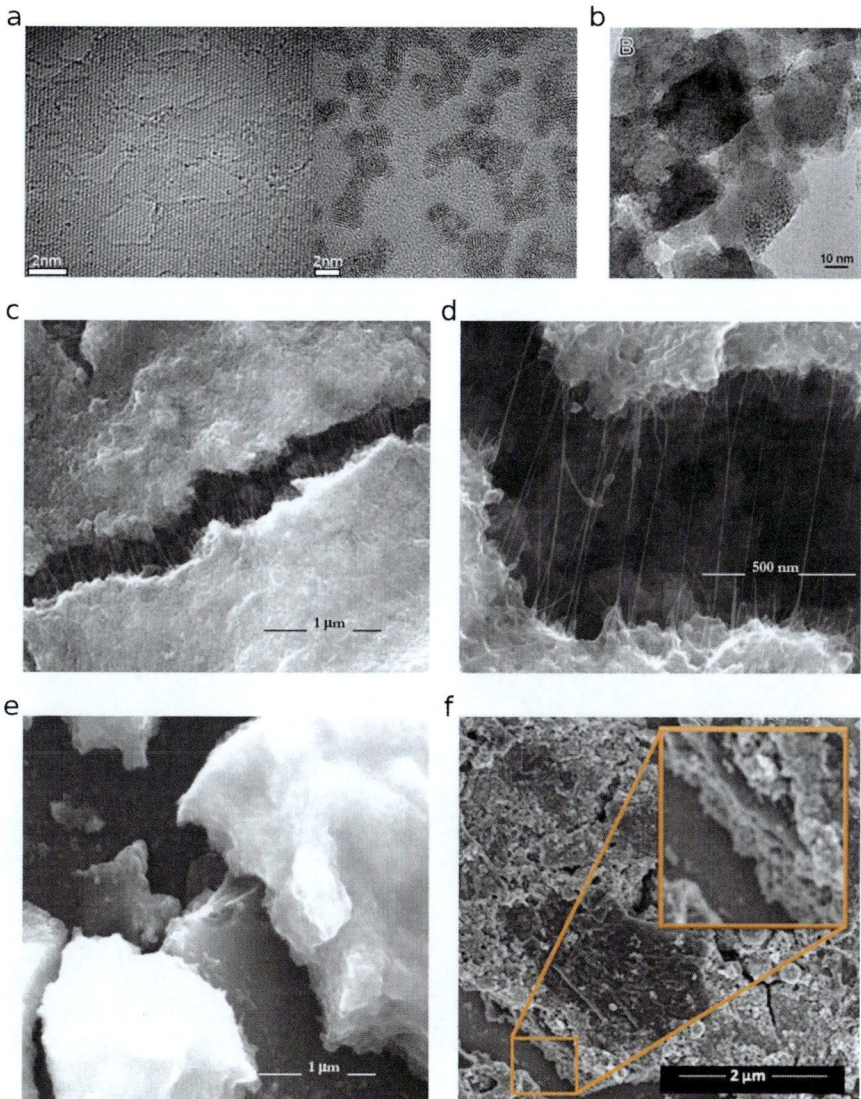

Fig. 5.6 (**a**) High-resolution TEM images of pristine graphene obtained by electrochemical exfoliation from pure graphite (left) and iridium oxo species (right) attached to it. Unpublished data (acknowledgement to Dr. Raul Arenal, ICMA-CSIC, Spain). (**b**) TEM images showing the adhesion of iridium precursors to the graphene surface. Reproduced with permission from reference [25]. Copyright © 2015 Materials Research Engineering C. Published Elsevier B.V. (**c**, **d**) SEM images of nanostructured electrodeposited IrOx-CNTs hybrids and IrOx-CNTs-PEDOT hybrids, respectively, showing the nanostructure where CNTs sustain IrOx as in reinforced concrete and how PEDOT encapsulates both IrOx and CNTs. (**c**) Reproduced with permission from reference [23]. Copyright © 2014 Acta Materialia Inc. Published Elsevier Ltd. (**d**) Reproduced from [39], Copyright (2018), with permission from Elsevier. (**e**, **f**) IrOx-pristine graphene and IrOx-graphene oxide hybrids, respectively, showing the millfeuille nanostructure. Cracks were developed only under the vacuum at the SEM microscope, allowing the view of the inner structure. (**f**) Reproduced with permission from reference [24]. Copyright © 2014 Elsevier B.V

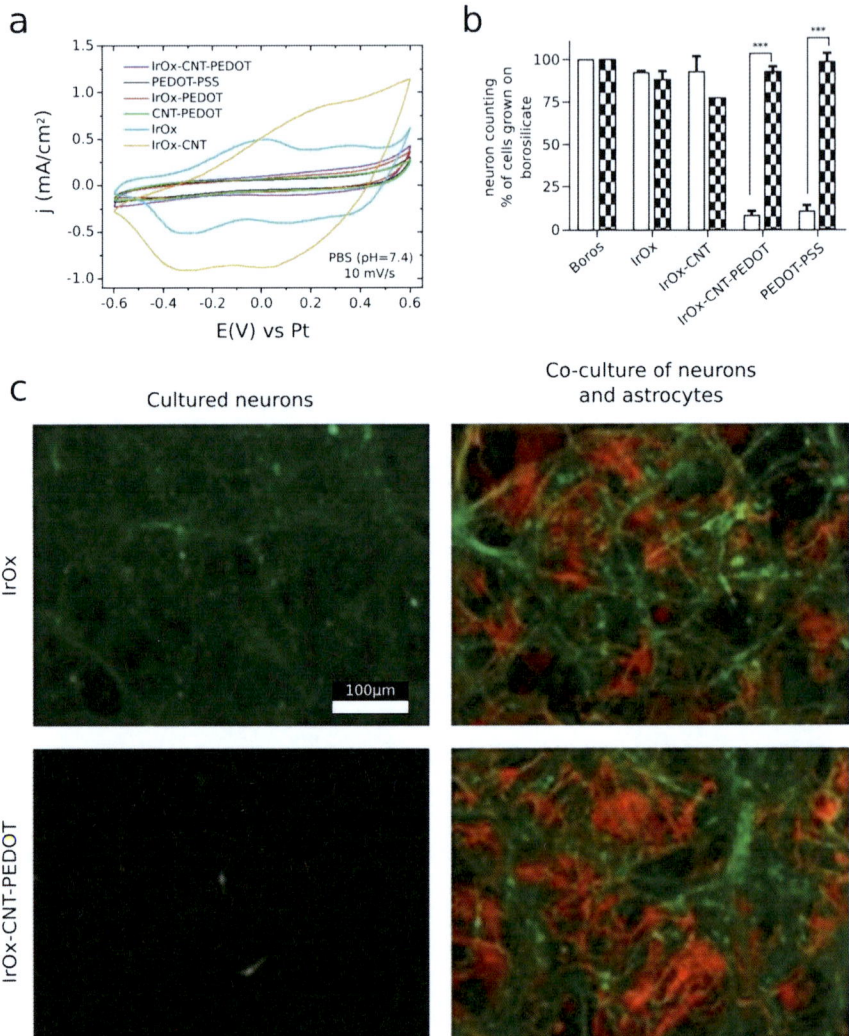

Fig. 5.7 (**a**) Comparison of cyclic voltammetries and derived charge capacities for all hybrids. (**b**) Quantitative data of neural cell growth on the diverse substrates (neural cells were seeded at 1.15×10^4 cells cm^{-2} on different materials coated with poly-L-lysine (white bars) or with a monolayer of confluent astrocytes (cross bars) and grown for 5 days in vitro). (**c**) Representative fluorescence images of neural cell cultures on the IrOx-based substrates. Cells were labeled for tau (green) and glial fibrillary acidic protein (red) to identify neurons and astrocytes, respectively. Reproduced from [39], Copyright (2018), with permission from Elsevier

nitrogen-doped graphene, neurons grow well. Again, hybrids containing PEDOT signify a different behavior. Neurons in cell culture with IrOx-CNT-PEDOT do not grow well, similar to that found for PEDOT-PSS (Fig. 5.7b, c), suggesting that these polymers may not sustain neural growth and that PEDOT is encapsulating the highly compatible IrOx and the nanocarbons. However, if co-culture of astrocytes and neurons is performed, both cell types grow and differentiate properly. Furthermore, neural cells grown in co-culture on these materials were more resistant against inflammatory insults [39]. On the other hand, *Xenopus* cells find an optimal substrate in PEDOT-PSS-based materials, with no surface adhesion phases like polylysine, but also on IrOx-based materials, with various behaviors in the presence of electric fields. Based on these observations, PEDOT encapsulation of other components in the trihybrid material is proven by the similarities in voltammetry, capacities, and neurons response with respect to single PEDOT. Such consideration, as well as the fact that IrOx-nanocarbon electrodes remain to date the optimal electrode material, may be useful in the design of future materials.

5.3 Biocompatibility Assessment of Electrode Materials with Neural Cell Cultures

Once materials with the desired characteristics have been prepared, it is necessary to first show that they are not harmful to cells. Initial biocompatibility tests typically involve cells cultured on the material surface. If destined for use in the nervous system, cells of selection include clonal cell lines that mimic neural cell phenotypes (e.g., PC-12 cells), neural progenitor cells, and primary neural cultures directly obtained from dissociated animal neural tissues (e.g., dorsal root ganglia, brain, and spinal cord). Although these mammalian cell culture systems are highly desirable for modeling the interactions of the neuron–material interface, such in vitro models have a significant limitation. Specifically, mammalian neurons (especially those from the central nervous system) require an adhesion layer to promote cell attachment to the substrate, without which mammalian neurons generally do not adhere or extend their neurites. Adhesion layers, such as polylysine, which extends 1 nm from the material surface, or collagen, which extends 100 nm or more, can also influence the outcome. Mammalian neurons growing 100 nm away from the material surface, as on collagen coatings, do not "see" the material chemistry. It is highly probable that this explains variable reports in the literature observed for cells grown on PEDOT-PSS coated in different ways. An attractive alternative in vitro model uses embryonic amphibian spinal neurons from *Xenopus laevis*, which do not require an adhesion layer for cell attachment or to initiate the sprouting of robust neurite arbors. Additionally, these neurons grow at room temperature in atmospheric conditions, so they do not require the use of a CO_2 incubator, axons extend much faster than those from mammalian neurons, and their cell bodies and growth cones are about twice as large as those of mammalian neurons. Individual cells are also observed easily. Collectively, these conditions make them ideal for time lapse observations of neuron growth and behavior. This is facilitated by using materials prepared as transparent films. Indeed, by using time lapse imaging, we have demonstrated that PEDOT-PSS and IrOx are favorable substrates for *Xenopus* axon outgrowth, which either matched or exceeded that on uncoated glass [27] and have positive outcomes from the above-described hybrid materials.

Although amphibian neurons are extremely useful as a first approach for testing novel material biocompatibility, it is decisive to test those with mammalian neural cell cultures as applications are predominantly intended for use in humans. This is often done using cultures of neurons from the embryonic rodent brain cortex, especially in the context of evaluating electrodes for nervous system repair, but co-culture systems of cortical neurons with astrocytes provide additional value as they more faithfully mimic the native neural tissue (Fig. 5.7).

When the main focus of the study is the use of those materials as electrodes for electrostimulation and repair, direct cell seeding and growth observation are far from enough. We focus below on two recent models, the scratch wound healing model in the presence of electrodes [26] and the remote control through unconnected electrodes [27], which have recently shown significant advantages for the study of electrode materials, their effects on living systems, and which suggest the design of new electric field protocols and devices.

5.4   In Vitro Wound Healing Models for Repair Effects Under Electric Field

A simple in vitro approach when studies are focused on repair is to induce a wound-like scratch on a cell culture and observe the replenishment of the void space over time. In these studies, a scratch is physically done with controlled dimensions in a monolayer of mammalian neurons previously grown on a variety of material surfaces. The rate of wound closure in the presence of electric fields is then monitored to model the rate of recovery of the nervous culture as compared with spontaneous repair.

Figure 5.8 shows the spontaneous neurite growth into the wounded area at different times after the scratch is made on cultures of mouse cortical neurons grown on PEDOT-PPy-lysine bilayers. A maximum rate of neurite growth was found between day 1 and day 3, allowing complete spontaneous healing (seen as connection between both sides of the wound) within 7 days [26]. By exploiting selective immunohistochemistry techniques that label distinct cell populations, it is possible to look at complex behaviors in such cultures.

Using this spontaneous repair as reference, we observe changes in wound recovery upon electrostimulation using a variety of electrode materials, evidencing one of the closest in vitro models to the in vivo scenario and helping to determine the specific electric field protocols for which repair is achieved, as a function of the electrode material properties.

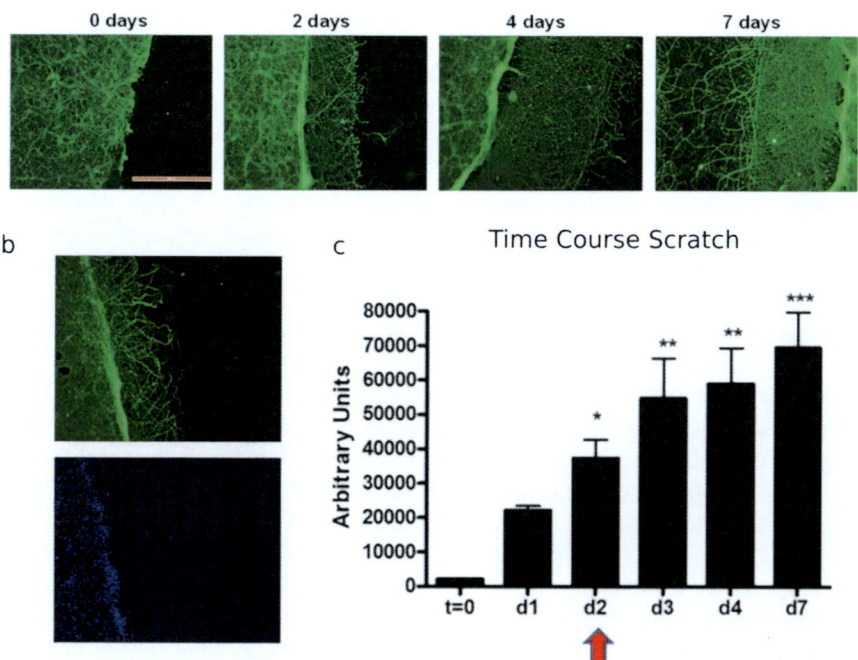

Fig. 5.8 (**a**) Spontaneous neural cell recovery after scratch wounding of monolayer cultures grown on PEDOT-PPy-lysine bilayers. Cells are immunolabeled for tau protein. Scale bar 500 μm. (**b**) Neurite and nuclei images for comparison (benzimide staining for nuclei). (**c**) Neurite counts at different time points, showing that 2 days in vitro after scratch is a time where 50% of recovery was obtained after 2 days and it is chosen as the best time to study electric field effects. Reproduced from [26], Copyright (2016), with permission from Elsevier

5.5 Establishing Safe Limits for Applying DC or Low-Frequency AC Electric Fields for Wound Repair Depending on Material Properties

When evaluating the possible use of electrostimulation for tissue repair, specific responses of neurons to electric fields reported to date are significant. A variety of responses have been described for neurons growing on insulating substrates (glass or plastic) and exposed to steady (DC) or oscillating (AC) electric fields in vitro. Responses include increased migration rates of neural progenitor cells, directed migration (usually to the cathode of a DC field), directional neurite outgrowth (biased toward the cathode, the anode, or orthogonal to the field vector, depending on the cell type), and increased branching [40]. Other than neurons, in vitro scratch wound cultures akin to those described in Fig. 5.8 have been used to model epithelial wound healing on insulating materials. Indirect electrical stimulation of such cultures has shown that the presence of an electric field in the culture medium influences the rate of wound closure for cultured cells including fibroblasts, epithelial cells, and keratinocytes [41]. Collectively, these observations suggest that the application of an electric field is therefore likely to influence the rate of wound closure in the neural culture scratch wound model, probably through directional neurite outgrowth. Typically, these studies [41] use conditions in which there are limitations to the electrode materials used. The low capacity of usual electrodes and the effects of surpassing those capacity values to deliver the charge are detrimental to cells if the electrodes are positioned directly in the culture medium. Therefore, such studies are made by separating cells from electrodes through salt bridges. On the other hand, it is not unusual to find reports where no separation is made (and perhaps it is not possible) especially for in vivo studies, thus surpassing the values of charge capacities.

Charge storage capacity (CSC) studies with various materials define CSC as a decisive point when establishing DC field effects, also in repair models. Other parameters, such as heating and geometry, are highly related to the material chosen, and therefore, the electric field protocol must be defined as a function of those properties, as clearly seen below. Figure 5.9a shows an example of pulsed potential electric field protocols with simultaneous current integration, and therefore charge control, that allow the application of constant polarity fields with relaxation times in the order of second to minutes to prevent heating effects [26]. A protocol of applied potential with alternating pulses (anodic pulses smaller than cathodic pulses) and final charge control is also included, similar to described previously [3], but in larger time scales and with larger net delivered charge to induce repair. Figure 5.9b, c shows a comparison of cell culture density and morphology for mouse cortical neurons, when values of charge delivery are below and above the previously measured charge capacity for a particular electrode (bilayer PEDOT-PPy-lysine in this case). Independently of the DC or AC mode, cell growth near the electrode drops dramatically once the charge capacity is surpassed (Fig. 5.9b). Not only the specific material charge capacity is significant, but also the global capacity from the full electrochemical cell. Such cell may be built either symmetrically, with the same material in both sides in the same oxidation state, or modifying the anode to achieve a previously reduced state to magnify the global electrochemically cell capacity. This possibility, which is not present in capacitive materials, shows here its full potential since the global capacity may double even for the same area and geometry [26].

When applied to the scratch model, evaluation of neurite extension into the wound after electrostimulation offers additional results. By comparing various electrode materials as anode and cathode, and using the same charge delivered (that of the material with lower capacity), significant results are obtained (Fig. 5.10). Also if anode and cathode compartments are separated by a salt bridge, differences are observed in each side. The result of neurite extension over the scratch wound depends clearly on the materials used as electrode (Fig. 5.10a). For the same charge delivered using various materials, those with the largest charge capacity in

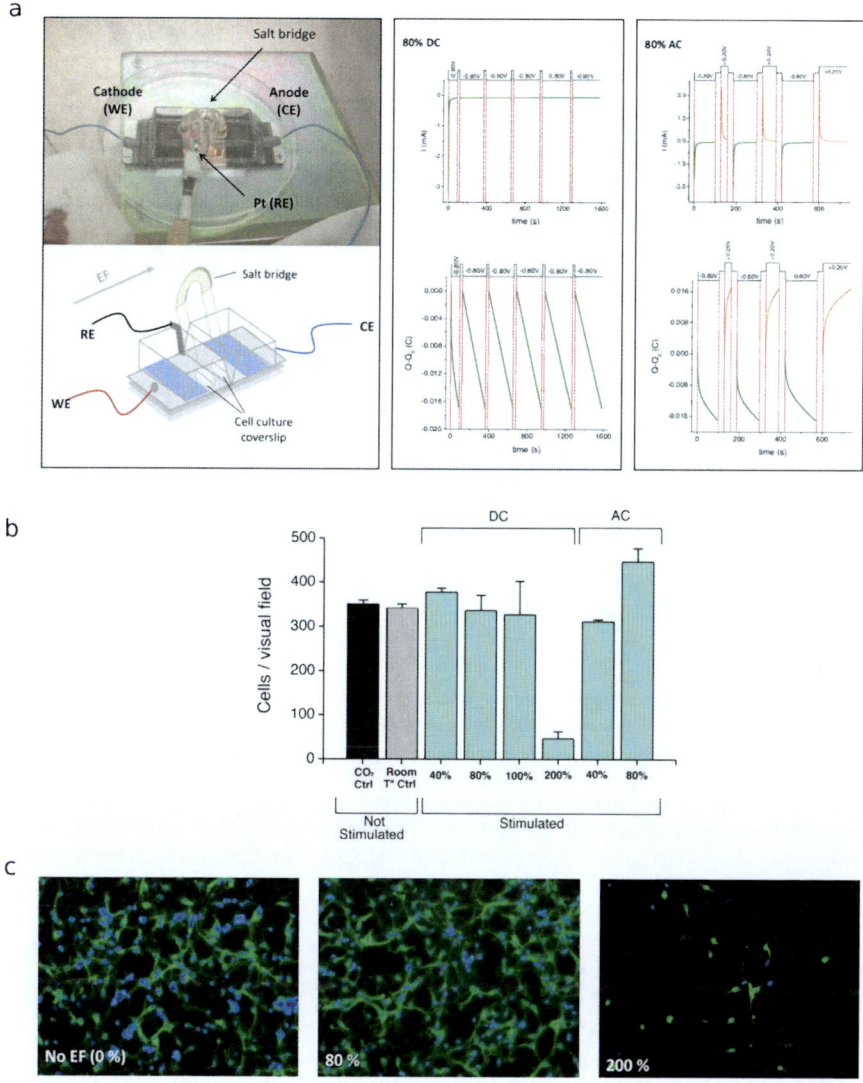

Fig. 5.9 (**a**) Electrochemical cell and electric field protocol applied in scratch wound cell cultures and current responses. (**b**) Cell counts after electric field being applied in a cell culture with a symmetrical electrochemical cell and Q (electrical charge) delivered below and above electrode CSC limits (0, 80, and 200% of CSC) of bilayer PEDOT-PPy-lysine. (**c**) Images of neuron cultures after charge delivery below and above charge capacities. Neurons were labeled for tau and cell nuclei with bis-benzimide staining. Reproduced from [26], Copyright (2016), with permission from Elsevier

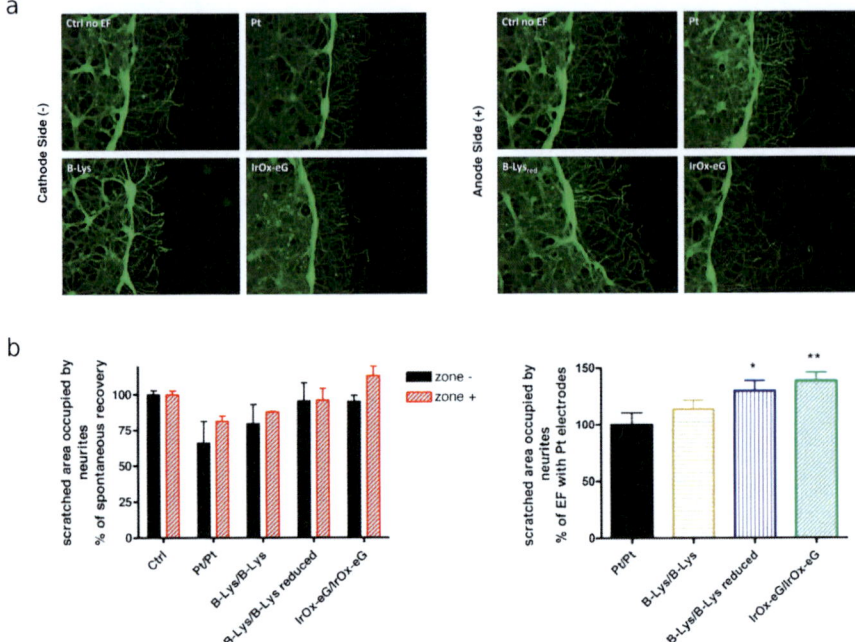

Fig. 5.10 Scratch model in neuron cultures under short periods of electrostimulation. (**a**) Neurite regrowth 2 days after the following protocol: Neuron culture 5 days in vitro, scratch wound and 40 min of electrostimulation. Charge delivered 80% of the lowest CSC intercalation material, PEDOT-PPy-lysine bilayer. Cathode and anode sides are shown for comparison. (**b**) Quantification of the area covered by neurites and relative comparison with Pt electrode for various electroactive materials (bilayered polymers and IrOx-pristine graphene hybrid coatings). Neurons were identified by tau immunofluorescence. Reproduced from [26], Copyright (2016), with permission from Elsevier

experiments where up to 80% of the maximum charge capacity is delivered show the best repair (of course, that charge would only be a small portion of their maximum charge capacity for the materials with larger CSC, but comparison among materials are made with the same charge delivered).

Very significant also is that recovery of the scratch space occurs having used very short electrostimulation time periods. After less than 40 min of stimulation, cell cultures were followed in vitro during 7 days in vitro, showing that an electric field induces neurite growth above spontaneous healing on the same material (Fig. 5.10b, 2 days after electrostimulation). It is worth comparing their response with Pt electrodes because they are so usual in clinical electrostimulation. In this particular case, charge delivered is necessarily above the limits of the capacity of the electrodes, and the scratch model shows that, for Pt electrodes, growth is inferior to the spontaneous response. Thus, electric fields in net charge mode would not help repair if Pt electrodes are positioned near the lesion, but they do with other large capacity electrodes. The largest repair is found for the hybrid IrOx-pristine graphene electrodes, even if the same time, and applied charge is used then for others, within the limits of capacity for all. Very significant is also that the behavior is slightly better at the anode, suggesting that O_2 reduction at the cathode may be the cause of radical formation in the biological media and the origin of harmful responses. For that particular IrOx-graphene material, charge delivery could still be expanded more than five times, with respect to bilayer PEDOT-PPy electrodes, before reaching its maximum capacity, working at 100 times more charge than the value of Pt. Thus, electrostimulation times could easily be enhanced before inducing detrimental effects.

The results above are relevant clinically as of today. Recently, electrostimulation during months, using AC fields and rehabilitation programs, yielded actual recovery in spinal cord injury patients [42, 43], although no electrodes or electric field protocols are being disclosed. Their results may also benefit from the added extra charge capacity described above for new electrodes and the corresponding protocols.

5.6 A Promising Alternative to Direct Electric Field Stimulation: Induced Dipolar Electrostimulation

Usually, electrodes work connected to an external power source. However, unconnected conducting materials immersed in an ionic medium will show a dipole between their edges, if a remote external field is applied with external driving electrodes. If the potential is sufficiently large, the induced dipole may yield electrochemical reactions at any of the poles, negative or positive. This is known in the literature as bipolar electrochemistry [44], recently rediscovered for materials preparation, but never studied in terms of materials–cells interactions until now.

Several electrode materials described above have been useful to test the influence of electric fields on the possible growth of neurites, in terms of speed and direction, using them as unconnected substrates. The *Xenopus* neuron culture system was exploited by us to ensure that cells were in direct contact with the material itself (not intervening any adhesion layer). A number of biocompatible materials, insulators like TiO_2, pure electronic conducting materials (such as Au and Pt), and mixed conducting materials (IrOx, PEDOT-PSS, and PPy-X), were compared in the absence and presence of the electric fields created by remote external electrodes. Without electrical stimulation, neurons extended neurites on Au, Pt, PEDOT-PSS, IrOx, and the mixed (Ir-Ti)Ox. As previous results showed that nano- and macro-scale topographies influence the rate and direction of neurite outgrowth of *Xenopus* neurons on fused quartz substrates [45], we expected optimal neural growth to be related to surface texture. However, there was no direct correlation in ranges from nanometers to microns. On the one hand, IrOx and PEDOT-PSS, which are both rougher than glass, improved neurite growth rates over glass. On the other hand, neurites also grew faster on Au, which is smoother than glass. Additionally, both the smoothest (Au) and the roughest (PEDOT-PSS) surfaces encouraged reliable neurite sprouting.

After identifying the most suitable materials that supported neural growth in the absence of electric fields, we used time lapse observation to test the ability of those materials to sustain growth during 3 h of constant electric field stimulation. Although external imposed electric fields always induce a dipole between the poles of the immersed conducting material, for magnitudes 100 to 150 mV/mm, they induced a large enough voltage between the poles to generate H_2 (and OH^-) and O_2 (and H^+) at the induced cathode and anode poles, respectively. This was visually evident by the gas bubble production and by the change in color of the pH indicator in the culture medium at the extremes of the material (Fig. 5.11) [27]. Furthermore, delamination of the Au and Pt coating occurred because of oxidation of the adhesion layer of titanium underneath, an oxidation that progressed as a wave along the material over time, as the conducting border moves. Indeed, those electric field values were demonstrating the effects of a dipole induced in the immersed conducting material and suggested that cells would be sensitive to it. Although 100 and 150 mV/mm electric fields are safe when used on insulating substrates, and no dipole appears, only 50 mV/mm fields were used for neural growth experiments on conducting materials to prevent undesired effects on the material and the electrolyte surrounding it.

The presence of the imposed external DC electric field in the culture medium indirectly induces an electric field in the conductive substrate material that has a polarity opposite to that in the culture medium (Fig. 5.11e–i). Therefore, a different behavior is expected between purely conducting Au or Pt and intercalation materials like IrOx and PEDOT-PSS because of the expected intercalation/deintercalation of M^+ ions (e.g., Na^+, H^+, K^+) and anions like OH^- at the anode, not possible in pure metals. We have used finite-element analysis to map the charge gradients that are likely to occur and have observed ionic gradients with spatial resolution by in situ XRay absorption spectroscopy in IrOx in bipolar conditions [32, 46] and have observed how the intercalation process occurs yielding a gradient material while softening the changes in electric field near the electrode border. However, to date, no model predicts the differences observed among materials.

Importantly, at lower external driving fields (50 mV/mm), the material maintained its integrity, no pH changes or gas bubbles were observed and neurons continued to grow. Results indicate that specific cell growth responses were induced depending on the conductive substrate material in which the field was induced. For example, on PEDOT-PSS, neurite growth was directed preferentially toward the external imposed cathode (Fig. 5.12). However, at the same external field conditions (50 mV/mm), neurite growth on IrOx and (Ir-Ti)Ox was not directional, but the neurites extended faster than those on PEDOT-PSS. This suggests that selected electrostimulation effects are possible through implanted unconnected electrodes, with defined behaviors for each material type.

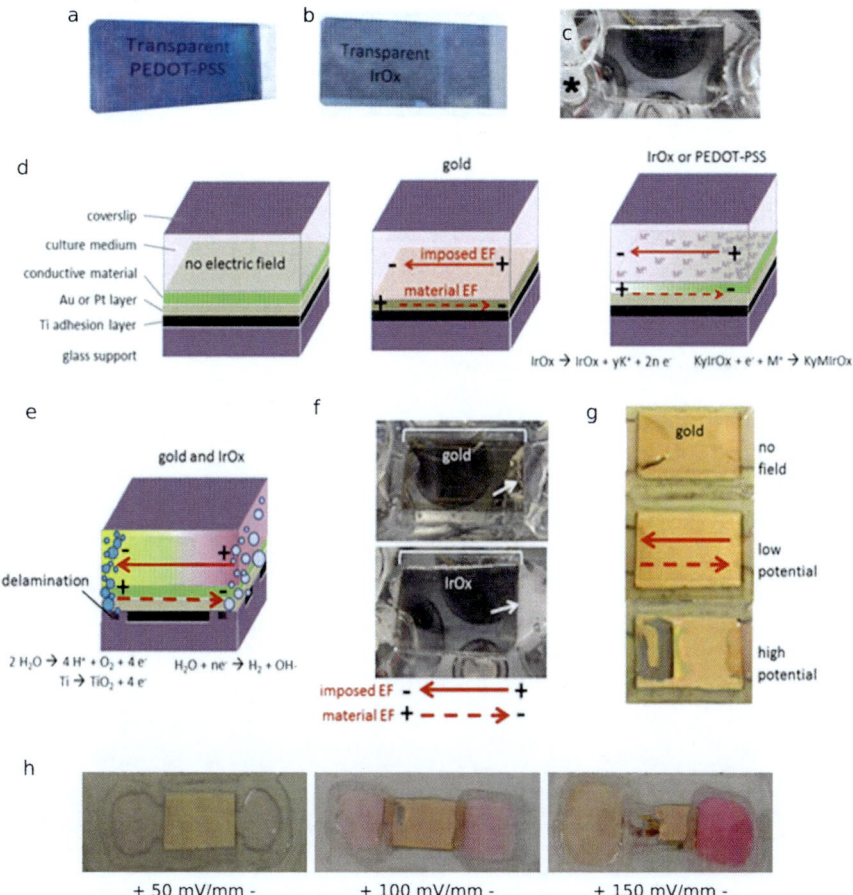

Fig. 5.11 (**a–c**) Transparent material coatings used in the study. (**d–g**) Scheme of reactions expected at low and high electric field application. (**h**) pH changes observed under external applied fields. pH increase changes the medium color to pinkish ($H_2 + OH^-$ formed). Reproduced from [27], with permission from John Wiley & Sons. © 2018 WILEY-VCH Verlag GmbH & Co. KGaA, Weinheim

5.7 Conclusions and Future Perspectives

The results summarized in this chapter evidence a wide perspective in the development of new electrode materials with a large span of electrochemical properties and nanostructures. It is clearly seen that a direct electric field stimulation modifies cell behavior not only because of ionic gradients created in the ionic media but also depending on the electrode surface chemistry. Short-term electrostimulations with net charge allow enhancement of neurite growth speed and angle modification. On the other hand, remote wireless stimulation is possible through the induced

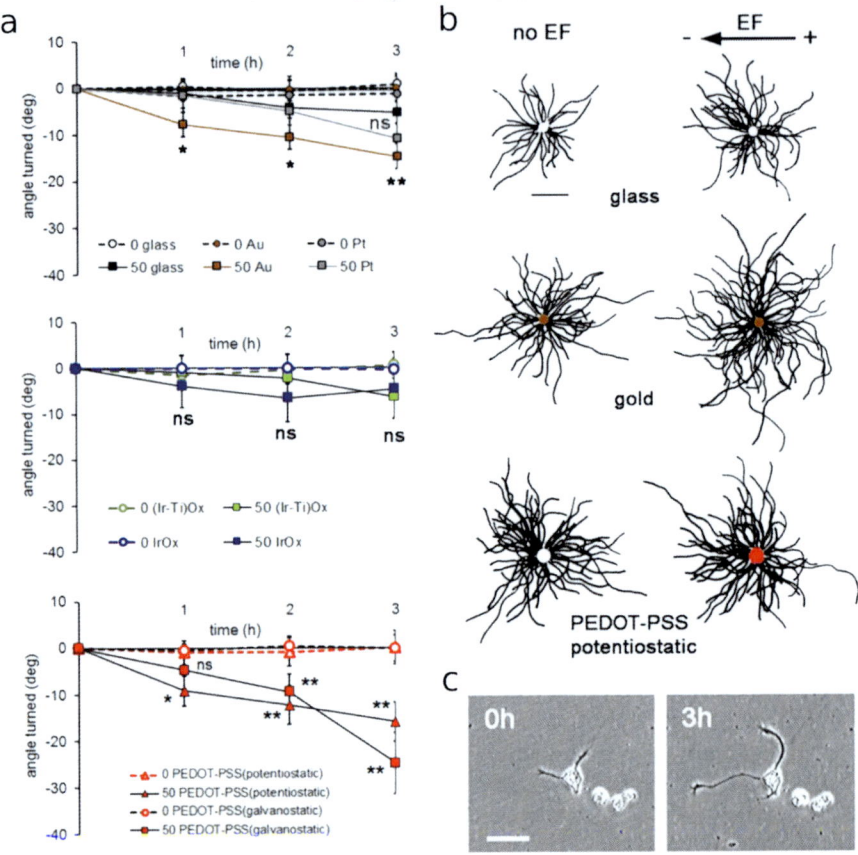

Fig. 5.12 Neuron growth during 50 mV mm^{-1} electric field (EF) stimulation on several materials. (**a**) Angle of growth cone migration. Negative values indicate migration toward the cathode, while zero indicates randomly directed migration. (**b**) Exact diagrams of individual *Xenopus* neurons growing in various materials in the absence and presence of an external electric field. (**c**) Time evolution of a neuron, neurites pointing the cathode. Reproduced from [27], with permission from John Wiley & Sons. © 2018 WILEY-VCH Verlag GmbH & Co. KGaA, Weinheim

dipoles in conducting immersed materials. Not only do these experiments suggest that remote electrostimulation is possible, but they also offer a wide perspective of what the cell–material interactions may be. Since neurons develop dynamic action potentials of high local intensity, a conducting material implanted near them could respond in the form of dipoles or multipoles to the electric signals in neurons that would, in turn, be dynamic. This material dipolar response may, in turn, influence in a feedback manner, cell behavior and growth. The two-way road through which interactions can occur opens a vast field of study. Even previous studies with the materials described herein, or others based on nanocarbons, for instance, may be reviewed differently thanks to the observations from bipolar electrodes.

The authors have been aware of recent work where a more active behavior of rats was observed upon the implantation of an electrode for DBS, prior to the electrode being connected [30, 31]. It is our opinion that such behavior may develop from the direct interaction between neurons and the dipoles induced in the material and that, being dynamic, could create a larger overlapping brain wave. If true, this would open the possibility of indirect stimulation in the brain through unconnected implants.

Another area ripe for exploration is the targeted synthesis of active materials in situ that could act later as bipolar electrodes. In this sense, a recent report proved that polymerization of conductive polyaniline (PANI) and PEDOT could be accomplished by using aqueous components that polymerize at physiological pH, genetically targeting polymer synthesis to neuronal cell membranes [47]. The potential to design a therapy that indirectly delivers electrical stimulation to implanted materials, customized to generate desired electrochemical reactions, or for an implanted conductive material to exploit the natural DC field present at every injury site (no external power source needed), is an exciting area for future exploration. These novel devices might bring useful and unprecedented therapeutic avenues for the repair of electrically active tissues such as brain, spinal cord, and muscles.

Acknowledgments The present work has been financed by the Ministry of Science of Spain (MAT2015-65192-R and RTI2018-097753-B-I00), Fundació Marató TV3 (110130/31), Severo Ochoa Program (SEV-2015-0496 and CEX2019-000917-S) and EU grant FP6-2004-NEST-C1 028473. The authors thank J. Oró and F. Sandiumenge (ICMAB) and R. Arenal (INA) for TEM and HRTEM studies.

References

1. Krebs R (1999) Scientific development and misconceptions through the ages: a reference guide. Greenwood Press, London
2. Catania K (2014) The shocking predatory strike of the electric eel. Science 346:1231–1234
3. Cogan SF (2008) Neural stimulation and recording electrodes. Annu Rev Biomed Eng 10:275–309
4. Adewole DO, Serruya MD, Wolf JA, Cullen DK (2019) Bioactive neuroelectronic interfaces. Front Neurosci 13:269
5. Merrill DR, Bikson M, Jefferys JG (2005) Electrical stimulation of excitable tissue: design of efficacious and safe protocols. J Neurosci Methods 141:171–198
6. Erskine L, Stewart R, McCaig CD (1995) Electric field directed growth and branching of cultured frog nerves: effects of aminoglycosides and polycations. J Neurobiol 26:523–536
7. Levin M (2007) Large-scale biophysics: ion flows and regeneration. Trends Cell Biol 17:261–270
8. McCaig CD, Rajnicek AM, Song B et al. (2005) Controlling cell behaviour electrically: current views and future potential. Physiol Rev 85:943–978
9. Nair HKR (2018) Microcurrents as an adjunct therapy to accelerate chronic wound healing and reduce patient pain. J Wound Care 27:296–306
10. Li N, Zhang Q, Gao S et al. (2013) Three-dimensional graphene foam as a biocompatible and conductive scaffold for neural stem cells. Sci Rep 3:1604

11. Serrano MC, Patiño J, García-Rama C et al. (2014) 3D free-standing porous scaffolds made of graphene oxide as substrates for neural cell growth. J Mater Chem B 2:5698–5706
12. Guitchounts G, Cox D (2020) 64-Channel carbon fiber electrode arrays for chronic electrophysiology. Sci Rep 10:3830
13. Vara H, Collazos-Castro JE (2019) Enhanced spinal cord microstimulation using conducting polymer-coated carbon microfibers. Acta Biomat 90:71–86
14. Balint R, Cassidy NJ, Cartmell SH (2014) Conductive polymers: towards a smart biomaterial for tissue engineering. Acta Biomat 10:2341–2353
15. Cruz AM, Abad Ll, Carretero NM (2012) Iridium oxohydroxide, a significant member in the family of iridium oxides. Stoichiometry, characterization, and implications in bioelectrodes. J Phys Chem C 116:5155–5168
16. Meyer RD, Cogan SF, Nguyen TH et al. (2001) Electrodeposited iridium oxide for neural stimulation and recording electrodes. IEEE Trans Neural Syst Rehabil Eng 9:2–11
17. Acton QA (2013) Oxides—advances in research and application: 2013 edition. ScholarlyBrief
18. Hoffman-Kim D, Mitchel JA, Bellamkonda RV (2010) Topography, cell response, and nerve regeneration. Annu Rev Biomed Eng 12:203–231
19. Marcus M, Baranes K, Park M et al. (2017) Interactions of neurons with physical environments. Adv Healthc Mat 6:1700267
20. Cruz AM, Casañ-Pastor N (2013) Graded conducting titanium-iridium oxide coatings for bioelectrodes in neural systems. Thin Solid Films 534:316–324
21. Moral-Vico J, Carretero NM, Pérez E et al. (2013) Dynamic electrodeposition of aminoacid-polypyrrole on aminoacid-PEDOT substrates: conducting polymer bilayers as electrodes in neural systems. Electrochim Acta 111:250–260
22. Mantione D, del Agua I, Sanchez-Sanchez A et al. (2017) Poly(3,4-ethylenedioxythiophene) (PEDOT) Derivatives: innovative conductive polymers for bioelectronics. Polymers 9:354
23. Carretero NM, Lichtenstein MP, Pérez E et al. (2014) IrOx-Carbon nanotubes hybrid: a nanostructured material for electrodes with increased charge capacity in neural systems. Acta Biomat 10:4548–4558
24. Carretero NM, Lichtenstein MP, Pérez E et al. (2015) Enhanced charge capacity in iridium oxide-graphene oxide hybrids. Electrochim Acta 157:369–377
25. Pérez E, Lichtenstein MP, Suñol C et al. (2015) Coatings of nanostructured pristine graphene-IrOx hybrids for neural electrodes: layered stacking and the role of non-oxygenated graphene. Mater Sci Eng C 55:218–226
26. Lichtenstein MP, Pérez E, Ballesteros L et al. (2017) Short term electrostimulation enhancing neural repair *in vitro* using large charge capacity intercalation electrodes. Appl Mater Today 6:29–43
27. Rajnicek AM, Zhao Z, Moral-Vico J (2018) Controlling nerve growth with an electric field induced indirectly in transparent conductive substrate materials. Adv Healthc Mater 7:1800473
28. Lichtenstein MP, Madrigal JL, Pujol A et al. (2012) JNK/ERK/FAK mediate promigratory actions of basic fibroblast growth factor in astrocytes via CCL2 and COX2. Neurosignals 20:86–102
29. Pampaloni NP, Lottner M, Giugliano M et al. (2018) Single-layer graphene modulates neuronal communication and augments membrane ion currents. Nature Nanotech 13:755–764
30. Perez-Caballero L, Soto-Montenegro ML, Hidalgo-Figueroa M et al. (2018) Deep brain stimulation electrode insertion and depression: Patterns of activity and modulation by analgesics. Brain Stimul 11:1348–1355
31. Perez-Caballero H, Carceller J, Nacher V, Teruel-Marti E, Pujades N, Casañ-Pastor N, Berrocoso E (2021) Induced dipoles and possible modulation of wireless effects in implanted electrodes. Effects of implanting insulated electrodes on an animal test to screen antidepressant activity. J Clin Med 10:4003
32. Abad Ll, Rajnicek A, Casañ-Pastor N (2019) Electric field gradients and bipolar electrochemistry effects on neural growth. A finite element study on immersed electroactive conducting electrode materials. Electrochim Acta 317:102–111

33. Yin ZF, Wu L, Yang HG et al. (2013) Recent progress in biomedical applications of titanium dioxide. Phys Chem Chem Phys 15:4844–4858
34. Casan-Pastor N (2021) Nanocarbon-iridium oxide nanostructured hybrids as large charge capacity electrostimulation electrodes for neural repair. Molecules 26:4236
35. Zhao Y, Hernandez-Pagan EA, Vargas-Barbosa NM et al. (2011) A high yield synthesis of ligand-free iridium oxide nanoparticles with high electrocatalytic activity. J Phys Chem Lett 2:402–406
36. Talanov A, Phelan WA, Kelly ZA et al. (2014) Control of the iridium oxidation state in the hollandite iridate solid solution $K_{1-x}Ir_4O_8$. Inorg Chem 53:4500–4507
37. Moral-Vico J, Sánchez-Redondo S, Lichtenstein MP et al. (2014) Nanocomposites of iridium oxide and conducting polymers as electroactive phases in biological media. Acta Biomat 10:2177–2186
38. Pérez E, Carretero NM, Sandoval S et al. (2020) Nitro-graphene oxide in iridium oxide hybrids: electrochemical modulation of N-graphene redox states and charge capacities. Mater Chem Front 4:1421–1433
39. Lichtenstein MP, Carretero NM, Pérez E et al. (2018) Biosafety assessment of conducting nanostructured materials by using co-cultures of neurons astrocytes. Neurotoxicology 68:115–125
40. Rajnicek AM (2011) Electric field effects on neuronal growth cone guidance. In Pullar CE (ed) The physiology of bioelectricity in development, tissue repair, tissue regeneration and cancer: weak electric field effects on cells, sub-cellular systems and tissues. CRC Press, Taylor and Francis Group, Boca Raton, pp 201–232
41. Yan T, Jiang X, Guo X et al. (2017) Electric field-induced suppression of PTEN drives epithelial-to-mesenchymal transition via mTORC1 activation. J Dermatol Sci 85:96–105
42. Wagner FB, Mignardot J, Le Goff-Mignardot CG et al. (2018) Targeted neurotechnology restores walking in humans with spinal cord injury. Nature 563:65–71
43. Formento E, Minassian K, Wagner F et al. (2018) Electrical spinal cord stimulation must preserve proprioception to enable locomotion in humans with spinal cord injury. Nat. Neurosci 21:1728–1741
44. Fosdick SE, Knust KN, Scida K et al. (2013) Bipolar electrochemistry. Angew Chem Int Ed 52:10438–10456
45. Rajnicek AM, Britland S, McCaig C (1997) Contact guidance of CNS neurites on grooved quartz: influence of groove dimensions, neuronal age and cell type. J Cell Sci 110:2905–2913
46. Fuentes-Rodriguez L, Llibertat A, Simonelli L, Tonti D, Casañ-Pastor N (2021) Iridium oxide redox gradient material: operando X ray absorption of Ir gradient oxidation states during IrOx bipolar electrochemistry. J Phys Chem C 125:16629–16642
47. Liu J, Kim YS, Richardson CE et al. (2020) Genetically targeted chemical assembly of functional materials in living cells, tissues, and animals. Science 367:1372–1376

Chapter 6
Magnetoelasticity and Magnetostriction for Implementing Biomedical Sensors

Jesús María González

Abstract In this chapter, the physical principles underlying the phenomenologies arising from the magnetoelastic coupling occurring in materials exhibiting ferromagnetic order are first reviewed. From those principles, several generic designs of magnetoelastic sensors are discussed and, with special detail, that corresponding to resonant sensors. Finally, exemplary biomedical applications of magnetoelastic sensors to the measurement of the mechanical stress in bone fractures healing plates, sutures, and laryngeal muscles, the monitoring of the local curvature of epithelial tissues, the control of cell growth, and the measurement of blood coagulation kinetic parameters are reviewed. We conclude with a brief vision on the future perspectives of the magnetoelastic sensing technology and its biomedical applications. Although unexplored to date in the context of traumatically injured neural tissues, we hope these biomedical approaches, secondarily benefitting patients with traumatic brain and spinal cord injuries, could inspire advances in the field toward their implementation for neural scenarios.

Keywords Blood coagulation sensor · Bone fracture plate stress sensor · Cell culture sensor · Laryngeal muscles sensor · Magnetoelastic resonance · Magnetoelastic sensors · Magnetoelasticity · Magnetostriction · Skin curvature sensor · Suture stress sensor

6.1 Introduction

Materials with magnetic order and spontaneous magnetization (ferro- and ferrimagnetic) are currently used in an immense variety of applications that involve (a) the production of forces and/or torques, (b) that of magnetic fields, and (c) the conversion of magnetic field fluxes into currents (and vice versa) through

J. M. González (✉)
Instituto de Ciencia de Materiales de Madrid (ICMM), Consejo Superior de Investigaciones Científicas (CSIC), Madrid, Spain
e-mail: jm.g@csic.es

© The Author(s), under exclusive license to Springer Nature Switzerland AG 2022
E. López-Dolado, M. Concepción Serrano (eds.), *Engineering Biomaterials for Neural Applications*, https://doi.org/10.1007/978-3-030-81400-7_6

induction mechanisms. In particular, an application based on the creation of fields that are maintained over time, the magnetic recording of information, has come to constitute, from the beginning of its development in the 1880s and its peak diffusion from the 1950s to the end of the twentieth century [1], one of the basic information process supporting technologies. Its associated commercial market is predicted to reach 6.5 billion $ US by 2022 [2], which is a remarkable size considering the current decrease of the use of the technology (being substituted by the solid-state recording). But, as of today, the application of the ordered magnetic materials with the greatest recognized growth potential is that linked to their use in the implementation of sensors. Magnetic materials, from the point of view of sensorization, react to variations in the different fields present in their environment (e.g., magnetic, temperature, stresses, electromagnetic) with detectable changes in their magnetization state, interact with either charge or spin currents [3], and can support controllable excitations such as the spin waves whose basic parameters can also be modified by the system environment [4].

The two main current application fields for magnetic sensors are the automotive industry (by far the largest one since a high-end car currently integrates hundreds of magnetic sensors and this number will certainly increase in the immediate future) and the magneto-resistive sensing (used extensively in magnetic recording reading). Along with these two fields, a third one, emerging but wide-ranging and of undoubted potential, is identified: the use of magnetic sensors in biomedicine.

In this chapter, we will focus on the latter sensors and, in particular, on those based on the coupling of a sensing magnetic material to a stress field relevant from the standpoint of monitoring physiological functions or medical treatments. Although magnetoelastic sensors have been unexplored to date in the context of traumatically injured neural tissues, some of these devices have proved applicability in biomedical applications that could secondarily serve to patients suffering traumatic brain and spinal cord injuries. Some of these approaches will be discussed with the aim to inspire advances in the field toward their implementation for neural scenarios.

6.2 Hysteresis in Ferromagnets

A macroscopic ferromagnetic material is divided, at the microscopic scale, into the so-called magnetic domains. Such domains are regions within which the atomic moments are parallel due to the exchange interaction (originating the magnetic order) and point along the direction globally minimizing the system free energy: that of the local internal field. The local internal field is determined by the externally applied field, the effective anisotropy easy axes (directions along which the effective anisotropy energy is minimized) and magnitudes, the dipolar and exchange interactions between the atomic moments and the coupling of those moments to the stress fields acting over the material. In a zero magnetic moment system, the magnetization at the different domains changes orientation in order to cancel out the material total moment. When that demagnetized material is submitted to an externally applied magnetic field, the free energy of the domains oriented at directions close to that of the applied field decreases, which favors the increase of the volume of those domains at the expense of those of the domains originally oriented close to antiparallel to the applied field. If the magnitude of the applied field continues to increase, the antiparallel domains eventually disappear to arrive at a state in which the material includes a single domain whose moments point in a direction close (but in general non-coincident) to that of the applied field. From that state, the material saturation (i.e., the parallelism of the field-induced single-domain magnetization and the applied field) is achieved by incrementing the applied field, which induces the rotation of the local moments toward the field direction. In general, the described first magnetization process is irreversible, so that when the field is reduced to zero the magnetization of the material does not cancel out (the material achieves its remanent magnetization) and it is necessary to apply and increase in magnitude a field opposite to the saturating one (demagnetizing field) to achieve null magnetization (at the coercivity or coercive force). If the demagnetizing field increases beyond coercivity, the saturation antiparallel to the initial one is reached and if, after that, the applied field is decreased in magnitude, switched in sense, and increased to achieve a magnitude equal in absolute value to that applied when achieving the negative saturation, the material completes a hysteresis (from Greek, $\upsilon\sigma\tau\acute{\varepsilon}\rho\eta\sigma\eta$, delay) loop (Fig. 6.1). The loop and its characteristic parameters (the main ones being the remanence, coercivity, and saturation field value) constitute the most important set of data to decide the use of a given material in a particular application. Precisely, the coercivity value allows to classify the ferromagnetic materials into three conventional (but widely customary) groups: the soft materials (the better of which exhibit coercivities as reduced as 1 A/m), the recording materials (which depending of the particular use can have coercivities of up to 10^5 A/m), and the hard materials or magnets (that, as of the present date, have a minimum coercivity of 10^6 A/m). Although quite frequently the coercivity value is not an explicit constraint for the materials used into the implementation of sensors, their magnetic softness (i.e., the possibility of magnetic reversal associated to small fields) can be considered a general basic requirement.

6.3   Magnetoelasticity and Magnetostriction

The two effects that underlie the functionalities of the sensors that are the subject of this chapter are magnetoelasticity and magnetostriction. Magnetoelasticity is the phenomenon by which a magnetically ordered material experiences some kind of modification in its magnetic state when it is submitted to a certain mechanical stress field. Reciprocally, magnetostriction corresponds to the observation of changes in the dimensions of the ordered material associated with the application of magnetic fields. The magnetoelastic effect was first reported by Villari in 1865, whereas the magnetostrictive one was first observed in iron by Joule in 1842 [5].

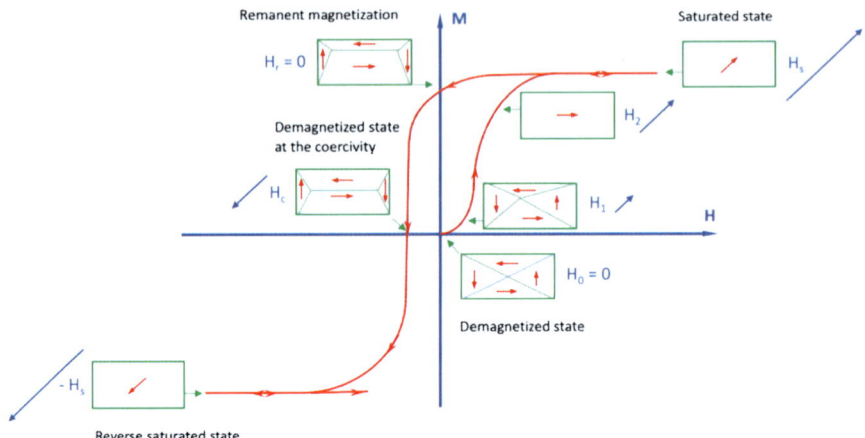

Fig. 6.1 Evolution from a demagnetized state corresponding to zero field ($H_0 = 0$) of the (schematic) domain structure present in a macroscopic ferromagnet. The red curve corresponds to the evolution of the component of the system magnetization, M, parallel to the applied field, H, when that field is varied following the path indicated by the arrows in the red curve, and corresponding to the sequence $H_0 (=0)$, H_1, H_2, H_s, $H_r (=0)$, H_c, and $-H_s$ (the field directions are shown by blue arrows). The saturation, associated to the occurrence of uniform magnetization oriented parallel to the applied field direction, is achieved at H_s. The remanent magnetization state (corresponding to $H_r = 0$) shows the delay (hysteresis) of the magnetization with respect to the applied field occurring upon saturation is achieved (the existence of hysteretic behavior does not require that of saturation). At the coercivity (H_c), a zero magnetization state is achieved but that state, in general, differs locally from the starting demagnetized one. Finally, the reversed saturation is achieved at $-H_s$ and, from that state, the behavior of the system corresponds to the curve from H_s to $-H_s$, symmetrically inverted with respect to the ($M = 0$, $H = 0$) point (not shown for the sake of clarity)

Both effects are originated by the interaction between the part of the magnetic moment linked to the atom orbital moment and the charge of the lattice ions creating the so-called crystalline field. From the occurrence of that interaction, it is clear that a local lattice strain, at the neighborhood of a magnetic moment including non-zero angular magnetic moment contribution, originates a modification of the local crystalline field and hence can generate a variation of the total magnetic moment minimum energy orientation. Conversely, if a change in the spatial distribution of the charge associated to an atomic moment is produced (for instance, by applicating a magnetic field to a material having large spin–orbit coupling as the rare earths are), that change could locally strain the lattice through the minimization of the crystalline field-orbital angular moment interaction energy. Both effects are illustrated at the atomic level in the sketches shown in Fig. 6.2.

At the macroscopic level, the magnetoelastic effect can be described by means of an additional term in the total free energy of the system, E_{mel}, that includes the applied stresses, σ, and the so-called saturation magnetostriction constant, λ_s, according to

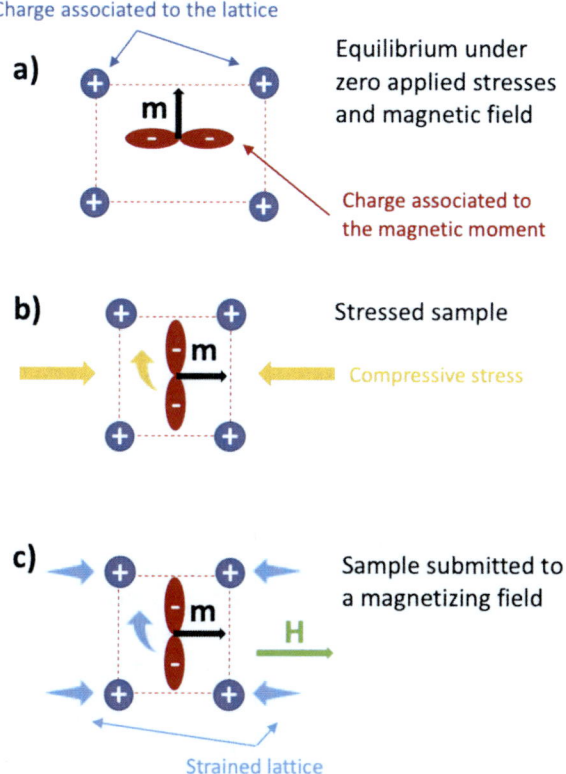

Fig. 6.2 Scheme illustrating the mechanisms underlying the magnetoelastic and magnetostrictive effects at the atomic level. (**a**) Equilibrium of a system incorporating elastic and magnetic contributions to its free energy under zero applied stresses and magnetic field. The (positive) charge associated to the lattice and that (negative) linked to the orbital part of the magnetic moment, m, of an atom situated at the center of the system are shown. The energy minimum originating the equilibrium state corresponds to the minimization of the distance between the negative and positive charges. (**b**) Magnetoelastic effect: Equilibrium configuration of the system in (**a**) when it is submitted to the shown compressive stresses. The applied stresses reduce the lattice parameter, and a new energy minimum achieved through a magnetization rotation occurs. (**c**) Magnetostrictive effect: Equilibrium configuration of the system shown in (**a**) when it is submitted to a magnetic field, H. The field-induced rotation of the magnetic moment originates a new energy minimum at which the lattice is strained in order to minimize the global free energy

$$E_{mel} = (3/2)\lambda_s\sigma \sin^2\theta \qquad (6.1)$$

(isotropic system, θ being the angle between the magnetization direction and that of the stress). When the magnetostriction is positive ($\lambda_s > 0$), the effect of a tensile (positive) stress is to favor the orientation of the magnetization parallel to the direction of the applied stress. A compressive (negative) stress applied along the same direction than the latter one will originate an easy plane perpendicular to the stress direction (any direction perpendicular to the stress direction has minimum magnetoelastic energy). The saturation magnetostriction λ_s is measured as the elongation per unit length experienced by a magnetostrictive sample when it is taken, by an externally applied field, from a demagnetized state to a saturated one, through an anhysteretic process exclusively involving magnetization rotations (i.e., excluding domain wall displacements). That process can be implemented by either applying the saturating field along a uniaxial hard axis (maximum effective anisotropy energy axis) or by polarizing the sample by means of a saturating field applied along a direction perpendicular to a unidirectional easy axis.

6.4   Basic Designs of Magnetoelastic Sensors

Non-exhaustive examples of the use of the effects originating from the magnetoelastic coupling are detailed in the following:

- Measurement of the strain (vertical displacement) involved on the oscillation of a micro-, nano-cantilever (either exhibiting magnetic order and positive magnetostriction or being coated with a film exhibiting those properties). The cantilever at rest is unstressed and exhibits perpendicular magnetization correlated to an out-of-plane easy axis. Upon periodically flexing, the upper (lower) region of the magnetoelastic material gets tensile (compressive) stressed, which modify the orientation and magnitude of the local easy axes (as illustrated in Fig. 6.3) and originate magnetization rotations that can be measured from the voltage induced at a suitable pick-up coil detecting the magnetization component along the long cantilever axis. The sensor output voltage depends, in addition to the magnitude of the magnetoelastic easy axes originated by the bending movement, on the volume of the magnetoelastic material that is under tensile (compressive) stress since that volume varies during the cantilever oscillations as a function of its elastic constants and the movement amplitude. Because of that, even if the system properties are known, the quantification of the vertical displacement is non-linear and requires calibration.
- Measurement of an identifiable DC magnetic field (spatially homogeneous and having known application direction and sense), acting over a magnetostrictive material with uniform anisotropy and easy axis perpendicular to the field to be measured, through the observation of a change of the dimensions of the sensing material (Fig. 6.4). The method yields a non-linear response requiring calibration and can be used with either DC or slowly varying AC fields (provided the availability of an adequate strain detection subsystem).
- Measurement of periodic stress acting over a saturated magnetoelastic sensing material located inside a pick-up coil through the measurement of the coil-induced voltage (Fig. 6.5). This sensor design requires, for $\lambda_s > 0$, an easy axis transverse to the measured stress. The time-varying magnetization is associated to the time evolution of the effective easy axis resulting from the composition

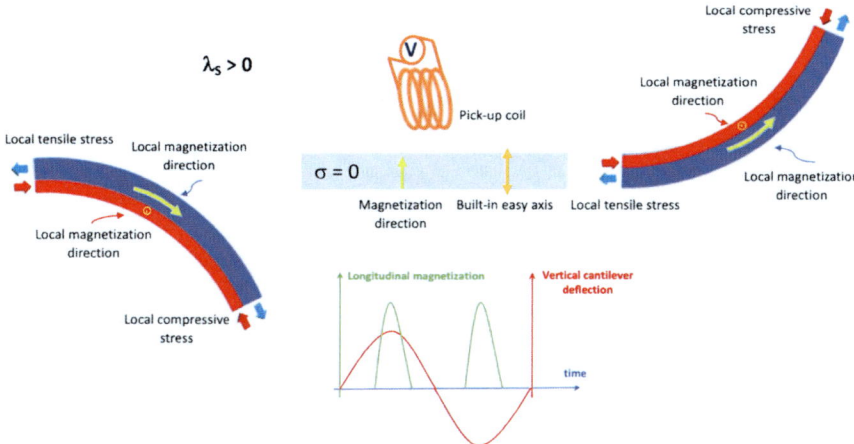

Fig. 6.3 Measurement of the vertical displacement of a magnetoelastic cantilever implemented through that of the voltage induced in a pick-up coil sensitive to the component of the magnetization along the cantilever axis (see text). In the strained cantilever, the regions under tensile (compressive) stresses are shown in blue (red), and the local magnetization direction is shown in yellow (in the compressed regions, it is perpendicular to the figure plane). The plot sketches the vertical bending and longitudinal magnetization evolutions with time (the times at which zero longitudinal magnetization is measured are associated to stresses not large enough so as to overcome the original perpendicular easy axis)

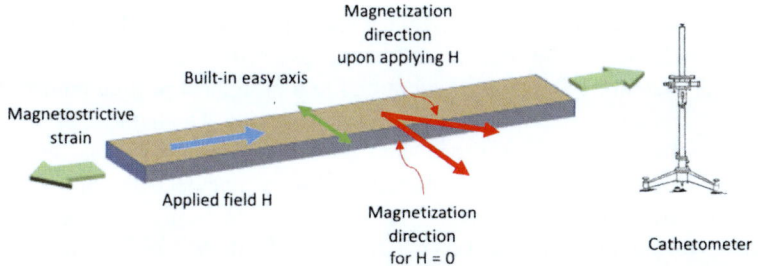

Fig. 6.4 Measurement of a field applied to a magnetostrictive material perpendicularly to its built-in transverse easy axis. The shown cathetometer stands for any device adequate to measure the magnetostrictive strain induced by the measured field

of the built-in and magnetoelastic easy axes. The measurable stress range has as upper bound the stress equivalent to the built-in easy axis magnitude. Therefore, for the implementation of the sensor, reduced magnetostriction and/or large transverse easy axes are preferred.

These magnetoelastic sensor implementation examples can be combined with other design elements to attain different sensing functionalities. The most widely used type of combined sensor is the presence detection sensor. In such sensor, for

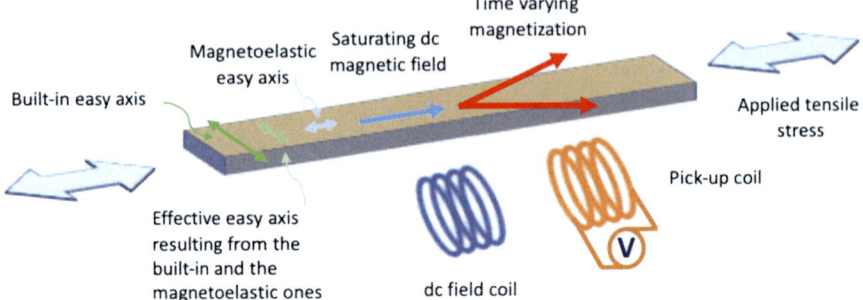

Fig. 6.5 Measurement of a time-varying tensile, longitudinal stress through that of the voltage induced in a pick-up coil. The stress is sensed on the basis of the variation of the magnitude of the effective transverse easy axis resulting from the built-in one and the magnetoelastic contribution induced by the applied stress

example, the sensed mechanical stress is exerted by the entity whose presence is intended to be detected.

6.5 Magnetoelastic Resonance

The most widely used principle for magnetoelastic sensing is the exploitation of the magnetoelastic resonances that will be discussed in detail in this section. Any mechanical system exhibits, when submitted to a periodical in time unidirectional stress, the so-called mechanical resonances, states at which the elastic energy provided to the system is absorbed with maximum efficiency, thus resulting in maximum strains along the stress direction. The magnetoelastic coupling linked to the magnetoelastic and magnetostrictive effects can be used to excite such a resonance by applying a time-varying magnetic field. Magnetoelastic materials react to that field by experiencing magnetostrictive strains that influence, through a magnetoelastic easy axis, the magnetic response of the materials to the applied field. Under particular boundary conditions, the field excited mechanical resonance can be associated to standing waves that are termed magnetoelastic resonance waves. For a given material shape, a set of boundary conditions, and applied field configurations (often including biasing DC fields in addition to the AC one), the magnetoelastic resonances can be easily observed by just varying the applied alternating field frequency [6].

The resonance frequency, f_r, for a free ends sample submitted to an applied sinusoidal magnetic field is given by

$$f_r^n = (n/2L)[E(H)/\rho(1 - v^2)]^{1/2},$$ (6.2)

where n is the standing waves mode number, L is the length of the sample, $E(H)$ is the Young modulus for an applied DC field, H, value (in many cases, this dependence is eliminated by saturating the sample under the bias field), ρ is the material density, and v is its Poisson ratio [7]. As it is evident from Eq. (6.2), the measurement of the magnetoelastic resonance frequency at a known mode number (for instance, the first one that is easily identifiable as that having the lower frequency) allows the evaluation of one of the elastic constants provided the other is known.

The detailed measurement of the resonance (i.e., that of the frequency dependence of the quantity used to detect the sample strains) allows the evaluation of two important parameters: First, that of the so-called quality factor, Q, which gives a measure of the energy dissipation in the resonating magnetoelastic system. The quality factor is defined as

$$Q = f_r/\Delta f, \tag{6.3}$$

where Δf is the full width (in frequency) at half maximum of the resonance peak. The larger the Q value (the narrower the resonance peak) is, the lower are the elastic energy losses. Conversely, the smaller the Q value is, the larger the elastic dissipation at resonance. Second, from the magnetoelastic resonance, it is possible to measure the magnetoelastic coupling coefficient, k, giving the percentage of magnetic energy that is transduced in elastic energy at the resonance [8]. That coefficient is defined as

$$k^2 = (\pi^2/8)[1 - (f_r/f_{ar})^2], \tag{6.4}$$

where f_{ar} is the so-called antiresonance frequency, characteristic of coupled oscillators, at which the introduction of magnetic energy in the system is $180°$ phase-shifted with respect to the elastic oscillations and, consequently, no magnetic elastic energy is stored into the system (in practice, the amplitude of the antiresonance peak gives the zero-amplitude reference). It is rather usual to measure, in soft high-magnetostriction materials, values for k^2 above 0.9, evidencing the high efficiency of the magnetic-to-elastic energy conversion occurring at magnetoelastic resonances [9].

6.6 Measuring the Magnetoelastic Resonance

Magnetoelastic resonances are measured in samples of simple shape and high aspect ratio (e.g., wires, ribbons, stripes) with the purpose of originating quasi-unidirectional strains and lowering the frequency of the first mode as much as possible (which makes it possible to use reduced excitation powers and simple AC field creation electronics).

From the standpoint of the magnetic properties of the material, it is crucial to reduce the inhomogeneity of the local anisotropies as much as possible since otherwise: (a) the magnitude of the variation of the Young's modulus with the applied field gets too high (up to 50% in some melt-spun materials [10]) and difficult to saturate and (b) the resonance quality factor is significantly reduced. Also, in the particular case of ribbons and stripes, an effective anisotropy having easy axis transverse to the sample long axis is desirable since it renders possible to observe the resonance by applying fields parallel to that long axis by means of a simple to design solenoid. The homogeneity of the sample can be improved by biasing the magnetization to a saturated state by means of a suitable DC magnetic field.

The resonance measurement is carried out by applying either sinusoidal fields of variable frequency in the range of interest of the frequency domain or controlled time width pulses whose Fourier transform covers the mentioned range in a dense way. The response of the resonant system can be obtained through three different approaches: (a) the inductive coupling of the measured sample to a receiving coil (this coupling is linked to the temporal variations of the magnetization at the resonance), (b) the detection of the acoustic waves emitted by the sample (and produced by its deformations at the resonance), which is carried out by means of a microphone of suitable bandwidth, and (c) the time monitoring of the light intensity reflected by the sample when it is illuminated by a laser beam (in this case, and depending on the frequencies to be measured, a photodiode or a phototransistor is used as detector device). Of the three methods, the inductive one is, by far, the preferable one since it does not require, like the other two, direct vision of the resonant system. Inductive detection is carried out from a suitable design of the pick-up coil (avoiding both resonances of the coil itself and excessive impedance in the range of frequencies of interest), but it must be implemented by maintaining reduced distances between the sample and the detector coil. Acoustic detection is preferred in continuous test environments as it requires very simple adjustments. Finally, light detection allows measurement at long distances from the resonant system and can be done through windows in cryostats, vacuum systems, or ovens. This last method can be optimized by exciting the first resonant mode and, from this, taking advantage of the local strain maximum associated with the center point along the axis of the measured system. A scheme of the discussed measuring methods is displayed in Fig. 6.6.

6.7 Magnetoelastic Resonance Modifications Originated by External Fields

The magnetoelastic resonance frequency, Eq. (6.2), can be modified due to the exposure of the resonating material of a variety of different ambient fields that include magnetic [11, 12], temperature [13], pressure [14], stress, and liquid flow velocity fields [15]. Also, the resonance frequency can be changed upon submitting

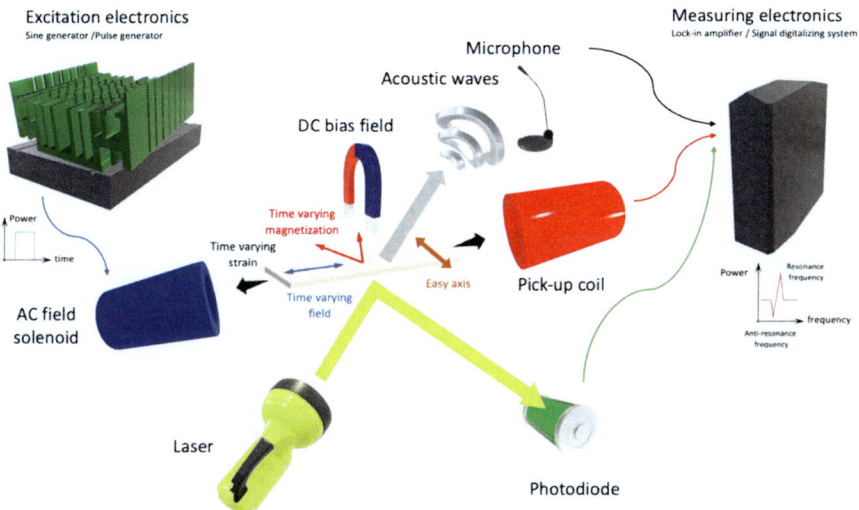

Fig. 6.6 Scheme of the measurement of the magnetoelastic resonances illustrating the three main detection methods: inductive, acoustic, and reflectometric. The figure also shows the exciting and measuring electronics, the homogeneous easy axis direction, and a biasing DC field used to optimize the quality factor. Finally, sketches of a typical time-domain excitation pattern and the frequency-domain system response are also displayed

the sensing material to different mass loads [16]. Some of the resonance frequency changes associated to the aforementioned quantities are linked to the variation of the elastic constants with the corresponding externally applied field (this includes, mainly, temperature and magnetic fields). But, in some other cases, being the pressure sensing the archetypical example, the sensing of a particular quantity is exclusively possible by recurring to particular excitations.

To implement a pressure magnetoelastic resonant sensor, it is necessary to induce strains perpendicular to the sensing material surface (that corresponding to the two larger material dimensions) [15]. In principle, those strains can be produced either by inducing rotations in domains whose local magnetization points perpendicular to the sensing surface or by suitably bending the sensing material. Since out-of-plane anisotropies are not typically associated to soft rotations, bending is the preferred design approach.

A flat, ribbon-shaped, stress-free sensing material submitted to a field originating a resonant behavior at frequency f_0 can be slightly bent to experience strains perpendicular to the flat sensor surface. The gas particles will interact with that surface (the largest lateral of the sensing ribbon) and its strains causing a shift, Δf, of the resonance frequency that will, under these conditions, be given by Kouzoudis and Grimes [15]:

$$\Delta f = f - f_0 = -(1/3)^{1/2}[\nu^2/(1 - \nu^2)](m_g u/k_B T d\rho_s)P, \tag{6.5}$$

where v is the Poisson ratio, m_g is the gas particle mass, u is the gas mean velocity, k_B is the Boltzmann constant, T is the absolute temperature, d is the sensor thickness, and ρ_s is the sensing material density. As it is clear from Eq. (6.5), the resonance frequency shift is linear with the measured pressure, which allows for a simple calibration in order to take into account the particular bending parameters.

When considering chemical or biological applications of magnetoelastic resonant sensors, coating of the sensing material is usually required, either for selective molecular detection or for providing an adequate substrate for cell growth. The coating modifies the flat sensor response by modifying its resonant frequency according to

$$f_0' = f_0[1/(1 + \Delta m/m_s)]^{1/2}, \tag{6.6}$$

where f_0' is the coated sensor resonance frequency and Δm is the, uniformly applied, coating mass [16]. Equation (6.6) can be used for the measurement of any small mass (with respect to the sensor mass) uniformly deposited on top of the sensing (either coated or uncoated) material [16].

In the case of the mass-measuring sensors, the sensitivity, S, is defined as the shift in resonance frequency per uniformly distributed unit mass over the sensing surface. It can be estimated as [17, 18]

$$S = -\Delta f/\Delta m = -[(f_0 \Delta m/2m_0)/\Delta m] = -f_0/2m_0, \tag{6.7}$$

where Δf is the variation of the resonance frequency associated to the load of the sensor with a mass Δm. Typical sensitivities obtained in tens of mm-sized magnetoelastic sensorization platform materials are in the range of a few units of Hz/μg [18].

6.8 Materials Used for the Implementation of Platforms for Magnetoelastic Sensorization

As already mentioned, an almost universal requirement for the implementation of magnetoelastic sensors is their soft hysteretic character (small coercive field). The requirement is based on the need to avoid significant power consumption for the supply or polarization of the materials with which the platforms are implemented. There are, however, exceptions to this requirement. These are the sensors implemented either with very high-magnetostriction materials such as Terfenol, $Tb_x Dy_{1-x} Fe_2$ ($x \approx 0.3$), having a saturation magnetostriction constant of up to $\lambda_s = 2 \times 10^{-3}$, or with intermediate magnetostriction ones such as Galfenol, with typical composition $Fe_{80}Ga_{20}$, exhibiting highly varying magnetostriction depending on their particular composition, fabrication and treatment, and reaching maxima of $\lambda_s = 2 \times 10^{-4}$ in bulk quenched materials. Although both materials were originally developed to implement submarine acoustic devices (e.g., ultrasonic transducers), magnetoelastic strain sensors designed for sensing large-size pieces have been implemented based on Terfenol [19, 20] and Galfenol [21] cores adequately polarized by means of permanent magnets (NdFeB alloys). Even a resonant sensor has been designed using Galfenol [22] for the in situ monitoring of the osteogenesis.

However, the materials mostly used in the implementation of magnetoelastic sensors are the amorphous transition metal/metalloid (TM/M) alloys. These materials are obtained in ribbon form (several tens of μm thick) by the melt spinning procedure (i.e., high cooling rate quenching of a liquid alloy in contact with a thermal focus, usually a massive high thermal conductivity and high thermal capacity, metallic, high-velocity spinning cylinder). The amorphous TM/M alloys are materials with a saturation magnetization higher than 8×10^5 A/m, coercivities lower than 40 A/m, and high-magnetoelastic coupling coefficients that allow both sensorization and transduction with energy yields in resonant applications of the order of 90% [23]. The order of magnitude of the saturation magnetostriction constant measured in TM/M amorphous alloys corresponds to 10^{-5}–10^{-6}. Zero-magnetostriction alloys, exhibiting λ_s values of the order of 10^{-7} or lower, are based on high-Co-containing CoFe alloys [23, 24].

Also these amorphous alloys exhibit remarkable mechanical properties with reduced Young's modulus [23] and giant, non-linear elastic regimes ending in fragile fracture without any intermediate plastic deformation or mechanical hysteresis [25]. But the property that adds the greatest potential to amorphous TM/M for their use in sensors is the fact that their metastable nature allows, upon anneals at temperatures below the crystallization one, both structural relaxation (i.e., removal of the free volume or residual stresses built-in during the amorphization process) and, primarily, induction of magnetic anisotropy with extensive control over the direction of the easy axes and the magnitude of their anisotropy constants [26]. The anisotropy induction is usually achieved by submitting the material during the thermal treatment to either a magnetic or a stress field performing as a polarizing agent for the structural disorder favored atomic diffusion. The symmetry and spatial variation of that field can be exploited to induce complex anisotropies as, for instance, the helical one, relevant to excite low-frequency torsion resonances [27, 28]. The possibility of including in the sensor design that of the sensing material easy axes characteristics largely favors the simplification of the sensing material polarization, response detection, sensitivity, and linearity [29, 30].

Finally, an important fact regarding amorphous TM/M ferromagnets is that corresponding to their commercial production, which started in the 1970s and has been continued to date by including the possibility of using ribbons having longitudinal dimensions of several km, from 1 mm up to 15 cm wide and different thicknesses, saturation magnetostrictions, saturation magnetizations, and coercive forces [31–33]. Most of the mass–pressure-measuring sensors incorporate a coating providing the response to the chemicals or biological entities being selectively detected. In the particular case of TM/M amorphous alloys, this active layer can be implemented with, typically, polymers or inorganic phases.

6.9 Force-Measuring Magnetoelastic Sensors for Use in Biomedicine

The need to control the forces (mainly the tensile or shear ones) acting on a bone fracture (or any surgically closed injury) during its healing process is an almost universal therapeutical requirement that, until recently, has only been covered with strategies not considering the in vivo measurement of the tension acting in the area of the fracture/injury, such as, typically, immobilization systems that, because of being heavy and bulky, cause a notable decrease in the comfort of patients without guaranteeing the safeguarding of the local tensions at levels appropriate to the welding or healing process.

This scenario began to evolve toward the use of a quantitative control of the involved mechanical stresses in the specific case of the healing of long bone fractures, for which metallic plates were used in order to provide the stiffness conditions required by the different fracture healing stages. One of the problems associated with this technique is caused by the fixation of the plates to the bone, which, being carried out using screws threaded in the bone itself, gave rise (for example, in the case of osteoporosis) to significant wear of the fixation that often resulted in an uncontrolled variation of the stresses applied to maintain the wound bone pieces in the optimal relative positions for the fracture welding.

Previous work reports a sensor that measures the tension acting on a plate that joined two fractured parts of a long bone (tibia) [34]. It is a magnetoelastic sensor that allows following (in real time) how the tension relaxes with the progress of the fracture healing. The sensor can also be used as a tool for the experimental characterization of the design of plates and fixations carried out through 3D models of different fractures. Similar bone fracture plate sensors have been reported elsewhere [35–39]. The operating principle of these sensors is that of the variation of the permeability of a magnetoelastic material subjected to different tensile stresses. The permeability of the sensor material varies from the modification of its effective anisotropy induced by the mechanical stress to be measured through the magnetoelastic term of the free energy shown in Eq. (6.1). In the particular case of work by Tan [34], an amorphous ribbon of the commercial material Metglas 2826 MB is used, by fixing it to the fracture welding plate. The sensing material is interrogated by an alternating low-frequency field created by an external surrounding solenoid. Its response to tensile stresses is followed from the harmonic analysis of the signal induced in an external pick-up coil that also surrounds the tape. The sensor can be calibrated ex situ by using a strain gauge.

The above-described sensor constitutes an example of an already wide family of force sensors used in biomedicine [34–39] and characterized by three essential advantages: (1) they are passive sensors, in the sense that the implanted components do not require any in-body power supply (power is exclusively used to excite the external interrogating currents and detection electronics); (2) they are wireless (no implanted connections are used) due to the almost non-existent attenuation of the low-frequency magnetic fields produced by the tissues, allowing to locate the used coils outside the body; and (3) they are little or non-intrusive, since the needed sensing material mass and dimensions are reduced enough so as to allow high stress sensitivities without interfering with the physiological functionalities.

Several articles have been published on passive, wireless force sensors designed to monitor the relevant stresses in skin surgical sutures [40, 41], a type of structure that, independently of its permanent or temporary nature, requires an implementation resulting on adequate transverse to the surgical incision (or scar) stress and that can be submitted to large efforts during rehabilitation following the surgery. The availability of the quantitative information on the stress state of the sutures along the wound healing process will help physicians to solve the clinical patient variability problems [42, 43] and those linked to the extreme low and high transverse stresses (i.e., insufficient mechanical support for the proper functionality of the sutured tissues and local ischemia, respectively). Also, when the sensors are integrated with the suture threads [40], a minimally invasive configuration, it is possible to obtain real-time information during the sewing process, which could help both to gain in systematicity and to design optimized suturing/stitching procedures. It has been already shown how that integration can be performed either to measure small forces by directly inserting the sensing material in between two thread sections or to monitor large stresses [40]. In the latter case, the magnetoelastic material is fixed to an intermediate rigid (non-magnetic) structure that is inserted in the suture, which helps to maintain the sensing material submitted to limited stresses since the total thread stress is supported by the rigid structure to a great extent. It is interesting to point out that in order to design coils as reduced in size as possible [40], the sensing material response was followed through the measurement of its harmonic response (third harmonic of the exciting field frequency), as detected by a pick-up coil almost coincident in dimensions and position with the interrogating coil.

Similar in basic design, force sensors have also been proposed to measure the force between limbs and prosthetic devices that often can lead to significant ulceration of the stump [39]. The proposed device is, in correspondence with the complexity of the stresses involved in the limb–prosthesis interaction, a multipoint one that, due to the dimensions of the used magnetoelastic materials, is not invasive at its position in the limb–prosthesis interface. It is clear that this sensor functionality will allow, if a correlation between the interfacial forces and the ulcers development is clinically established, enhancement of prosthesis design and adaptation to the patients and prevention of degenerative process.

Kaniusas and colleagues have proposed an original application of the force magnetoelastic sensing principle: the detection of skin curvature changes [44]. The device is characterized by the use of a bilayer formed by a large magnetostriction layer affixed to a non-magnetic one (Al and steel are used). This provides a thickness to the bilayer sufficient so as to maintain the whole volume of the magnetoelastic layer under either tensile or compressive stress when the bilayer is bent to globally exhibit either positive or negative curvatures, respectively (see Fig. 6.6, where a similar bending cantilever and its response to different curvatures are shown). That mechanical design results in large permeability variations with the bilayer curvature (up to 15% increase with respect to the flat bilayer) that can be detected inductively by using a coil surrounding the bilayer. The authors show how the sensor is able to follow the curvature of the carotid artery in time, providing information about the propagation of blood pressure waves. Also, it is proposed to use the sensor to monitor the respiratory activity through the chest volume modifications associated to breathing (thus simplifying with a single device the apnea monitoring setups), limb unvoluntary movements, and eyelids muscular activity.

The latter application has a resulting use of clear interest: the functionalization of symmetric muscular pairs, one of whose elements is denervated. The application can be implemented on the basis of the sensing of the movement of the active muscle, the electronic conditioning of the output of the sensor, and the use of the conditioned signal to excite the denervated muscle through an actuator (that, in the case of low-volume muscles, can simply be a suitable contacts network injecting a modulated current). A test of concept of this application was published by Moreno Iglesias et al. [45] and results corresponding to the use of the sensor described and tested by Pina et al. [46]. The work was carried out on the thyroarytenoid and cricothyroid muscles pairs in an in vivo animal model (beagle breed dogs) by collecting statistics (*ca.* 600 samples) on the exerted force by the considered larynx muscles during swallowing and phonation. To validate the forces measurement by means of the magnetoelastic sensor, an electromyographic signal was recorded simultaneously with the sensor reading, which evidenced a complete time correlation between the events detected using both techniques. The magnetoelastic detection of the deglutition and phonation processes was characteristic. Whereas in deglutition the sensor signal output corresponded to a symmetric and highly repetitive peak associated to the decrease of the potential outputted by the sensor, in the phonation case the sensor signal was linked to a lower magnitude output voltage decrease describable as a square pulse of varying time width. Since the same animal was medically intervened four times (allowing for full recovery between surgeries), it was possible to observe how the sensor response was not affected by the succeeding surgical procedures. From the sensor calibration, it was possible to conclude that the forces ranged from 0.03 N up to 0.9 N for swallowing and from 0.005 N up to 0.48 N for phonation, in correspondence with sensor outputs from the tenths up to the tens of mV. As a whole, the sensorization of the laryngeal muscles was concluded to be adequate both to quantitatively investigate the laryngeal biomechanics and to implement actuation over the denervated muscle of the pair, at least to the level corresponding to the recovery of the sphincter function of the larynx.

In in vitro systems, the in situ monitoring of the proceeding of a cell culture has been implemented through the measurement of its mass by means of a magnetoelastic sensor [47, 48]. The added value of this quantitative measure is clear if we consider that the best alternative available is filming the evolution of the culture, which could yield errors with respect to the mass measurement of almost an order of magnitude. Also, the use of sensors makes possible the automation of the modification of some growth parameters and the supervision of a large number of simultaneous growths. The used sensor is a differential one, concurrently following the measured cell culture and an identical mass of culture bath without cells on it. The corresponding plates are maintained in close proximity so as to compensate any possible temperature drift occurring during the measurement. The sensing materials, working in resonance conditions, were two identical magnetoelastic ribbons, each one of them placed in one of the culture plates and operating according to Eq. (6.6). The use of a highly sensitive detection electronics allowed for a relative mass resolution better than 1% in a cervical cancer cell line.

Alternatively, Rivero et al. described a differential non-resonant device based on the variation in the magnetic permeability of a magnetoelastic microwire induced by the change on blood viscosity occurring during coagulation [47]. The permeability variation of a microwire in contact with the clotting blood is compared to that of an identical microwire kept immersed in a reference liquid. The device allows the measurement of the prothrombin time and the values that derive from it, such as the international normalized ratio.

Finally, Hernando and colleagues have reported an implanted sensor designed to follow the physiological deterioration of heart valves [49, 50]. The sensor is built using a glass-coated Co-based magnetoelastic microwire having 45 μm in diameter and 10 mm in length. Due to their reduced dimensions, the sensing material can be directly integrated in the flaps without obstructing in any way the valve normal functioning. The sensing material exhibits a high permeability resulting from its shape anisotropy and allowing very little power investment for the interrogation and detection processes. The design principle of this sensor is significantly different from those discussed previously. Specifically, it works by being interrogated by an AC magnetic field having a frequency of a few units of kHz. The moment of the sensing material switches sense, following the interrogating field frequency in a non-sinusoidal way due to the high material permeability. Thus, the moment creates a detectable magnetic field outside the body that varies in time as the exciting field. That field is amplitude modulated according to the movements of the valve to which the sensing material is attached due to both the variations of the relative orientation of the microwire with respect to the interrogating field direction and the stress-induced magnetoelastic contribution to the effective anisotropy originated by the moving flaps. The modulation allows to identify the heart rate, and more importantly, the carrier signal envelope can be associated to the onsets of the valve malfunctioning (e.g., regurgitation, stenosis), which allows for the early detection of valvular diseases and, consequentially, permits to plan valve replacement before its complete failure.

6.10   Future Perspectives

Although unexplored to date for their application in traumatically injured neural tissues, magnetoelastic sensors are being already investigated, as extensively discussed in this chapter, for several biomedical applications that can indirectly assist some aspects of the reality of patients suffering traumatic brain and spinal cord injuries. That is the case, for instance, of sensing at bone fractures, denervated muscle pairs, skin sutures, and cardiovascular elements. Importantly, the field of the biomedical application of these sensors is clearly growing, as exposed above, with several trends identified. The most remarkable ones are outlined as follows:

- Reduction of sensor dimensions: This trend is obvious in what concerns the reduction of the interference of the sensors with the normal body functioning. Sensor reduction can be achieved through the use of higher frequency fields in the interrogation and detection processes (the fundamental resonance frequency corresponds to a wavelength twice the maximum sensing material dimension). The reduction is, nevertheless, limited by the penetration of the high-frequency fields in the water-rich human body, which corresponds to attenuation by factor of 10^3 for electromagnetic waves, having wavelengths below $100\,\mu m$ (3 GHz) [51]. The use of devices with maximum dimensions in this range will consequently require large electromagnetic powers and the confirmation that no excessive heat dissipation is produced in the body while the sensor is operated.

- Enhancement of the response of the sensing material: From the standpoint of the saturation magnetostriction values, no new phases are anticipated to be used on the implementation of sensors. This is so due to the limited range of variation of that quantity in the binary, ternary, and pseudo-ternary known alloys. Despite this, two envisaged possibilities can be mentioned: (a) reduction of the coercivity of high-magnetostriction alloys (and, particularly, that of Galfenol), which would render an operational alloy having a magnetostriction of $\lambda_s = 2 \times 10^{-4}$, ten times larger than that of the highest magnetostriction material currently used in magnetoelastic sensors, (b) increase of the availability of the perpendicular-to-the-largest-sensor-surface magnetic anisotropy, which would render unnecessarily the need for sensor bending in mass–pressure-detecting sensors.

- Enhancement of the detectability of the sensing materials: This is a clearly open point since the use of either ultra-high sensitivity devices (as the lock-in amplifiers) or dedicated digital electronics (e.g., FPGAs, PLAs) designed and optimized for particular applications is just a market size question linked to the growth of the sensor applicability.

- Improvement of the sensor design: Although it is not possible to make concrete predictions about this point, it is clear that new sensor morphologies and, specially, local anisotropies distributions still constitute a field in which the performance of the magnetoelastic sensors used in biomedicine can be improved. In particular, it can be anticipated that the performance of bilayer and multilayer sensors can significantly overcome that of single-layer sensors using bending designs.

- Identification of new applications: This is by far the potentially most relevant driver for the growth of the use of magnetoelastic sensors in biomedicine. A priori, the most interesting fields are those associated to the permanent implant of sensors: (a) as part of sensing–actuating (transducing) devices as those discussed in the case of muscle pairs with lateral paralysis; (b) in situ continuously measuring different specific molecular (e.g., proteins, antibodies, glucose) concentrations in order to automatically deliver their associated drugs or chemicals; (c) in situ measuring stresses with the purpose of controlling, by means of adequate actuators, bones welding, wound healing, and rehabilitation processes; (d) in situ detection of wear and malfunction of a broad range of prostheses.

Improvements in most of these trends would necessarily benefit the application of magnetoelastic sensors in the context of traumatic brain and spinal cord injuries. The development of specific devices to sense mechanical forces at fixed skull bones and backbones after traumatic injuries, to monitor the effect of rehabilitation programs on denervated muscles, and to control the affectation of vital functions such as breathing and cardiovascular pumping are just a few examples of the potentiality that magnetoelastic sensors could exert in the field. Although the promise is high, rare advances, if any, have been done to date on the particular context of injured neural tissues.

In summary, magnetoelastic sensors applications will continue to grow mainly due to the increase of the demand of new functionalities (in many cases, making part of transducing systems) and to the optimization of the sensing elements and electronics detection systems design. We hope this chapter serves to inspire researchers in the field to explore the design and utility of magnetoelastic sensors in neural environments in order to improve their functional recovery after injury.

Acknowledgments JMG acknowledges project MAT2016-80394-R, financed by the Spanish Research Agency (AEI/FEDER, UE) for providing funds for the preparation of the chapter. JMG would also like to dedicate this work to the memory of Dr. Guillermo Rivero.

References

1. Kimizuka M (2012) Historical development of magnetic recording and tape recorder. Survey reports on the systematization of technologies, vol. 17. National Museum of Nature and Science, Tokyo. http://sts.kahaku.go.jp/diversity/document/system/pdf/073_e.pdf. Cited 30 Sep 2021
2. Infoholic Research. https://www.infoholicresearch.com/report/magnetic-tape-market-trends-and-forecast-to-2022/#:~:text=AccordingtoInfoholicResearch%2Cthe,reach%246. 5billionby2022. Cited 30 Sep 2021
3. Rousseau O, Viret M (2012) Interaction between ferromagnetic resonance and spin currents in nanostructures. Phys Rev B 85:144413
4. Urdiroz U, Gómez A, Magaz M et al. (2021) Antiphase resonance at X-ray irradiated microregions in amorphous $Fe_{80}B_{20}$ stripes. J Magn Magn Mater 520:167017
5. Joule JP (1842) On a new class of magnetic forces. Ann Electric Magn Chem 8:219–224
6. Garcia-Arribas A, Gutiérrez J, Kurlyandskaya GV et al. (2014) Sensor applications of soft magnetic materials based on magneto-impedance, magneto-elastic resonance and magneto-electricity. Sensors 14:7602–7624
7. Lacheisserie EDTD (1993) Magnetostriction: theory and applications of magnetoelasticity. CRC Press, Boca Raton, London
8. Lacheisserie EDTD (2002) Magnetostrictive materials. Kluwer Academic Publishers, Boston
9. del Moral A (2007) Magnetostriction and magnetoelasticity theory: a modern view. In: Handbook of magnetism and advanced magnetic materials. Wiley, London

10. Baghdasaryan G, Danoyan Z (2018) Basics of the theory of magnetoelasticity. In Magnetoelastic waves. Springer Nature
11. Grimes CA, Mungle CS, Zeng K et al. (2002) Wireless magnetoelastic resonance sensors: a critical review. Sensors 2:294–313
12. Livinston JD (1982) Magnetomechanical properties of amorphous metals. Phys Status Solidi A 70:591–596
13. Mungle CS, Grimes CA Dreschel WR (2002) Magnetic field tuning of the frequency temperature response of a magnetoelastic sensor. Sensor Actuat A-Phys 101:143–149
14. Kouzoudis D, Grimes CA (2000) The frequency response of magnetoelastic sensor to stress and atmospheric pressure. Smart Mater Struct 9:885–889
15. Kouzoudis D, Grimes CA (2000) Remote query fluid-flow velocity measurement using magnetoelastic thick-film sensor. J Appl Phys 87:6301–6303
16. Tormes CD, Beltrami M, Cruz RCD, Missell FP (2014) Characterization of drying behavior of granular materials using magnetoelastic sensors. NDT&E Int 66:67–71
17. Zhang K, Zhang L, Fu L et al. (2013) Magnetostrictive resonators as sensors and actuators. Sensor Actuat A-Phys 200:2–10
18. Sagasti A (2018) Functionalized magnetoelastic resonant platforms for chemical and biological detection purposes. PhD Thesis (EHU/UPV, Leioa)
19. Zheng Y, Xie D, Ma J et al. (2011) Micro magnetic field sensor based on Terfenol-D/PZT/Terfenol-D magnetoelectric composites. Key Eng Mat 483:190
20. Lasassmeh SM, Lynch E, Law CT (2017) Fiber optical current sensor based on tapered Terfenol-D composite. Front Optics, OSA Technical Digest (online) (Optical Society of America, 2017), paper JTu2A.29
21. Swartz A, Singh C (2016) Automated scour detection arrays using bio-inspired magnetostrictive flow sensors. Final technical report, Michigan Technological University—USDOT Cooperative Agreement No. RITARS-12-H-MTU
22. Fischer WJ, Sauer S, Marschner U et al. (2009) Galfenol resonant sensor for indirect wireless osteosynthesis plate bending measurements. Sensors 2009 IEEE, Christchurch, New Zealand p 611–616
23. Petrakovski GA (1981) Amorphous magnetic materials. Sov Phys Usp 24:511
24. Niinomi M (2010) Metals for biomedical devices. CRC Press, Boca Raton, Boston, New York, Washington, Oxford, Cambridge, New Delhi
25. González JM, Madurga V, Poza M, Hernando A (1981) Torsional elastic behavior in MET-GLASS $Fe_{40}Ni_{40}P_{14}B_6$. J Phys D App Phys 14:2243
26. Mizoguchi T (1991) Amorphous magnetic materials, in Physics and engineering applications of magnetism. Springer Series in Solid-State Sciences, vol 92. Springer, Berlin, Heidelberg
27. Hernando A, Madurga V, González JM, Cebollada F (1983) Helical anisotropy induced by annealing in METGLAS-2826. J Magn Magn Mater 1553–1554
28. Liniers M, Madurga V, Vázquez M, Hernando A (1985) Magnetostrictive torsional strain in transverse-field-annealed Metglas® 2605. Phys Rev B 31:4425
29. Nielsen O, Hernando A, Madurga V, Gonzalez JM (1985) Experiments concerning the origin of stress anneals induced magnetic anisotropy in metallic glass ribbons. J Magn Magn Mater 46:341–349
30. Murillo N, Blanco JM, González J et al. (1994) Stress annealing in $Fe_{73.5}Cu_1Ta_3Si_{13.5}B_9$ amorphous alloy: induced magnetic anisotropy and variation of the magnetostriction constant. J App Phys 76:1131–1134
31. China Amorphous Technology Co. Amorphous Core & Nanocrystalline Core Factory. http://www.catech-china.cn/Fe-based-Amorphous-Ribbon.html. Cited 30 Sep 2021
32. Vacuumschmelze.   https://vacuumschmelze.com/products/soft-magnetic-materials-and-stamped-parts/Amorphous-Material---VITROVAC. Cited 30 Sep 2021
33. Liquid Metal. https://www.liquidmetal.com. Cited 30 Sep 2021
34. Tan Y (2017) A passive and wireless sensor for bone plate strain monitoring. Sensors 17:2635
35. Oess NP, Weisse B, Nelson BJ (2009) Magnetoelastic strain sensor for optimized assessment of bone fracture fixation. IEEE Sens J 9:961

36. Yu K, Ren L, Tan Y, Wang J (2019) Wireless magnetoelasticity-based sensor for monitoring the degradation behavior of polylactic acid artificial bone *in vitro*. Appl Sci 9:739
37. Klosterhoff BS, Tsang M, She D et al. (2017) Implantable sensors for regenerative medicine. J Biomech Eng 139:021009-1
38. Ren L, Yu K, Tan Y (2019) Applications and advances of magnetoelastic sensors in biomedical engineering: a review. Materials 12:1135
39. Pereles BD, DeRouin AJ, Ong KG (2014) A Wireless, passive magnetoelastic force-mapping system for biomedical applications. J Biomech Eng 136:011010
40. DeRouin A, Pacella N, Zhao C et al. (2016) A wireless sensor for real-time monitoring of tensile force on sutured wound sites. IEEE T Biomed Eng 63:1665–1171
41. Horeman T, Meijer E, Harlaar J et al. (2013) Force sensing in surgical sutures. PLoS ONE 8:0084466
42. Schreiber J (2005) A review of the literature on evidence-based practice in physical therapy. Int J Allied Health Sci Practice 3:1–10
43. Jette DU, Bacon K, Batty C et al. (2003) Evidence-based practice: beliefs, attitudes, knowledge, and behaviors of physical therapists. Phys Ther 83:786–805
44. Kaniusas E, Pfützner H, Mehnen L et al. (2004) Magnetoelastic skin curvature sensor for biomedical applications. In: Sensors 2004 IEEE, Vienna, Austria, vol 3, p 1484–1487
45. Moreno MCI, Gil-Loyzaga P, Gómez ER et al. (2007) Magnetoelastic sensors as a new tool for laryngeal research. Acta Oto-Laryngol 127:1182–1187
46. Pina E, Burgos E, Prados C et al. (2001) Magnetoelastic sensor as a probe for muscular activity: an *in vivo* experiment. Sensor Actuat A 91:99
47. Rivero G, Multigner M, Spottorno J (2012) Magnetic sensors for biomedical applications. In: Magnetic Sensors—Principles and Applications, Kevin Kuang, IntechOpen, https://doi.org/10.5772/37285. Available from: https://www.intechopen.com/chapters/30947. Cited 30 Sep 2021
48. Rivero G, Crespo P, Spottorno J et al. (2008) Sensor system for continuously monitoring the cellular growth *in situ* based in magnetoelastic sensor. Patent P200801973, Spain
49. Rivero G, García-Páez JM, Álvarez L et al. (2007) Magnetic sensor for early detection of heart valve bioprostheses failure. Sensor Lett 5:263
50. Rivero G, García-Páez JM, Alvarez L et al. (2008) Sensor system for early detection of heart valve bioprostheses failure. Sensor Actuat A 142:511–519
51. Fuse T, Taki M, Minohara S (1984) Submillimeter-wave penetration into biological tissues. In: Int Symposium on Electromagnetic Compatibility, p. 376

Chapter 7
Nanoparticles for Neural Applications

Jesús García Ovejero, Edina Wang, Sabino Veintemillas-Verdaguer, María del Puerto Morales, and Anabel Sorolla

Abstract Nanotechnology represents a novel powerful technology with the potential to overcome some of the problems related to brain research. In particular, nanoparticles can be used as platforms for drug delivery, contrast agents for imaging diagnosis, or therapeutic probes. The small scale of these materials can be exploited to interact with individual neurons and even with target-specific molecules of the neuron. Moreover, some nanoparticles can be transcranially activated with external stimuli such as infrared light, ultrasounds, or magnetic fields. Nanoparticles are able to transduce such stimuli in a variety of physicochemical responses such as mechanical forces, local heat, catalytic activity, and drug release that modify the internal physiology of the neurons. Despite the enormous potential of nanoparticles, some issues still need to be solved such as the biodistribution through the blood–brain barrier and the minimization of potential toxicological hazards.

Keywords Anticancer drugs · Blood–brain barrier disruption · Central nervous system · Colloidal synthesis · Engineered nanoparticles · Imaging diagnosis · Nanotoxicity · Neuronal actuators · Regenerative Medicine · Targeted delivery

7.1 Introduction: Nanoparticles and the Central Nervous System

The study of the brain and its neuronal activity has historically been fostered by the development of new technologies that open new windows to the exploration of this

J. G. Ovejero · S. Veintemillas-Verdaguer · M. Puerto Morales
Instituto de Ciencia de Materiales de Madrid (ICMM), Consejo Superior de Investigaciones Científicas (CSIC), Madrid, Spain
e-mail: jesus.g.ovejero@csic.es; sabino@icmm.csic.es; puerto@icmm.csic.es

E. Wang · A. Sorolla (✉)
Harry Perkins Institute of Medical Research, ,, QEII Medical Centre, Nedlands

Centre for Medical Research, The University of Western Australia, Perth, WA, Australia
e-mail: edina.wang@perkins.org.au; asorolla@irblleida.cat

© The Author(s), under exclusive license to Springer Nature Switzerland AG 2022
E. López-Dolado, M. Concepción Serrano (eds.), *Engineering Biomaterials for Neural Applications*, https://doi.org/10.1007/978-3-030-81400-7_7

fascinating black box [1]. Disruptive technologies such as magnetic resonance imaging (MRI) and transcranial magnetic stimulation (TMS), among many others, have modified the established paradigms of neuroscience and created new exploration pathways in neurophysiology. In this sense, nanomaterials and nanotechnology facilities are instrumental in the development of technologies for the study of the central nervous system (CNS). The possibility of interacting in the scale of single neurons, or even single molecules, creates a technological interface with unprecedented spatiotemporal resolution [2]. The small size and higher compliance of the nanosystems also reduce undesirable immune responses as opposed to other kinds of bulky neuroimplants [3, 4]. For these reasons, the use of nanotechnology in neuroscience is an expanding field of research (Fig. 7.1).

Among the multiple nanostructures used in neuroscience, nanoparticles (NP)—materials with the three dimensions smaller than 100 nm [5]—possess suitable features for application in neuroimaging [6–8], drug delivery [9], and transcranial interaction [10, 11]. On the one hand, the high surface-to-volume ratio of the NPs can be used to attach a plethora of biocompatible molecules, such as peptides,

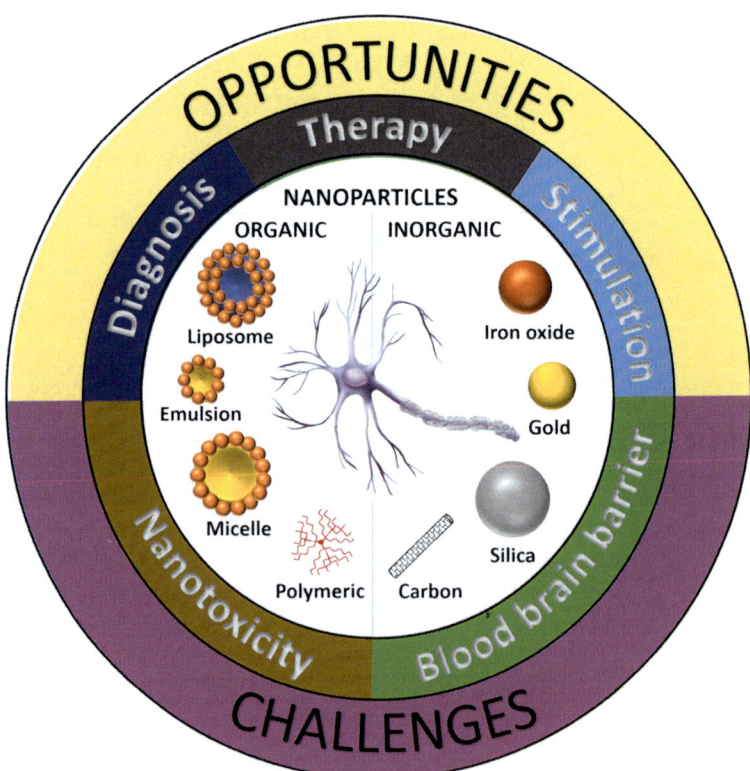

Fig. 7.1 Scheme illustrating the major opportunities and challenges of the use of organic and inorganic nanoparticles in neural applications

oligonucleotides, and antibodies [9], which provide interesting functionalities for a selective interaction with CNS neurons such as an immune shield [12], neuronal targeting [13], and antitumoral drug delivery [14, 15]. On the other hand, thanks to their small dimensions, they can be administrated *via* respiratory inhalation or, in the case of stable colloids of NPs, *via* intravenous injection avoiding a surgical intervention. In addition, by using the appropriate coatings, NPs can penetrate inside the neuron and behave as local wireless actuators or as vehicles for other molecules [16].

7.2 Designing Nanoparticles for Neuroscience

Regarding traumatic brain injury (TBI) and spinal cord injury (SCI), there is not yet any NP clinically approved as a therapy, presumably due to their insufficient delivery to the CNS and target binding. Nevertheless, the assessment of new nanoformulations for their applicability in both imaging and therapy of traumatic brain and spinal cord injuries is under intensive investigation. Currently, there is a myriad of NPs being tested in preclinical studies based on organic and inorganic materials.

7.2.1 Organic vs. Inorganic Nanoparticles

The development of advanced synthetic routes has provided to neuroscientists a wide variety of nanomaterials with different composition, geometry, and functionality (Fig. 7.2). Depending on the nature of the materials, NPs are usually classified into two large families: Organic and inorganic NPs [17].

Organic NPs can be defined as organic materials structured at the nanoscale. The most common geometries are liposomes, nanomicelles, nanoemulsions, and polymeric NPs [18]. They represent approximately 60% of the Food and Drug Administration (FDA)-approved nanodrugs [19] and provide interesting properties for their use in humans such as high biocompatibility, flexibility, and biodegradability. In addition, some of these geometries present internal pockets that can be filled with drugs and other functional elements such as luminescent tracers [20] and cavitation liquids to create advanced contrast agents [21]. The most important organic carriers and present in large of formulations accepted by the FDA are liposomes, vesicles formed by phospholipid bilayers with an aqueous internal cavity. In fact, Doxil® (Doxorubicin liposomes) is the first example of a nanoformulation approved by the FDA for the treatment of Kaposi's sarcoma in HIV patients in 1995 [22]. The use of Doxil® is encouraged due to its safer profile compared to free doxorubicin [22]. However, it possesses some disadvantages such as lack of uniformity between batches, fast degradation, and short circulation time

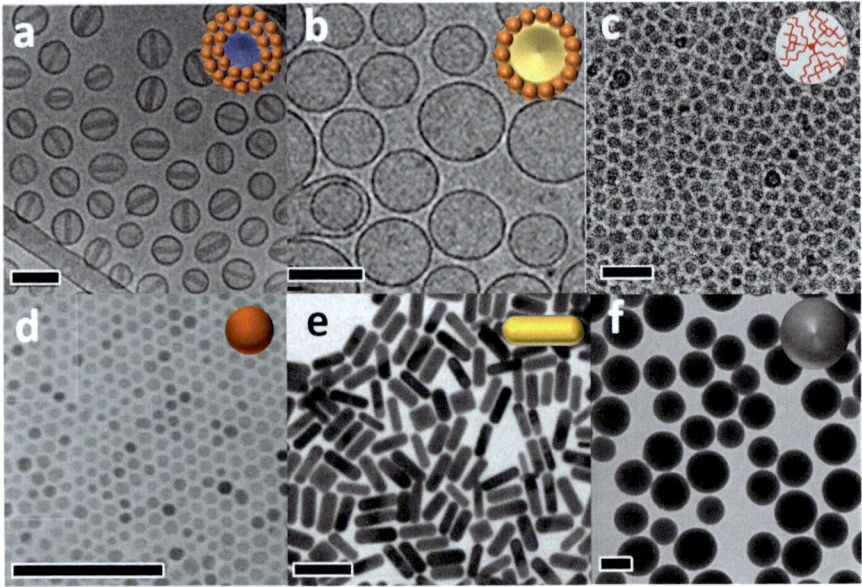

Fig. 7.2 Transmission electron microscopy images of the most commonly used nanoparticles in neurological studies: (**a**) Doxil® liposomes. Reprinted from [22], Copyright (2012), with permission from Elsevier. (**b**) Fluorescent bolaamphiphiles vernolic acid vesicles. Reprinted from [30], Copyright (2012), with permission from Elsevier. (**c**) PAMAM nanodendrimers. Adapted with permission from [31]. Copyright (1998) American Chemical Society. (**d**) Iron oxide nanoparticles. Republished with permission of Future Medicine Ltd., from [32]; permission conveyed through Copyright Clearance Center, Inc. (**e**) Gold nanorods and (**f**) mesoporous silica (SiO_2) nanoparticles. Copyright © 2015 Josef Jampilek et al. from [33]. This is an open access article distributed under the Creative Commons Attribution License. Scale bars correspond to 100 nm

[23]. Despite these drawbacks, Doxil® has been used frequently in clinical therapies [22].

Second in relevance are synthetic vesicles, known as polymeric NPs. They are colloidal particles usually made from biodegradable polymers such as poly(lactic-co-glycolic acid) (PLGA) [24] and poly(glycidyl methacrylate) (PGMA) [25]. Depending on their preparation method, they can be nanostructured as either nanocapsules or nanospheres. They are highly versatile systems that allow for a controlled drug release, shape, size, and charge control and varied surface functionalizations. Polymeric NPs are the most utilized vehicles for therapeutic purposes. However, they also posses some similar disadvantages such as lack of uniformity between batches and short circulation time [23]. Nevertheless, they have been extensively used in the nervous system. For instance, Xu et al. engineered polymeric nanocapsules made of methacryloyloxyethyl phosphorylcholine (MPC) as the monomer and polylactic acid diacrylate (PLA) as the crosslinker. These nanocapsules enclosed and delivered nerve growth factor (NGF) in the CNS of mice and non-human primates [26]. Polymeric dendrimers are a variant of polymeric

NPs, although they possess a branched structure instead that allow for many drug and ligand conjugations on both the backbone and the ramifications. They are often made of poly(amidoamine) (PAMAM), poly(propylene imine) (PPI) and polyether-co-polyester (PEPE) [27]. Dendrimers are however complex to synthesize and present certain toxicity issues [28]. They have been recently studied as drug delivery systems for the CNS. For example, PAMAM dendrimers carrying the anti-inflammatory agent N-acetyl cysteine and targeting to glial mitochondrias with triphenyl-phosphonium successfully targeted injured glial mitochondrias in a TBI model in rabbits [29].

Less employed organic formulations are micelles, similar to liposomes in biodegradability but composed of amphiphilic molecules, other than phospholipids, forming a massive hydrophobic core and a hydrophilic shell that provides stability in circulation and the opportunity to deliver high payloads of hydrophobic drugs. They accept neural-specific targeting components [34]. Nanomicelles can be synthesized to make them sensitive to external stimuli, allowing controlled drug release under temperature and pH changes, ultrasounds, or enzymes [35]. This is particularly interesting in the case of brain tumors such as glioma, characterized for presenting an acidic microenvironment. For example, Zeng et al. reported the use of worm-like micelles made of a pH responsive polymer loaded with a bioreducible peptide-drug conjugate (cyclic RGD-emtansine) as a successful therapeutic strategy for gliomas [36]. Another example is the intranasal delivery of mRNA of the brain-derived neurotrophic factor (BDNF) encapsulated into polyethylene glycol (PEG)-polyamino acid block copolymer nanomicelles in mice with impaired olfactory function [37]. Alternatively, nanoemulsions, smaller in size, are transparent dispersions of oil droplets in water (or the inverse) stabilized by surfactants. They allow the encapsulation of hydrophobic drugs in the dispersed oil phase and a tight control of drug release. Nanoemulsions have been utilized as drug delivery vectors to the CNS through intranasal administration, thus bypassing the blood–brain barrier [23]. Finally, it is worth mentioning the use of solid lipid NPs for brain delivery as an alternative to polymeric NPs. They are a particular class of nanoemulsions, with the hydrophobic phase in solid state [38].

On their turn, inorganic NPs such as iron oxide, gold (Au), and silica (SiO_2) are a large group of nanomaterials that also raised great interest in neurological studies due to their unique optical, electrical, and magnetic properties [23, 39]. Two examples of the physical properties that characterize inorganic NPs and have been extensively used in biomedical applications are the superparamagnetism of magnetic NPs and the localized surface plasmon of metallic NPs. Due to their relevance, these two phenomena are analyzed more in detail in the next paragraphs.

Ferromagnetic materials such as iron oxide suffer a drastic change in their magnetic response when the size is reduced to the nanoscale. Their magnetic moment is not able to preserve a fixed position, and the thermal fluctuations make this moment to switch between two bistable positions. Such a microscopic behavior, known as superparamagnetism, implies a lack of remanence magnetization and high susceptibility to static or quasistatic magnetic fields. For these reasons, superparamagnetic NPs can be prepared as a colloid avoiding limitations such as magnetic aggregation in the absence of a magnetic field but preserving a strong magnetic response when a magnetic field is applied. Superparamagnetic colloids based on iron oxide NPs have been exploited for biosensing [40], cell manipulation [41], and bioaccumulation [42]. At high-frequency magnetic fields, the superparamagnetic NPs recover the ferromagnetic response and follow the magnetic field, dissipating energy as heat. This characteristic behavior makes magnetic NPs excellent candidates for contactless local nanoheating in applications such as thermal cancer therapies [43], selective drug release [44], or triggering neuronal activation [11]. Numerous uses have been proposed for iron oxide NPs in the CNS given their target specificity and ability to surpass the BBB. Some of them include delivery of drugs, inhibition of microglial cells (e.g., to prevent inflammation responses associated with Parkinson's and Alzheimer's diseases and multiple sclerosis), targeting of beta amyloid plaques in brain arterioles in Alzheimer's disease, and increasing contrast for MRI diagnosis [45].

Other interesting property arising from metallic NPs is the localized surface plasmon. It can be defined as a collective oscillation of the electrons that take place in NPs made of highly conductive metals (generally noble metals) when illuminated with the appropriate light [46]. This oscillation enhances the electric field in the local environment of the NP, increasing the absorbance and scattering of the incident light [47]. Neuroscience can take advantage of these unique properties. The near-field enhancement generated close to the surface of the NPs has been used to increase the sensitivity of imaging techniques like Raman scattering [48] and to induce other chemical routes by local activation [49]. The strong light absorption is used to heat up specific tissues in hyperthermia cancer treatments [50, 51] and to create photoacoustic contrast agents with NPs that convert the energy of the light irradiation in ultrasound waves [52] (see Sect. 7.5.1). The scattering of the light can also be used to obtain high-resolution images using metallic NPs as contrast agents [53].

However, inorganic NPs possess some drawbacks such as lack of biodegradability, poor clearance, and toxicity, including neurotoxicity. Currently, few inorganic NPs have been approved for clinical trials: iron oxide NPs (Feraheme, AMAG Pharmaceuticals), silica NPs (Cornell Dots), silica-Au NPs (AuroLase, Nanospectra Biosciences), and hafnium oxide NPs (NBTXR3, Nanobiotix) [19, 54], to cite a few. Among them, only iron oxide NPs have been approved by the FDA for their use in patients. Other inorganic materials, such as quantum dots (semiconductor materials) and up-converting NPs (generally lanthanide compounds), have also been frequently used for in vitro and in vivo studies. However, they have not yet been approved for their use in the clinical arena. Finally, carbon nanotubes are commonly considered inorganic materials despite their carbon composition. They present very interesting optical and electrical properties for neurological studies. However, the biocompatibility of these nanomaterials is still controversial [55, 56], especially due to the possible asbestos-like effect produced by their long aspect ratio. So, their use to date is restricted to preclinical in vitro and in vivo studies.

7.2.2 Administration Through Blood–Brain and Blood–Spinal Cord Barriers Challenge

The main routes for NPs delivery to treat TBI and SCI are either systemic [57] or nasal administrations as non-invasive options, or invasive administration during surgery [58]. However, both the brain and the spinal cord present an additional firewall protection that hinders these routes, called the blood–brain barrier (BBB) and the blood–spinal cord barrier (BSCB), respectively. These barriers are mainly composed by a densely populated endothelium, tightly connected by proteins like claudin-5 and occludin, among others [59]. They avoid the penetration of large molecules, and almost 98% of small molecules present in the blood stream, protecting the CNS from circulating toxins and pathogens [60]. However, a varied range of strategies enabling different NPs designs has been proposed to overcome this restriction [61, 62]. The simplest method consists of coating the NPs with positively charged compounds that present high affinity to the negatively charged BBB to promote its permeability [63]. More complex strategies use ligands with high affinity to receptors overexpressed in the BBB such as transferrin (Tf) receptor, insulin receptor, low-density lipoprotein receptor (LDLR)-related protein, nicotinic acetylcholine receptor, insulin-like growth factor receptor, diphtheria toxin receptor, scavenger receptor call B type, leptin receptor, and the neonatal Fc receptor [61]. The chemical routes used to target these receptors are discussed in Sect. 7.3 of this chapter.

An alternative to the biochemical labelling for the penetration of NPs through the BBB and the BSCB is the use of physical stimulus like focused ultrasounds and alternating magnetic fields. In the former case, the acoustic energy of ultrasound waves is concentrated in a targeted region of the brain to disturb the endothelium integrity, enhancing the penetration of large drugs and nanoscaled systems [64]. Different organic [65] and inorganic nanoformulations have been used to create cavitation centers that enhance the effect of focused ultrasounds [66, 67]. Studies of antitumoral drug delivery with NPs have shown that focused ultrasounds increase the efficiency of the drug lengthening the life span of rats with glioma tumors from 23 to more than 35 days [65]. Magnetic fields can also be used to force the diffusion of magnetic NPs through the BBB with magnetic gradients and to disrupt the BBB by magnetic field-induced hyperthermia [68, 69].

Intranasal administration or nose-to-brain administration is an alternative strategy to overcome the BBB limitation. Pharmaceutical agents can easily enter to the CNS through olfactory nerve fibers since they are in direct contact with the nasal cavity [70]. Consequently, olfactory mucosa has been largely used as a bypass of the BBB [71–73]. However, recent studies have shown that nose-to-brain administrated NPs do not reach neither olfactory nor forebrain neurons but get trapped by microglia cells instead [74], indicating that further research is needed to determine the suitability of this approach. Another limitation of this method is the small volume that can be administrated intranasally (25–200 μL), imposing concentrated colloids of NPs [61]. The intranasal access route has also been explored by nanotoxicological studies to evaluate the effect of micro- and nanopollutants on neural activity [75, 76]. Less common routes of NPs administration such as ocular, vaginal, rectal, and cutaneous have also been studied although scarcely and generally restricted to the administration of organic NPs like dendrimers [77].

7.2.3 Nanoparticles as Neuronal Actuators

External stimulus such as infrared light, ultrasounds, and static and alternating magnetic fields are able to penetrate the skull and activate deeply allocated NPs to generate a specific response such as mechanical movement, heat dissipation, or production of chemical species.

Biological tissues present two optical windows of minimum absorbance in the near and far infrared. By using wavelengths inside these windows, it is possible to reach deep regions of the CNS and interact with photosensitive NPs. For example, by activating the localized surface plasmon of metallic NPs, it is possible to carry out photothermal treatments of glioblastomas [78] and to generate reactive oxygen species (ROS) that activate biochemical mechanisms controlling neuronal differentiation [79] or cell apoptosis [80]. By using irradiation of lower intensity, it is also possible to stimulate the electrical activity of dorsal ganglion neurons through a localized mild heating [81, 82] and to promote the differentiation and growth of NG108-15 neuroblastoma cells [83].

Magnetic fields have traditionally been used in neuroscience to obtain brain and spinal cord images (by MRI) and activate specific regions (by TMS) as they can penetrate into biological tissues with minimal biological processes interference. Magnetic NPs are able to convert the magnetic fields applied in mechanical stimuli or even induce neuronal signaling by mild local heating [41, 84]. The position and orientation of magnetic NPs can be modified by applying a magnetic gradient or a rotating magnetic field, respectively. Mechanical stimulation with magnetic NPs has already been used to induce axonal growth [85], enhance microglial cell transfection [86], and destroy glioma cells by mechanical oscillation [87], but the biorheology of neurons is still an open field of research [84, 88].

In the case of magneto-thermal activation, works by Chen and colleagues are particularly noteworthy. They were able to remotely activate the heat-sensitive capsaicin receptor TRPV1 using iron oxide NPs and low-radiofrequency alternating magnetic fields (0.1–1 MHz) [89]. This approach has lately been used to stimulate motor behavior in awake mice [11], creating a new field of exploration in magneto-thermal stimulation.

There are other examples of inorganic NPs that can be used to exploit well-established stimulation techniques like optogenetics. Up-converting NPs are a special family of lanthanide NPs that convert infrared light into visible light. Such conversion can be used for optogenetic-triggered activation [90, 91] or inhibition [92] of neuron activity in deep neural tissues by infrared stimulation. Quantum dots, another family of fluorescent inorganic NPs, have been used to track membrane potential activation in neurons [93].

7.3 Synthesis and Surface Engineering

The number of synthesis routes for the production of organic and inorganic NPs is enormous and constantly growing. Liposomes and iron oxide NPs are the most promising nanoformulations for clinical neural applications due to their high safety and scalable production [94]. For this reason, the emphasis of this section is the synthesis and surface engineering of those formulations. We will first review liposomes, which can be used as cargoes for the transport of drugs to the CNS, and enhance their retention. Second, we will discuss about iron oxide NPs, which can be targeted to the CNS and be locally activated by non-invasive magnetic fields to manipulate cellular events from distance. These two types of NPs constitute highly versatile tools with numerous applications in neuroengineering and regenerative medicine [95].

7.3.1 Synthesis of Organic Nanovesicles

As mentioned before, liposomes are biodegradable and biocompatible spherical vesicles made of phospholipid bilayers, with a hydrophilic core and a lipophilic surface. Due to their dual nature, they can be loaded with hydrophilic drugs in the core or within the lipidic bilayer. Liposomes can be functionalized with different coatings and ligands to increase their half-life in circulation and to make them targeted systems. Moreover, drug release and flexibility of liposomes can be controlled by changing their lipidic composition. Interestingly, they have the ability to cross the BBB *via* receptor-mediated entry [96], as discussed in Sect. 7.2.2.

Liposomes can be prepared using the thin-layer evaporation method. In short, lipids are dissolved in chloroform, the solvent is removed by rotary evaporation under vacuum, and then the dried lipid film is hydrated with water. The product of this hydration is a large multilamellar vesicle (LMV) that is next broken by sonication at 55 °C, producing small unilamellar vesicles of around 200 nm [97]. Other methods for the preparation of liposomes are ethanol injection and freeze–thaw sonication. For the ethanol injection method, the lipid is dissolved in ethanol and very slowly injected into distilled water at 55 °C [98]. For the freeze–thaw sonication method, the lipid is directly hydrated with the aqueous solution and rapidly frozen with liquid nitrogen, followed by bath sonication at 55 °C to break the LMV [99]. Liposomes can be decorated with Au and magnetic NPs that can be incorporated during the synthesis either in the organic or in the aqueous phase. Taking into account the experimental procedure and the surface charge of the particles, these NPs can be spatially located inside, outside the liposome, or within the lipid bilayers [100]. Liposomes are versatile carrier systems that can simultaneously encapsulate hydrophilic drugs in their core (e.g., doxorubicin) and hydrophobic materials in their membrane (e.g., Zinc phthalocyanine). As the drugs may be photosensitive, the synthesis process in both cases should be done in the dark.

An alternative to liposomes are the polymersomes, vesicles in which membrane is composed of a bilayer of amphiphilic block-co-polymers. Polymersomes exhibit enhanced stability and lower permeability compared to liposomes due to their thicker and more robust membrane. Moreover, by contrast to liposomes, the polymersome membranes can be tailored by tuning the length of the amphiphilic block-co-polymers [101]. As the typical hydration method to produce such vesicles results in low encapsulation efficiency and high polydispersity, a microfluidic approach has been proposed to solve these problems [102]. The improved stability of polymersome membrane facilitates the funtionalization with polymers such as PEG and poly(N-isopropylacrylamide) (PNIPAM) on the outside, as well as Au and iron oxide NPs in the hydrophobic part of the bilayer.

7.3.2   Synthesis of Inorganic Nanoparticles

Among iron oxide NPs, magnetite and maghemite are of particular interest because of their potential in different biomedical areas such as MRI, targeted drug delivery, hyperthermia, gene therapy, and tissue regeneration [103]. In general, iron oxide NPs can be classified according to their size (i.e., superparamagnetic and ferromagnetic), structure (i.e., single core and multicore), and shape (e.g., sphere, rod, disk) [104, 105]. Size of the particles should be kept within the superparamagnetic range (<70 nm) to minimize the magnetic interactions responsible for aggregation and loss of colloidal stability. Shape is also important and has recently been shown to be a crucial factor for some biological applications. For example, nanodisks and nanorods of magnetite present suitable features to induce mechanical damage of cancer cells [106], while cubes, nanorings, and nanoflowers seem to be ideal as heat mediators for hyperthermia [107], recently applied to neuronal cells [43].

There are two different approaches for the direct synthesis of magnetite NPs: (i) starting from iron(II) and/or iron(III) inorganic salts such as nitrates, sulfates, or chloride or (ii) starting from organic precursors such as iron oleates, acetates, acetylacetonates, or pentacarbonyl [105]. In most cases, a mild reducer or oxidizer is added to the reaction to get the final Fe_3O_4 stoichiometry. The presence of different ligands may promote the formation of anisometric particles. For example, elongated magnetic iron oxide NPs can be synthesized directly in organic media, starting with iron pentacarbonyl, hexadecylamine, oleic acid, and 1-octanol, heated at 200 °C for 6 h, rendering rods of around 100 nm in length and $\sim$10 in axial ratio [108]. Both length and axial ratio can be tuned by varying the amount of hexadecylamine and the heater reactor.

Indirect methods involve the initial synthesis of iron oxides (hematite) or oxohy-droxides (e.g., goethite, akaganeite, lepidocrocite) NPs that act as a shape template, setting the final dimensions and shape of the nanostructure. Those templates are thermally transformed into magnetite and then into maghemite through oxidation at 240 °C in air. However, when the size is reduced to nanoscale, the oxidation temperature can be reduced down to 50 °C and even to room temperature, for a long period of time (e.g., 1 year for sizes below 10 nm). In rare cases, lepidocrocite can evolve to maghemite by simple dehydration.

Multicore structures and flower-like magnetite NPs can be synthesized either following the polyol route or by thermal decomposition in organic media [109]. The polyol method consists of the alkaline hydrolysis of iron salts at high temperature (usually 210–220 °C) and slow rates in the presence of N-methyldiethanolamine (NMDEA). This leads to the generation of primary units of 4–6 nm that quickly agglomerate to 11–16 nm cores that aggregate up to 55 nm particles depending on the amount of NaOH and the time of reaction at 220 °C. Magnetite nanoflowers can also be synthesized by thermal decomposition of iron(III) oleate in a surfactant mixture of trioctylphosphine oxide (TOPO) and oleic acid (5:1 in molar ratio). It renders aggregates of 5 nm cores in final 20 nm NPs. Also, calixarene molecules have been shown to stabilize intermediate reaction stages, leading to flower-like structures before magnetite particles are transformed into octahedrons [110]. This synthesis method yields hydrophobic particles stabilized by surfactants, while polyol-mediated process yields hydrophilic NPs, with a versatile surface chemistry depending on the reagents used in the synthesis.

It is worth mentioning that Au NPs are also involved in numerous preclinical settings as contrast agents for computed tomography (CT) for brain imaging and therapeutic agents. These particles present different morphologies such as nanospheres, nanocages, nanorods, and nanoshells. Available reports in the litera-ture describing the synthesis of colloidal Au NPs are abundant, with various degrees of reproducibility and simplicity [111]. However, the preparation of high-quality inorganic NPs is still challenging and needs fine-tuning. For example, traditional protocols fail to prepare large citrate-capped Au NPs (>50 nm) with acceptable monodispersity [112]. Recent modifications of the seed-mediated approach allowed for the synthesis of citrate-capped Au NPs with excellent monodispersity and larger diameters (up to 100 nm) [113]. In the case of nanorods, traditional protocols based on seeding use a high concentration of cetyltrimethylammonium bromide (CTAB, 0.1 M), which is cytotoxic and thus limits its use in biomedical applications. In 2012, Murray and colleagues proved that the presence of aromatic additives in the growth solution allows reducing CTAB concentration to 0.037 M. This methodology employs a binary surfactant mixture, CTAB, and sodium oleate (NaOL) [114, 115].

7.3.3 Surface Engineering

One of the most relevant advantages of nanoscale systems is their interaction with injured tissues at the molecular and cellular level as no other systems can do. Such interaction can be customized to achieve particular physiological effects, while preventing unwanted secondary side effects in the organism [116]. For targeting injured brain and spinal cord tissues, NPs have to accomplish, in addition to biocompatibility and biodegradability, a prolonged half-life in blood circulation and ability to cross the BBB and the BSCB [116, 117]. If utilized as a drug-delivery platform for the CNS, apart from the above-mentioned essential properties, the encapsulation efficiency and their drug release mechanisms are very important. All of these properties are affected in varying degrees by the carrier nanostructure (e.g., size, shape, elasticity) and its surface charge [116, 117].

Engineering NPs coating is a popular strategy to increase their half-life in circulation. PEGylation and surfactant coatings are the most common approaches to ensure long half-life in circulation by preventing the degradation and removal of NPs by the reticuloendothelial system. For example, PEGylated polystyrene NPs remained longer in circulation and subsequently crossed the BBB in rat brains more efficiently than their non-PEGylated counterparts [118]. Importantly, surface charge greatly affects solubility, clearance, biodistribution, stability, cell uptake, tissue diffusion, and cytotoxicity of NPs [118]. Achieving a neutral charge on the NPs surface is perhaps the best strategy to achieve a good balance between good BBB penetrability and a low chance of NPs elimination by the body and adsorption by random serum proteins [119]. In this sense, PEG coating represents a useful strategy since it neutralizes the zeta potential in NPs keeping the dispersions stable by steric effects [120]. In regard to NPs surface, other strategies that facilitate BBB crossing are the coating with surfactants and surface functionalization with biomolecules, the latter being extensively discussed in the next section of this chapter. Besides, surfactants such as polysorbate 80, Pluronic P85, and Poloxamer 188 adsorb the apolipoproteins ApoA and ApoE from the blood. As ApoA and ApoE receptors are highly expressed in the endothelial cells conforming the brain microvessels, crossing of NPs through the BBB is facilitated when coated with surfactants *via* receptor-mediated endocytosis followed by transcytosis [116, 121]. This phenomenon has been exploited for polymeric NPs, in which polysorbate coating facilitated the delivery of molecular imaging probes into the brain by ApoE adsorption exclusively [122, 123].

7.4 Functionalization and Loading

As mentioned before, increasing therapeutics transport across the BBB and BSCB represents one of the unmet needs for the treatment of neuronal injuries. Some functionalization strategies used to overcome this limitation are the surface linkage

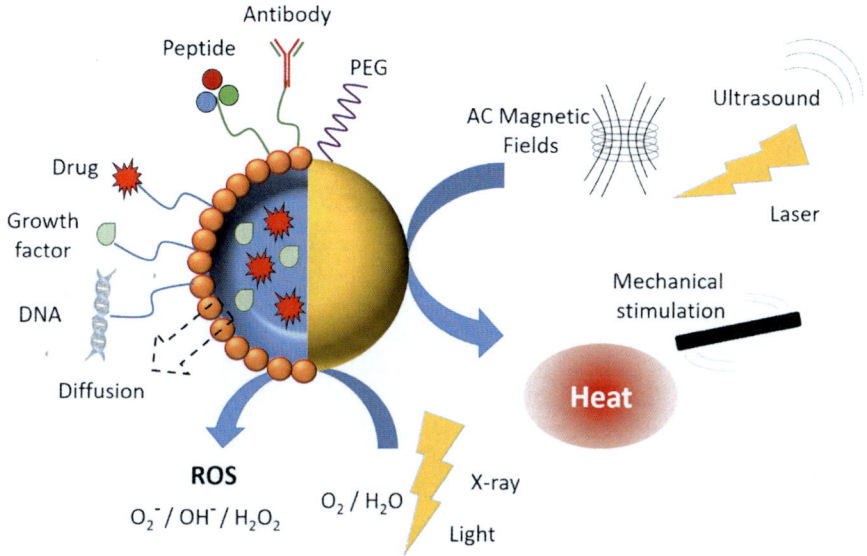

Fig. 7.3 Schematic representation of the most common functionalities of nanoparticles: Delivery cargoes of drugs, growth factors, and nucleic acids (left); delivery vectors using peptides, antibodies, or stabilizers like polyethylene glycol (up); contactless nanoheaters and nanoactuators (right); opto- and radiosensitizers (down)

with antibodies, aptamers, and peptides, while common cargoes to be transported include growth factors, drugs, and genome engineering tools (Fig. 7.3). In this context, organic and porous NPs offer the advantage of delivering biomolecules in their native structure, hidden in their interior, which is crucial for their successful therapeutic performance. Such structures have to be effectively targeted in order to enter the CNS due to their impermeant nature.

7.4.1 Antibodies

Antibodies can be used to enhance BBB targeting of nanovehicles [124]. As mentioned before, only very small lipophilic molecules (<400 Da) and very small iron oxide NPs (<1 nm) can cross the BBB by passive diffusion [125]. The rest enter through more sophisticated mechanisms than that, including endocytosis and receptor-mediated transport [124], as this is the case for insulin and transferrin receptors. Importantly, these receptors are among the most abundantly expressed by the endothelial cells present in the BBB [126, 127], fact that has been exploited by nanotechnology. For instance, anti-transferrin receptor antibodies have served as functionalization elements to enhance brain targeting for Au NPs [128–131], liposomes [128] (tested in a mouse model of glioblastoma [132]), PEGylated PLGA

NPs tackling an in vitro model of Alzheimer's disease [24], rod-shape polymeric NPs [133], and PEGylated human serum albumin NPs [134] were used in vivo for the same purpose. Similarly, anti-insulin receptor antibodies have been utilized as BBB-targeting ligands for PEGylated human serum albumin NPs loaded with loperamide, a nociceptive drug unable to cross the BBB by itself, in mice [135].

7.4.2 Peptides

The use of peptides is another strategy utilized to facilitate NPs transport to the sites of CNS injury. For example, Kwon et al. engineered NPs with a targeting peptide from the rabies virus (i.e., rabies virus glycoprotein, RVG) to target the injured site and with an siRNA against caspase-3, the executor caspase of apoptosis, to prevent neuronal cell death after injury [136]. Such nanoformulation successfully accumulated to the site of injury and downregulated caspase-3 after systemic administration in mice. Also, Bao et al. used ceria NPs loaded with edaravone, a drug to treat stroke, and targeted with Angiopep-2, which is a peptide that triggers transcytosis and crosses the BBB by recognizing the low-density lipoprotein-related protein-1 (LRP-1) present in the brain capillary endothelial cells [137]. These NPs eliminated ROS induced by brain injury and presented an excellent intracephalic uptake in rats treated systemically. Another peptide, CAQK, targeting a proteoglycan complex upregulated in brain injuries, induced homing of PEGylated silver NPs and porous silicon NPs carrying a siRNA against the green fluorescent protein (GFP, as a proof of concept) into the mice brain after severe TBI and systemic administration of the nanoformulations [138].

7.4.3 Neural Growth Factors

Regarding the loading of neural growth factors, the work done with neurotrophins is specially relevant. Neurotrophins are a family of growth factors that support survival, differentiation, and neuronal growth. One member of this family, BDNF, has been found to increase its correspondent mRNA levels after acute TBI in adult rats [139], suggesting the existence of endogenous neuroprotection events triggered by BDNF. However, Oyesiku et al. reported no clinical benefit in terms of survival in patients with amyotrophic lateral sclerosis after BDNF administration, either subcutaneously or via the spinal canal, in a phase III clinical trial [140]. This lack of clinical efficacy was likely due to the low ability of BDNF to cross the BBB and its short half-life in circulation as reported by others [141]. Interestingly, NPs have the potential to solve this problem [142]. Multiple preclinical studies have demonstrated signs of axonal regeneration and/or motor function recovery after TBI in mice and rats by treatment with hydrogels [143, 144], poly ε-caprolactone (PCL) scaffolds [145] and PLGA microspheres [146] loaded with BDNF. Similarly, other neurotrophins, such as the nerve growth factor (NGF) and neurotrophin-3 (NT-3), elicited axonal regeneration when encapsulated in PLGA microspheres [146], PCL-based conduits [147], chitosan-based nerve conduits [148], PLGA-based conduits [149], and hydrogels [150, 151] in rat models. In non-human primates, Xu et al. achieved functional recovery after SCI with 2-methacryloyloxyethyl phosphorylcholine and polylactic acid diacrylate nanocapsules loaded with NGF [26].

7.4.4 Genome Engineering Tools

Apart from delivering growth factors and drugs in the CNS, NPs have been recently deployed for delivering the groundbreaking genome editing tool clustered regularly interspaced short palindromic repeats (CRISPR)/Cas9. Since its discovery in 2012 [152], CRISPR/Cas9 technology opened the avenue to modern gene therapy by precisely modulating gene expression to resolve multiple diseases. The use of nanomaterials for the delivery of CRISPR/Cas9 to tissues including nervous tissues is preferable to classic virus-mediated delivery due to the toxicity, immunogenicity, and prolonged CRISPR/Cas9 expression associated with viral gene transfer [153, 154]. Lee et al. followed this nanomaterial approach and used Au NPs to deliver CRISPR into the brain by intracranial injection in a mice model of fragile X syndrome. The developed therapy, namely CRISPR-Gold and targeting metabotropic glutamate receptor 5 (mGluR5), successfully rescued animals from the exaggerated repetitive behavior caused by this neurological disorder [155].

7.5 Applications

Some of the most interesting NPs applications in neuronal diseases include their use as contrast agents for imaging diagnosis (e.g., CT, MRI, and others) (Fig. 7.4), platforms for drug delivery, and therapeutic probes in cancer detection and treatment, cell labelling, and regenerative medicine. Further details of each of these applications are discussed below.

7.5.1 Imaging Diagnosis

Among all NPs, those of lipid nature and those incorporating metal derivates or metals such as Au are the most used for CT purposes. The need for increasing the sensitivity of CT has motivated the use of dense NPs with high payloads. Also, small NPs with a few nm in size are easier to be excreted by the kidneys than

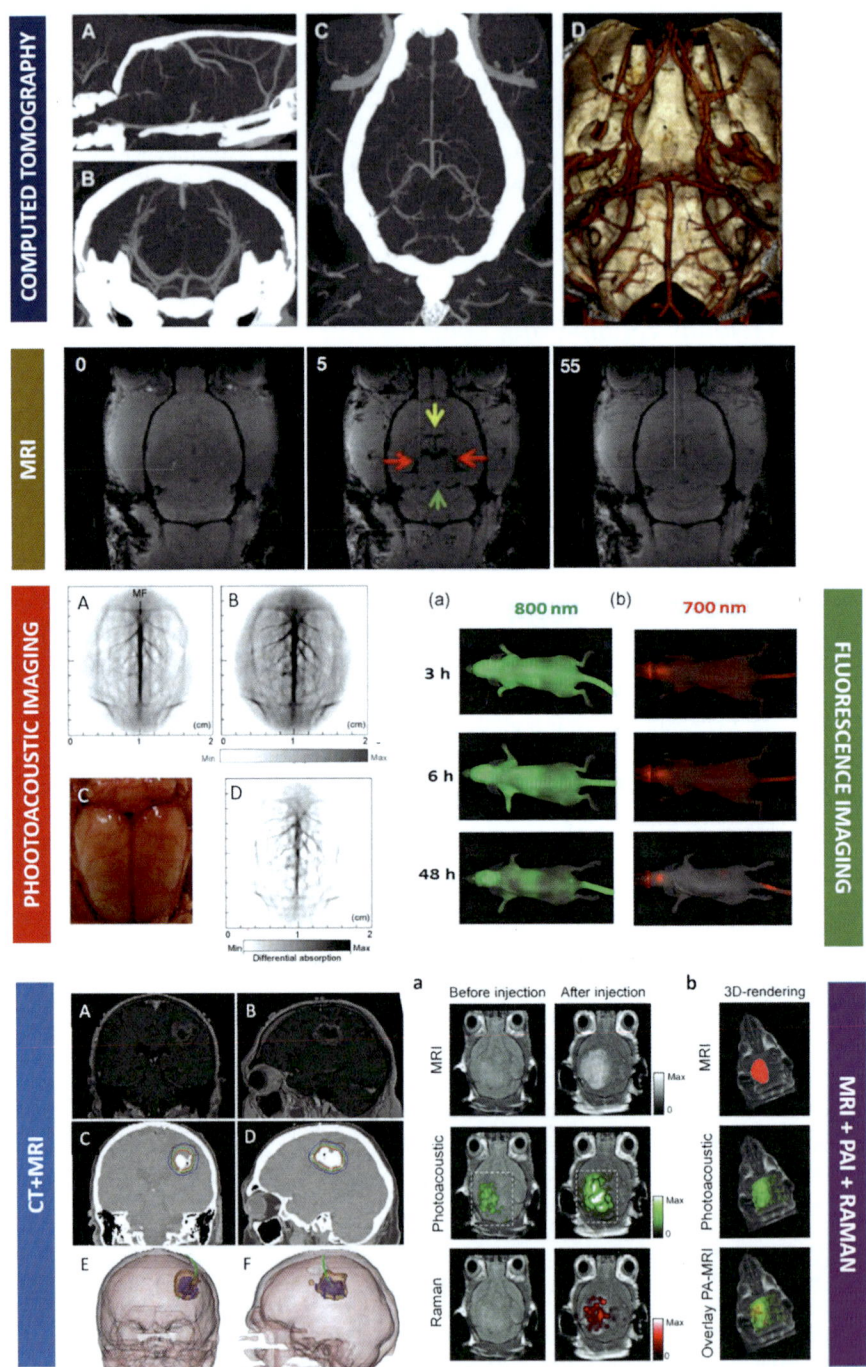

Fig. 7.4 Nanoparticles used as contrast agents for biomedical imaging. X-ray computed tomography (CT): (**a**) sagittal, (**b**) axial, and (**c**) coronal imaging of a rat brain after intravenous injection of iodine liposomes, and (**d**) 3D reconstruction of brain vascularization. Reprinted from [157],

commonly used CT contrast agents in patients with renal insufficiency [156]. One of the advantages of using NPs as contrast agents is the possibility to functionalize their surface to target brain tumors and cerebrovascular accidents, among others, and to add different coatings to increase their half-life in blood circulation, ideal for vascular imaging [156, 157]. NPs used for CT imaging can also have therapeutic applications (i.e., theranostic NPs), if loaded with drugs or use for their intrinsic therapeutic properties such as the ability of photothermal ablation [158] and high X-ray absorption. This latter property, displayed by Au NPs, leads to radiosensitization effects in mice models of glioma [159].

MRI is another imaging modality that can benefit from the use of NPs. Magnetic NPs, for instance, can increase the contrast of MRI and optimize the imaging of specific pathological phenomena such as tumors, inflammation, arteriosclerotic plaques, vascular stenosis, thrombus, and neoangiogenesis. This is possible by passive accumulation of NPs (e.g., enhanced permeability and retention effect, EPR) or by active targeting of NPs facilitated by surface functionalization with antibodies, peptides, and ligands like in the case of tumor imaging [32, 162]. Different nanoformulations have been investigated for MRI imaging, being micelles, liposomes, and dendrimers with magnetic material (such as gadolinium or iron oxide) the most used. For example, Starmans et al. compared iron oxide micelles functionalized with fibrin-binding peptides and non-functionalized ones and observed an enhanced uptake in the thrombi by MRI in an in vivo carotid artery thrombosis mouse model [163]. Also, Faucher et al. synthesized ultra-small gadolinium oxide NPs for MRI of glioblastoma cells implanted into a chicken embryo, suggesting their utility for the diagnosis of glioblastoma [164]. These NPs successfully accumulated in the

---

Fig. 7.4 (continued) Copyright (2016), with permission from Elsevier. Magnetic resonance imaging (MRI): Transversal section of a mouse brain at different times (0, 5, and 55 min) after intravenous injection of iron oxide nanoparticles for drug delivery. Republished with permission of Future Medicine Ltd., from [32]; permission conveyed through Copyright Clearance Center, Inc. Photoacoustic imaging (PAI): Photoacoustic signal of rat brain (**a**) before and (**b**) 20 min after gold nanoshell i.v. injection, (**c**) optical picture of the rat brain, (**d**) differential image (before–after injection) of brain rat vascularization. Reprinted with permission from [7]. Copyright (2004) American Chemical Society. Fluorescence imaging: Whole body fluorescence imaging at 3, 6, and 48 h post-induction of a cryolesion and subsequent intravenous injection of PLGA nanoparticles excited with (**a**) 800 nm and (**b**) 700 nm light. Reproduced by permission from Springer Nature Customer Service Centre GmbH: Springer Nature, Nano Research [160], COPYRIGHT (2016). Multimodal CT+MRI images: MRI (**a** and **b**) and CT (**c** and **d**) images of a human brain after direct injection of iron oxide nanoparticles in a glioblastoma. 3D reconstruction shows tumor (brown), magnetic fluid (blue), and thermometry catheter (green) distribution from posterior (**e**) and lateral (**f**) views. Reproduced by permission from Springer Nature Customer Service Centre GmbH: Springer Nature, Journal of Neuro-Oncology [161], COPYRIGHT (2011). Multimodal MRI+PAI+ Raman scattering imaging for brain tumor labelling: Coronal MRI images of a rat brain (**a**) before and after injection of hybrid (Au-Raman label-Silica-Gd) nanoparticles (upper row); PAI signal overimposed on MRI image (middle row) and Raman scattering signal overimposed on MRI image (bottom row). (**b**) 3D reconstruction of brain tumor labelling in each case. Reproduced by permission from Springer Nature Customer Service Centre GmbH: Springer Nature, Nature Medicine [8], COPYRIGHT (2012)

infarct region induced by intraluminal occlusion of the proximal left middle cerebral artery in mice [164]. Moreover, Jin et al. developed aminated dextran iron oxide NPs targeted with a P-selectin-binding peptide [165]. P-selectin is a marker of acute neuroinflammation and microthrombi formation [166]. Another approach for MRI imaging is the one reported by Frias et al., in which high-density lipoprotein (HDL)-based NPs efficiently imaged arteriosclerotic plaques in mice [167].

Apart from CT and MRI, NPs can be deployed for non-invasive optical imaging of TBI. This would allow the selective detection of necrotic areas found after the injury [168]. While optical imaging is very useful for the diagnosis of damage in tissue sections and cells, optical imaging of deep layer of tissue can be obtained by using dyes that emit in the near-infrared (NIR) biological window mentioned before. Such probes would enable the expansion of the imaging coverage up to a few centimeters [169]. For example, Cruz et al. have developed bimodal PEGylated PLGA NPs incorporating the NIR 700 nm dye and the perfluoro-15-crown-5-ether (PFCE) dye, together with the dead cell-specific ligand 800CW. Such NPs were able to recognize in vivo TBI areas in mice [160].

An alternative strategy to obtain high-resolution imaging of brain cortex is the photoacoustic imaging (PAI). It is an emergent technology for brain scanning based on the conversion of laser light pulses into ultrasound waves by the thermal expansion of photoabsorbing compounds [170]. The small dispersion of ultrasound waves in biological tissues, combined with the deep penetration of NIR light, is used to obtain real-time images with a resolution of approximately 70 μm for tissues within 2–3 mm in depth [171]. Furthermore, this technique represents a novel strategy for the detection of neuronal activity in the mesoscale by indirect measurement of oxygen consumption. Specifically, hemoglobin (Hb) is one of the strongest optical absorbers. In biological samples, Hb can be used to generate a PA image of brain vascularization. In addition, the different absorption coefficients between oxy-Hb (HbO$_2$) and deoxy-Hb (HbR) allow for the detection of the hypoxia generated by the accelerated activity of tumors in the brain [172, 173]. The combination of PAI with conventional ultrasounds provides a structural–functional image of the brain with unprecedented resolution. NPs are the contrast agents most commonly used for this technique. Some examples of organic NPs used in PAI include porphysomes [174], perfluorocarbon-based (PFC) nanodroplets [175], polymer NPs [176], and hydrogel NPs [177]. However, the absorption results stronger in the case of inorganic NPs with plasmonic response. Au NPs are the most frequent choice for PAI because the NIR absorption peak can be tuned by adjusting their geometry [178]. They have been used in PAI in neural tissues to enhance the vascular and tumor contrast (Fig. 7.4) [179].

Experimental studies have demonstrated that Au NPs are versatile, persistent, and safe contrast agents for multimodality imaging, thereby enhancing the precision of the microscopic determination of the brain tumor edges pre-, intra-, and post-operation. Several studies were conducted in vivo in rodent models, using mainly nanospheres with sizes ranging from 20 to 120 nm. The most commonly used imaging modalities were MRI, surface-enhanced Raman scattering, and fluorescent microscopy (Fig. 7.4) [180].

Finally, hybrid NPs made of materials with complementary response can be employed as dual contrast agents to obtain highly resolved dual images of a targeted region. The multimodal image obtained by complementary techniques is an extremely useful diagnostic tool. For instance, the combination of MRI and X-ray is clinically used to obtain a dual image of soft and hard tissues of the patients [181, 182]. One of the first examples of dual contrast agents are the Au-iron oxide NPs used for CT-MRI dual imaging [183]. The heavy nucleus of Au absorbs X-ray radiation, while iron oxide NPs modify MRI contrast. Other examples of multimodal imaging that use Au-iron oxide NPs as dual contrast agents include MRI-optical coherent tomography [184], MRI-photoluminescence [185], and MRI-PAI [186].

7.5.2   Cancer Detection and Treatment

Brain and spinal cord tumors are responsible for an important portion of CNS injuries. A study reviewing the cause of SCI in the region of Aragón (Spain) between 1991–2008 revealed that 32.5% of SCIs were due to tumor formation [187]. There is a wide diversity of tumors affecting the CNS. They can be either primary tumors (i.e., tumors that start in the CNS) or secondary tumors (i.e., tumors that start in other organs and then metastasize into the CNS) [188]. The primary tumors of the brain and spinal cord are gliomas, astrocytomas, oligodendrogliomas, ependymomas, meningiomas, medulloblastomas, gangliogliomas, schwannomas, and craniopharyngiomas [188]. Other tumors like chordomas, primary CNS non-Hodgkin lymphomas, and pituitary tumors are not originated from brain or spinal cord cells but can start in or near the brain, thus still affecting its functionality [188]. The secondary tumors correspond to metastatic disease located in the CNS but originated in other tissues, with special predilection to metastasize in the CNS, such as breast, lung, and prostate cancers and melanoma [187, 188].

Nanotechnology has the potential to reduce the morbidity and mortality of patients affected by cancer-related CNS injuries by providing an early diagnosis and more effective targeted therapies specifically directed to each of these cancers. Successful NPs have to be selectively recognized by the tumor, carry suitable drugs, and have appropriate biodegradability. There has been extensive research on how to target specifically CNS tumors, from gliomas to brain metastasis. For gliomas, Qian et al. used oxidized nanocrystalline mesoporous carbon particles linked to the Pep22 polypeptide targeting the LDLR and encapsulating doxorubicin. These NPs efficiently targeted LDLR-overexpressing gliomas and killed glioma cells with both doxorubicin and NIR light [189]. For glioblastomas, gliomas with the highest grade, Sun et al. employed polymalic acid-based nano-biopolymers to deliver CRISPR/Cas9 to target and suppress the expression of tumor vascular laminin-411 ($\alpha4\beta1\gamma1$), significantly inhibiting glioblastoma growth in vivo and extending mice life span [190]. For brain breast cancer metastasis, Patil et al. have successfully engineered polymeric poly(β-L-malic acid) (PMLA) NPs for sensitive MRI containing the MRI contrast agent gadolinium-DOTA and targeted to breast cancer cells with monoclonal antibodies anti-human epidermal growth factor receptor 2 (HER2) (i.e., trastuzumab) and anti-epidermal growth factor receptor (EGFR) (i.e., cetuximab). NPs were also targeted with anti-transferrin receptor to cross the BBB *via* transcytosis. As for treatment, trastuzumab-targeted NPs incorporating antisense oligonucleotides (AON) for the inhibition of HER2 increased survival in mice harboring HER2$^+$ intracranial breast tumors by 57% when animals were treated intravenously with these NPs, compared to the PBS control treatment [191]. Also, 66% and 114% increases in survival were achieved in mice bearing EGFR$^+$ brain and lung cancer tumors and EGFR$^+$ triple negative breast cancer tumors, respectively, when animals were treated with EGFR-targeted NPs carrying AON anti-EGFR [191].

The active properties of inorganic NPs can also be exploited for the treatment of brain tumors. Radiotherapy seeks to maximize the damage on the tumor while minimizing the radiation dose. Targeting heavy metal NPs in the tumor increases the X-ray absorption in this region promoting the production of ROS and amplifying DNA damage [192]. For instance, Hainfeld et al. reported the accumulation of Au NPs into brain gliomas with a 19:1 tumor-to-healthy parenchyma ratio after intravenous injection (4 mg Au/g) and observed a 50% long-term (N1 year) survival with a single dose of 30–35 Gy at 100 kVp compared to mice receiving only radiation [159]. The clinical trials for radiosensitization with gadolinium-doped Au NPs (AGuIX$^®$) have already completed phase I in patients with multiple brain metastases (NANORAD, NCT02820454) [193].

Other modalities of nanotherapy on brain tumors are based on the local heating of cancer cells (i.e., hyperthermia) and the photocatalytic generation of ROS (i.e., photodynamic). Hyperthermia, also known as thermotherapy, is a clinical procedure in which the temperature of the patient is increased. Hyperthermia can be local, regional, or whole body, depending on the type of cancer, its location, and its stage. Nanotechnology allows the focalization of the effect of temperature rise by concentrating the local heating in a desired site of the tumor by irradiating magnetic NPs with high-frequency magnetic fields (i.e., magnetic hyperthermia) or with NIR light (i.e., optical hyperthermia or photothermia). Magnetic hyperthermia has been successfully tested in animal models and is currently being evaluated in clinical trials to treat glioblastoma (NanoTherm$^®$) [161]. Clinical trials employing optical hyperthermia with silica-Au NPs (AuroLase, Nanospectra Biosciences) for the treatment of refractory and/or recurrent head and neck tumors are currently ongoing (NCT00848042) [194].

Photodynamic therapy is an alternative approach to radiotherapy that substitutes the X-ray and γ-ray for visible-NIR light. To increase the effect of the visible light on tumor cells, the tissue needs to be labeled with strong photosensitizes such as porfimer sodium (Photofrin$^®$) and 5-aminolevulinic acid (5–5-ALA, Gliolan$^®$). Metallic NPs combine the strong light absorption of their plasmonic response with the catalytic activity of their surface for an enhanced production of ROS

[195]. Moreover, metallic NPs present suitable properties for the combination of photodynamic therapy with drug delivery and tumor labelling [196].

7.5.3   Cell Labelling and Regenerative Medicine

NPs can be used for cell labelling, and the most prominent application of this type of labelling is cell tracking. Cell tracking is a non-invasive method that provides real-time visualization of the cells of interest. We have previously described how NPs can improve CT, MRI, and optical imaging. Likewise, in cell-based therapies, NPs represent a powerful tool for labelling cells and tracking them. In the context of brain and spinal cord treatments, cell tracking allows for monitoring migration, distribution, and functionality of transplanted cells during the process of neuronal regeneration [156]. Such implanted cells can be neural stem cells, which have pluripotent features and are able to differentiate into neurons or glia in the right context. Guzman et al. showed the applicability of iron oxide nanocomposites to track neural stem cells by MRI in vivo and analyzed their migration patterns into newborn and adult rodent brains [197]. Such labelled neural stem cells migrated to the site of injury in the situation of ischemic stroke (distal middle cerebral artery occlusion) being their functionality not affected by the NPs [197]. More recently, Duan et al. studied the long-term fate of transplanted mesenchymal stem cells labelled with iron oxide NPs-loaded polymersomes by MRI after induced cerebral ischemia in rats [198]. Immunotherapy is another type of cell-based therapy that uses NPs for cell tracking. In particular, the introduction of "activated" autologous T cells to the patient able to recognize cancer cell antigens is currently a rapid growing oncologic discipline. Such discipline has not been excepted from the use of nanotechnology. As a proof of concept, Smith et al. achieved in vivo T cell reprogramming in mice by the introduction of leukemia-targeting chimeric antigen receptors (CAR) genes mounted in NPs. These NPs were tracked by bioluminescence imaging thanks to the luciferase reporter gene that they were carrying [199]. To our knowledge, no studies employing CAR T cells and NPs have yet been performed in brain tumors. However, it represents a very attractive strategy to defeat this aggressive type of cancers.

An additional functionality of labelling cells with magnetic NPs is the contactless manipulation with magnetic fields. By using the magnetic gradient of strong magnets (generally, NdFeB-type), it is possible to accumulate T cells in a tumor region [200], as well as concentrating circulating tumor cells in the blood stream allowing their detection despite their very low concentrations [42, 201].

Finally, another exciting application of NPs is neural regeneration and neural tissue engineering after TBI and SCI. Magnetic NPs have been largely studied for this purpose. The migratory ability of magnetic NPs by themselves and when engulfed by cells and subcellular structures has been exploited for guided axonal and neurite growth of induced neuronal cells [202], and filopodia of retinal ganglion cells [203]. Moreover, magnetic NPs have been used to induce movement of subcellular compartments such as endosomes, to control growth cone behavior and subsequent axon growth [204], and neuron synaptosomes isolated from glutamatergic neurons without affecting their intracellular signaling [205]. Regarding peripheral nerve reconstruction, the only clinical alternative to autologous nerve grafts to treat extensive nerve injuries is the usage of synthetic nerve guidance conduits. In this sense, combining such nerve conduits with magnetic NPs encapsulating growth factors to control their spatiotemporal release represents a great strategy to favor nerve regeneration. For example, Giannaccini et al. demonstrated an accelerated the regeneration process in rats with a peripheral nerve lesion with the use of polyethylenimine (PEI)-coated iron oxide NPs encapsulating NGF and vascular endothelial growth factor receptor [206]. The effect was enhanced if the NGF-magnetic NPs were incorporated in aligned poly-L-lactic acid microfibers generating a NGF gradient in the direction of the lesion [207].

7.6 Major Concerns in Nanotoxicity

While NPs own the potential to improve the diagnosis and treatment of diseases affecting the CNS, the use of NPs-mediated selective drug delivery may trigger adverse effects in patients resulting in exacerbated reactivity and other hazorous biological effects. Due to their unique physicochemical properties, the interaction of NPs with living cells and tissues is often unpredictable [208]. For this reason, the characteristics that are responsible for the unique properties of NPs that account for their beneficial effects in medical applications often contribute to their toxicity leading to concerns about human and environmental health safety [209]. Therefore, understanding these safety health issues is critical for the successful application of NPs in the industry of biomedicine and biotechnology. Deliberately, many examples of toxic NPs showed below were excluded in this review, or directly not used in the CNS.

The ability of nanometric systems to cross the BBB and BSCB makes them potentially hazardous [210]. NPs are prone to accumulate in various brain regions after crossing the BBB, where they may enter into astrocytes, neurons, and microglia [211]. The exposure of neuronal cells to NPs can then lead to a direct alteration of the CNS structure and/or functioning, which may result in subsequent effects due to glial stimulation and glial–neuronal interactions. In this way, the neurotoxic effect of NPs that occur through various mechanisms, including oxidative stress, immune responses, and neuroinflammation, could permanently damage the CNS or trigger irreversible alterations to the barrier that control its access [212, 213].

The neurotoxicity of NPs is usually caused by substantial production of ROS leading to oxidative stress and chronic diseases. ROS targets neurons and glial cells, which are post-mitotic cells and thus particularly vulnerable to free radicals that inflict neuronal harm. The release of cytokines then triggers neuroinflammation and, eventually, neuronal death by apoptotic mechanisms [214]. Some structures of carbon-based NPs and phases of metal and metal-oxide NPs such as silica, iron oxide, titanium dioxide, Au, and silver have been identified as being potentially neurotoxic [215]. In contrast, organic NPs such as ferritin NPs, dendrimers, micelles, liposomes, and polymeric NPs are generally considered as non-toxic and biodegradable.

Due to their unique structural characteristics and diversity, carbon-based NPs have drawn considerable interest in numerous fields including the delivery of therapeutic molecules for disease treatment and tissue repair at a cellular level [216]. Carbon-based NPs such as carbon nanotubes, C_{60} fullerenes, graphene-derived materials, and carbon quantum dots produce ROS when exposed to ultraviolet light or transition metals. In vivo studies have shown that inhalation exposure to carbon black NPs results in accumulation in the olfactory bulb that induces an inflammatory response by activating microglia and enhances the risk of brain disorders [217]. More recently, Samiei et al., by examining studied toxicity effects on brains in respiratory contact with multiwalled carbon nanotubes, found an increased production of mitochondrial ROS in different parts of the rat brain (e.g., hippocampus, frontal cortex, and cerebellum) and disruption of the mitochondrial respiratory chain in varying degrees [218].

The catalytic activity of iron oxide NPs in the production of ROS by the Fenton reaction is also responsible for potential neurotoxicity *via* oxidative stress. The accumulation of iron ions released from iron oxide NPs, together with oxidative stress, may induce the aggregation of proteins such as the amyloid beta peptides and α-synuclein, which lead to the onset of Alzheimer's and Parkinson's diseases, respectively [219, 220]. Some studies have also suggested that regular uptake of these NPs has an effect on the synaptic transmission, causing apoptosis, inflammation of neurons, and infiltration of immune cells [221].

The toxicological effects on microglia-mediated activity of various inorganic NPs remain unknown. Microglia, accounting for the 12% of the CNS cells, represents macrophage-like cells regarded as important mediators of the initial inflammation within the CNS [222]. When activated, microglia releases nitric oxide, ROS, and inflammatory cytokines (e.g., TNFα, IL1β, and IL6), which are critical for neuronal damage and apoptosis [223]. Two studies from the same group showed that titanium dioxide [224] and silica [225] NPs, in the 20 nm range in size, could directly induce neuronal dysfunction and damage in PC12 cells through activation of the p53 and JNK pathways. Another study reported that silica NPs increased intracellular ROS and reactive nitrogen species production decreasing tumor necrosis factor alpha (TNFα), while increasing interleukin 1 beta (IL1β) and cyclooxygenase-2 expression in microglial cells [226].

The main mechanisms of neurotoxicity induced by titanium dioxide NPs are the impairment of antioxidative capacity, oxidative stress, increased production of ROS, inflammatory responses, shrinkage of nuclear envelopes, apoptosis, changes in the cellular components, and disruption of signaling pathways [227]. Many in vivo studies have demonstrated different toxicological effects triggered by this type of NPs depending on the intake period and dosage. For instance, it was shown that apoptosis was triggered after 96 h of exposure to titanium dioxide NPs on human U373 and rat C6 glial cells resulting in brain damage [228]. Two doses of titanium oxide NPs, 0.25 and 0.5 mg/ml, over 24 h of exposure increased apoptosis and release of lactate dehydrogenase while decreasing cell viability in neuroendocrine cells [229]. Besides that, maternal exposure of mice to titanium oxide NPs was shown to alter the expression of genes regulating apoptosis and oxidative stress, causing brain impairment in newborn pups [230].

Recent studies have shown the possible role of silver NPs, in the form of release rates of silver ion, surface coatings, shape, size, and interactions with different cells and proteins, in affecting neurotoxicity through intracellular ROS generation and cell death [231]. Another toxicity mechanism is the one triggered by the direct release of silver ions in cells, which harms the integrity of the cell membrane and causes cell necrosis. Intriguingly, even low doses of silver NPs caused a rise in the release of cytokines from astrocytes and caspase activation, which resulted in neuroinflammation and apoptotic cell death [232]. In regard to Au NPs, bulk Au is known to be non-toxic, chemically inert, and has actually been used in the clinic as anti-inflammatory agent to treat rheumatoid arthritis. However, the chemical reactivity of Au NPs varies depending on the size. Au particles larger than 5 nm are considered to be inert [233], while those with less than 3 nm are not [234]. Moreover, long-term exposure to Au NPs induced astrogliosis in rats [235]. Astrogliosis consists of the exacerbated production of reactive astrocytes, which has been associated to epilepsy, hypoxia, and ischemic events in neurological disorders. Finally, it should be mentioned that, in polymer-coated Au NPs, a partial separation of the organic shell from the inorganic core caused by proteolytic digestion upon cell uptake has been reported, thus having potential harmful consequences for drug targeting [236].

Unfortunately, no simple conclusions have emerged on the mechanism behind NPs-induced toxicity, and it has yet to be uncovered. Many studies attribute this to oxidative stress. However, neurotoxicity induced by NPs remains a novel area that requires further exploration, which is essential for enhancing the applicability of NPs. Importantly, there are ways to reduce the neurotoxicity of NPs. For instance, carbon NPs can be functionalized with hydrophilic polymers to improve their water solubility and dispersion. More recently, green chemistry employing non-hazardous and safe substances has been proposed for the development of NPs with reduced toxicity [237]. New routes of drug administration are also being explored to facilitate side-specific targeting effects, thereby increasing their bioavailability and reducing their toxicity such as ocular delivery [238]. Modifications to the size, shape, and surface of nanostructures can further enhance the bioactivity of these nanomaterials. In any case, the benefits of nanomaterials must be weighed against their potential harmful effects. Therefore, recognizing and understanding the neurotoxic effects of manufactured and engineered NPs will help to establish health standards to support the clinical implementation of nanotechnology for treating TBI, CSI, and other CNS diseases.

7.7 Conclusions

Despite numerous advantages in the use of NPs for neural applications in the last decades, there are some challenges that need further studies for practical application: (1) long blood circulation times, (2) BBB and BSCB penetration, (3) concentration at a specific target, (4) high drug/molecule payload, and (5) control of the release process. In this chapter, we have discussed different ways to address the above-mentioned challenges using a diversity of NPs and coatings. However, the design and production of some of them are still complicated.

Among all the nanoparticulated systems studied for neural applications, organic NPs (mostly liposomes and polymeric vesicles) are the most employed due to their versatility and degradability. Inorganic NPs are considered to be as one component more of the system to obtain additional functionalities as nanoheaters and contrast agents. The complexity of the formulations is a challenge itself for the regulatory point of view, since each component has to be approved individually as pharmaceutical excipient before its use. Therefore, we consider that simple formulations based on already approved components might be a clever strategy to boost the use of NPs in neural applications.

Finally, due to the delicate nature of the CNS in terms of structure and function, safety is one of the most important issues to take into account for the application of NPs in the neural tissue. To this regard, more exhaustive in vitro and in vivo toxicity studies are still necessary prior to the full exploitation of the great potential of these nanoparticulated systems.

Acknowledgments This work was funded by the Spanish Ministry of Economy and Competitiveness (COMANCHE, MAT2017-88148-R, AEI/ FEDER, UE) and by CSIC (201960E062). A.S. acknowledges the Healy Research Collaboration Award (HRCA05-19) from the Raine Medical Research Foundation.

References

1. Frank RG (1994) Instruments, nerve action, and the all-or-none principle. Osiris 9:208–235
2. Ledesma HA, Li X, Carvalho-de-Souza JL et al. (2019) An atlas of nano-enabled neural interfaces. Nat Nanotechnol 14:645–657
3. Yang X, Zhou T, Zwang TJ et al. (2019) Bioinspired neuron-like electronics. Nat Mater 18:510–517
4. Zhou T, Hong G, Fu TM et al. (2017) Syringe-injectable mesh electronics integrate seamlessly with minimal chronic immune response in the brain. Proc Natl Acad Sci USA 114:5894–5899
5. ISO/TC 229 Nanotechnologies. (2005) International Organization of Standardization. https://www.iso.org/committee/381983.html. Cited 27 Sep 2021
6. Wu LC, Zhang Y, Steinberg G et al. (2019) A review of magnetic particle imaging and perspectives on neuroimaging. AJNR Am J Neuroradiol 40:206–212
7. Wang Y, Xie X, Wang X et al. (2004) Photoacoustic tomography of a nanoshell contrast agent in the *in vivo* rat brain. Nano Lett. 4:1689–1692

8. Kircher MF, de la Zerda A, Jokerst JV et al. (2012) A brain tumor molecular imaging strategy using a new triple-modality MRI-photoacoustic-Raman nanoparticle. Nat Med 18:829–834

9. Nam L, Coll C, Erthal LCS et al. (2018) Drug delivery nanosystems for the localized treatment of glioblastoma multiforme. Materials (Basel) 11:779

10. Le Floc'h J, Lu HD, Lim TL et al. (2020) Transcranial photoacoustic detection of blood-brain barrier disruption following focused ultrasound-mediated nanoparticle delivery. Mol Imaging Biol 22:324–334

11. Munshi R, Qadri SM, Zhang Q et al. (2017) Magnetothermal genetic deep brain stimulation of motor behaviors in awake, freely moving mice. Elife 6:e27069

12. Fadeel B (2019) Hide and seek: nanomaterial interactions with the immune system. Front Immunol 10:133

13. Han D, Zhang B, Chong C et al. (2020) A strategy for iron oxide nanoparticles to adhere to the neuronal membrane in the substantia nigra of mice. J Mater Chem B 8:758–766

14. Michael JS, Lee BS, Zhang M et al. (2018) Nanotechnology for treatment of glioblastoma multiforme. J Transl Int Med 6:128–133

15. Paka GD, Ramassamy C (2017) Optimization of curcumin-loaded PEG-PLGA nanoparticles by GSH functionalization: investigation of the internalization pathway in neuronal cells. Mol Pharm 14:93–106

16. Marcus M, Karni M, Baranes K et al. (2016) Iron oxide nanoparticles for neuronal cell applications: uptake study and magnetic manipulations. J Nanobiotechnology 14:37

17. de la Fuente JMG, Grazu V (2012) Nanobiotechnology. Inorganic nanoparticles vs organic nanoparticles. In: Frontiers of neuroscience, vol 4. Elsevier, Amsterdam, p 2–520

18. Jia F, Liu X, Li L et al. (2013) Multifunctional nanoparticles for targeted delivery of immune activating and cancer therapeutic agents. J Control Release 172:1020–1034.

19. Ventola CL (2017) Progress in nanomedicine: approved and investigational nanodrugs. P T 42:742–755

20. Li S, Goins B, Zhang L et al. (2012) Novel multifunctional theranostic liposome drug delivery system: construction, characterization, and multimodality MR, near-infrared fluorescent, and nuclear imaging. Bioconjug Chem 23:1322–1332

21. Guvener N, Appold L, de Lorenzi F et al. (2017) Recent advances in ultrasound-based diagnosis and therapy with micro- and nanometer-sized formulations. Methods 130:4–13

22. Barenholz Y (2012) Doxil(R)-the first FDA-approved nano-drug: lessons learned. J Control Release 160:117–134

23. Agrahari V, Burnouf PA, Burnouf T et al. (2019) Nanoformulation properties, characterization, and behavior in complex biological matrices: challenges and opportunities for brain-targeted drug delivery applications and enhanced translational potential. Adv Drug Deliv Rev 148:146–180

24. Cui N, Lu H, Li M (2018) Magnetic nanoparticles associated PEG/PLGA block copolymer targeted with anti-transferrin receptor antibodies for Alzheimer's disease. J Biomed Nanotechnol 14:1017–1024

25. Clemons TD, Singh R, Sorolla A et al. (2018) Distinction between active and passive targeting of nanoparticles dictate their overall therapeutic efficacy. Langmuir 34:15343–15349

26. Xu D, Wu D, Qin M et al. (2019) Efficient delivery of nerve growth factors to the central nervous system for neural regeneration. Adv Mater 31:e1900727

27. Beg S, Samad A, Alam MI et al. (2011) Dendrimers as novel systems for delivery of neuropharmaceuticals to the brain. CNS Neurol Disord Drug Targets 10:576–588

28. Janaszewska A, Lazniewska J, Trzepinski P et al. (2019) Cytotoxicity of dendrimers. Biomolecules 9:330

29. Sharma A, Liaw K, Sharma R et al. (2018) Targeting mitochondrial dysfunction and oxidative stress in activated microglia using dendrimer-based therapeutics. Theranostics 8:5529–5547

30. Dakwar GR, Abu Hammad I, Popov M et al. (2012) Delivery of proteins to the brain by bolaamphiphilic nano-sized vesicles. J Control Release 160:315–321

31. Jackson CL, Chanzy HD, Booy FP et al. (1998) Visualization of dendrimer molecules by transmission electron microscopy (TEM): staining methods and cryo-TEM of vitrified solutions. Macromolecules 31:6259–6265
32. Mejías R, Pérez-Yagüe S, Roca AG et al. (2010) Liver and brain imaging through dimercaptosuccinic acid-coated iron oxide nanoparticles. Nanomedicine 5:397–408
33. Jampilek J, Zaruba K, Oravec M et al. (2015) Preparation of silica nanoparticles loaded with nootropics and their *in vivo* permeation through blood-brain barrier. BioMed Res Int 2015:812673
34. Zou D, Wang W, Lei D (2017) Penetration of blood-brain barrier and antitumor activity and nerve repair in glioma by doxorubicin-loaded monosialoganglioside micelles system. Int J Nanomed 12:4879–4889
35. Hanafy NAN, El-Kemary M, Leporatti S (2018) Micelles structure development as a strategy to improve smart cancer therapy. Cancers (Basel) 10:238
36. Zeng L, Zou L, Yu H et al. (2016) Treatment of malignant brain tumor by tumor-triggered programmed wormlike micelles with precise targeting and deep penetration. Adv Func Mater 2:4201–4212
37. Baba M, Itaka K, Kondo K et al. (2015) Treatment of neurological disorders by introducing mRNA *in vivo* using polyplex nanomicelles. J Control Release 201:41–48
38. Kaur IP, Bhandari R, Bhandari S et al. (2008) Potential of solid lipid nanoparticles in brain targeting. J Control Release 127:97–109
39. Teleanu DM, Chircov C, Grumezescu AM et al. (2019) Neuronanomedicine: An up-to-date overview. Pharmaceutics 11:101
40. Rong G, Corrie SR, Clark HA (2017) *In vivo* biosensing: progress and perspectives. ACS Sens 2:327–338
41. Gahl TJ, Kunze A (2018) Force-mediating magnetic nanoparticles to engineer neuronal cell function. Front Neurosci 12:299
42. Ovejero JG, Yoon SJ, Li J et al. (2018) Synthesis of hybrid magneto-plasmonic nanoparticles with potential use in photoacoustic detection of circulating tumor cells. Mikrochim Acta 185:130
43. Del Sol-Fernandez S, Portilla-Tundidor Y, Gutierrez L et al. (2019) Flower-like Mn-doped magnetic nanoparticles functionalized with alphavbeta3-integrin-ligand to efficiently induce intracellular heat after alternating magnetic field exposition, triggering glioma cell death. ACS Appl Mater Interfaces 11:26648–26663
44. Minaei SE, Khoei S, Khoee S et al. (2019) *In vitro* anti-cancer efficacy of multi-functionalized magnetite nanoparticles combining alternating magnetic hyperthermia in glioblastoma cancer cells. Mater Sci Eng C Mater Biol Appl 101:575–587
45. Yarjanli Z, Ghaedi K, Esmaeili A et al. (2017) Iron oxide nanoparticles may damage to the neural tissue through iron accumulation, oxidative stress, and protein aggregation. BMC Neurosci 18:51
46. Horvath H (2009) Gustav Mie and the scattering and absorption of light by particles: historic developments and basics. J Quant Spectrosc Radiat Transfer 110:787–799
47. Garcia MA (2011) Surface plasmons in metallic nanoparticles: fundamentals and applications. J Phys D Appl Phys 44:283001
48. Huefner A, Kuan WL, Barker RA et al. (2013) Intracellular SERS nanoprobes for distinction of different neuronal cell types. Nano Lett 13:2463–2470
49. Youssef Z, Yesmurzayeva N, Larue L et al. (2019) New targeted gold nanorods for the treatment of glioblastoma by photodynamic therapy. J Clin Med 8:2205
50. Fernandez Cabada T, Sanchez Lopez de Pablo C, Martinez Serrano A et al. (2012) Induction of cell death in a glioblastoma line by hyperthermic therapy based on gold nanorods. Int J Nanomedicine 7:1511–1523
51. Jang Y, Lee N, Kim JH et al. (2018) Shape-controlled synthesis of Au nanostructures using EDTA tetrasodium salt and their photothermal therapy applications. Nanomaterials (Basel) 8:252

52. Hartman RK, Hallam KA, Donnelly EM et al. (2019) Photoacoustic imaging of gold nanorods in the brain delivered *via* microbubble-assisted focused ultrasound: a tool for *in vivo* molecular neuroimaging. Laser Phys Lett 16:025603

53. Olesiak-Banska J, Waszkielewicz M, Obstarczyk P et al. (2019) Two-photon absorption and photoluminescence of colloidal gold nanoparticles and nanoclusters. Chem Soc Rev 48:4087–4117

54. Anselmo AC, Mitragotri S (2019) Nanoparticles in the clinic: an update. Bioeng Transl Med 4:e10143

55. Knudsen KB, Berthing T, Jackson P et al. (2019) Physicochemical predictors of Multi-Walled Carbon Nanotube-induced pulmonary histopathology and toxicity one year after pulmonary deposition of 11 different Multi-Walled Carbon Nanotubes in mice. Basic Clin Pharmacol Toxicol 124:211–227

56. Mohanta D, Patnaik S, Sood S et al. (2019) Carbon nanotubes: evaluation of toxicity at biointerfaces. J Pharm Anal 9:293–300

57. Vauthier C (2012) Formulating nanoparticles to achieve oral and intravenous delivery of challenging drugs. In: NanoFormulation, p 2–19

58. Teleanu DM, Chircov C, Grumezescu AM et al. (2018) Blood-brain delivery methods using nanotechnology. Pharmaceutics 10:269

59. Quintana FJ (2017) Astrocytes to the rescue! Glia limitans astrocytic endfeet control CNS inflammation. J Clin Invest 127:2897–2899

60. Alexander JJ (2018) Blood-brain barrier (BBB) and the complement landscape. Mol Immunol 102:26–31

61. Gao H (2016) Progress and perspectives on targeting nanoparticles for brain drug delivery. Acta Pharm Sin B 6:268–286

62. D'Agata F, Ruffinatti FA, Boschi S et al. (2017) Magnetic nanoparticles in the central nervous system: targeting principles, applications and safety issues. Molecules 23:9

63. Jallouli Y, Paillard A, Chang J et al. (2007) Influence of surface charge and inner composition of porous nanoparticles to cross blood-brain barrier *in vitro*. Int J Pharm 344:103–109

64. Konofagou EE, Tung YS, Choi J et al. (2012) Ultrasound-induced blood-brain barrier opening. Curr Pharm Biotechnol 13:1332–1345

65. Aryal M, Vykhodtseva N, Zhang YZ et al. (2013) Multiple treatments with liposomal doxorubicin and ultrasound-induced disruption of blood-tumor and blood-brain barriers improve outcomes in a rat glioma model. J Control Release 169:103–111

66. Etame AB, Diaz RJ, O'Reilly MA et al. (2012) Enhanced delivery of gold nanoparticles with therapeutic potential into the brain using MRI-guided focused ultrasound. Nanomedicine 8:1133–1142

67. Liu HL, Hua MY, Yang HW et al. (2010) Magnetic resonance monitoring of focused ultrasound/magnetic nanoparticle targeting delivery of therapeutic agents to the brain. Proc Natl Acad Sci USA 107:15205–15210

68. Huang Y, Zhang B, Xie S et al. (2016) Superparamagnetic iron oxide nanoparticles modified with Tween 80 pass through the intact blood-brain barrier in rats under magnetic field. ACS Appl Mater Interfaces 8:11336–11341

69. Tabatabaei SN, Girouard H, Carret AS et al. (2015) Remote control of the permeability of the blood-brain barrier by magnetic heating of nanoparticles: A proof of concept for brain drug delivery. J Control Release 206:49–57

70. Jansson B, Bjork E (2002) Visualization of *in vivo* olfactory uptake and transfer using fluorescein dextran. J Drug Target 10:379–386

71. Bonferoni MC, Rossi S, Sandri G et al. (2019) Nanoemulsions for "nose-to-brain" drug delivery. Pharmaceutics 11:84

72. Ruigrok MJ, de Lange EC (2015) Emerging insights for translational pharmacokinetic and pharmacokinetic-pharmacodynamic studies: towards prediction of nose-to-brain transport in humans. AAPS J 17:493–505

73. Sonvico F, Clementino A, Buttini F et al. (2018) Surface-modified nanocarriers for nose-to-brain delivery: from bioadhesion to targeting. Pharmaceutics 10:34

74. Kumarasamy M, Sosnik A (2019) The nose-to-brain transport of polymeric nanoparticles is mediated by immune sentinels and not by olfactory sensory neurons. Adv Biosystems 3:1900123

75. Ruan J, Qin B, Huang J (2018) Controlling measures of micro-plastic and nano pollutants: a short review of disposing waste toners. Environ Int 118:92–96

76. Haghani A, Johnson R, Safi N et al. (2020) Toxicity of urban air pollution particulate matter in developing and adult mouse brain: Comparison of total and filter-eluted nanoparticles. Environ Int 136:105510

77. Caster JM, Patel AN, Zhang T et al. (2017) Investigational nanomedicines in 2016: a review of nanotherapeutics currently undergoing clinical trials. Wiley Interdiscip Rev Nanomed Nanobiotechnol 9:1

78. Baek SK, Makkouk AR, Krasieva T et al. (2011) Photothermal treatment of glioma: an *in vitro* study of macrophage-mediated delivery of gold nanoshells. J Neurooncol 104:439–448

79. Abdal Dayem A, Lee SB, Choi HY et al. (2018) Silver nanoparticles: two-faced neuronal differentiation-inducing material in neuroblastoma (SH-SY5Y) cells. Int J Mol Sci 19:1470

80. Cramer SW, Chen CC (2019) Photodynamic therapy for the treatment of glioblastoma. Front Surg 6:81

81. Carvalho-de-Souza JL, Treger JS, Dang B et al. (2015) Photosensitivity of neurons enabled by cell-targeted gold nanoparticles. Neuron 86:207–217

82. Miyako E, Russier J, Mauro M et al. (2014) Photofunctional nanomodulators for bioexcitation. Angew Chem Int Ed Engl 53:13121–13125

83. Paviolo C, Haycock JW, Yong J et al. (2013) Laser exposure of gold nanorods can increase neuronal cell outgrowth. Biotechnol Bioeng 110:2277–2291

84. Christiansen MG, Senko AW, Anikeeva P (2019) Magnetic strategies for nervous system control. Annu Rev Neurosci 42:271–293

85. Falconieri A, De Vincentiis S, Raffa V (2019) Recent advances in the use of magnetic nanoparticles to promote neuroregeneration. Nanomedicine (Lond) 14:1073–1076

86. Smolders S, Kessels S, Smolders SM et al. (2018) Magnetofection is superior to other chemical transfection methods in a microglial cell line. J Neurosci Methods 293:169–173

87. Cheng Y, Muroski ME, Petit D et al. (2016) Rotating magnetic field induced oscillation of magnetic particles for in vivo mechanical destruction of malignant glioma. J Control Release 223:75–84

88. Grevesse T, Dabiri BE, Parker KK et al. (2015) Opposite rheological properties of neuronal microcompartments predict axonal vulnerability in brain injury. Sci Rep 5:9475

89. Chen R, Romero G, Christiansen MG et al. (2015) Wireless magnetothermal deep brain stimulation. Science 347:1477–1480

90. Chen S, Weitemier AZ, Zeng X et al. (2018) Near-infrared deep brain stimulation via upconversion nanoparticle-mediated optogenetics. Science 359:679–684

91. Lin X, Wang Y, Chen X et al. (2017) Multiplexed optogenetic stimulation of neurons with spectrum-selective upconversion nanoparticles. Adv Healthc Mater 6:1700446

92. Raimondo JV, Kay L, Ellender TJ et al. (2012) Optogenetic silencing strategies differ in their effects on inhibitory synaptic transmission. Nat Neurosci 15:1102–1104

93. Chen G, Zhang Y, Peng Z et al. (2019) Glutathione-capped quantum dots for plasma membrane labeling and membrane potential imaging. Nano Res 12:1321–1326

94. Khan FA, Almohazey D, Alomari M et al. (2018) Impact of nanoparticles on neuron biology: current research trends. Int J Nanomedicine 13:2767–2776

95. Monzel C, Vicario C, Piehler J (2017) Magnetic control of cellular processes using biofunctional nanoparticles. Chem Sci 8:7330–7338

96. Zhao Y, Jiang Y, Lv W et al. (2016) Dual targeted nanocarrier for brain ischemic stroke treatment. J Control Release 233:64–71

97. Maestrelli F, Gonzalez-Rodriguez ML, Rabasco AM et al. (2006) Effect of preparation technique on the properties of liposomes encapsulating ketoprofen-cyclodextrin complexes aimed for transdermal delivery. Int J Pharm 312:53–60

98. Jaafar-Maalej C, Diab R, Andrieu V et al. (2010) Ethanol injection method for hydrophilic and lipophilic drug-loaded liposome preparation. J Liposome Res 20:228–243

99. Costa AP, Xu X, Burgess DJ (2014) Freeze-anneal-thaw cycling of unilamellar liposomes: effect on encapsulation efficiency. Pharm Res 31:97–103

100. Fortes Brollo ME, Dominguez-Bajo A, Tabero A et al. (2020) Combined magnetoliposome formation and drug loading in one step for efficient alternating current-magnetic field remote-controlled drug release. ACS Appl Mater Interfaces 12:4295–4307

101. Amstad E, Kim S-H, Weitz DA (2012) Photo- and thermoresponsive polymersomes for triggered release. Angew Chem Int Edition 51:12499–12503

102. Zhu Y, Wu N, East CJ (2013) Micro segmented flow-functional elements and biotechnical applications. Front Biosci (Schol Ed) 5:284–304

103. Faivre D (2016) Iron oxides: from nature to applications. Wiley-VCH Verlag GmbH & Co. KGaA

104. Ling D, Lee N, Hyeon T (2015) Chemical synthesis and assembly of uniformly sized iron oxide nanoparticles for medical applications. Acc Chem Res 48:1276–1285

105. Roca AG, Gutierrez L, Gavilan H et al. (2019) Design strategies for shape-controlled magnetic iron oxide nanoparticles. Adv Drug Deliv Rev 138:68–104

106. Cheng D, Li X, Zhang G et al. (2014) Morphological effect of oscillating magnetic nanoparticles in killing tumor cells. Nanoscale Res Lett 9:195

107. Dias CSB, Hanchuk TDM, Wender H et al. (2017) Shape tailored magnetic nanorings for intracellular hyperthermia cancer therapy. Sci Rep 7:14843

108. Sun H, Chen B, Jiao X et al. (2012) Solvothermal synthesis of tunable electroactive magnetite nanorods by controlling the side reaction. J Phys Chem C 116:5476–5481

109. Gavilan H, Sanchez EH, Brollo MEF et al. (2017) Formation mechanism of maghemite nanoflowers synthesized by a polyol-mediated process. ACS Omega 2:7172–7184

110. Vita F, Gavilán H, Rossi F et al. (2016) Tuning morphology and magnetism of magnetite nanoparticles by calix[8]arene-induced oriented aggregation. CrystEngComm 18:8591–8598

111. Huhn J, Carrillo-Carrion C, Soliman MG et al. (2017) Selected standard protocols for the synthesis, phase transfer, and characterization of inorganic colloidal nanoparticles. Chem Mater 29:399–461

112. Frens G (1973) Controlled nucleation for the regulation of the particle size in monodisperse gold suspensions. Nat Phys Sci 241:20

113. Bastus NG, Merkoci F, Piella J et al. (2014) Synthesis of highly monodisperse citrate-stabilized silver nanoparticles of up to 200 nm: kinetic control and catalytic properties. Chem Mater 26:2836–2846

114. Murphy CJ, Thompson LB, Chernak DJ et al. (2011) Gold nanorod crystal growth: from seed-mediated synthesis to nanoscale sculpting. Curr Opin Colloid In 16:128–134

115. Alkilany AM, Nagaria PK, Hexel CR et al. (2009) Cellular uptake and cytotoxicity of gold nanorods: molecular origin of cytotoxicity and surface effects. Small 5:701–708

116. Bharadwaj VN, Nguyen DT, Kodibagkar VD et al. (2018) Nanoparticle-based therapeutics for brain injury. Adv Healthc Mater 7:1700668

117. Agarwal A, Lariya N, Saraogi G et al. (2009) Nanoparticles as novel carrier for brain delivery: a review. Curr Pharm Des 15:917–925

118. Nance EA, Woodworth GF, Sailor KA et al. (2012) A dense poly(ethylene glycol) coating improves penetration of large polymeric nanoparticles within brain tissue. Sci Transl Med 4:149ra119

119. Albanese A, Tang PS, Chan WC (2012) The effect of nanoparticle size, shape, and surface chemistry on biological systems. Annu Rev Biomed Eng 14:1–16

120. Yang M, Lai SK, Wang YY et al. (2011) Biodegradable nanoparticles composed entirely of safe materials that rapidly penetrate human mucus. Angew Chem Int Ed Engl 50:2597–2600

121. Wohlfart S, Gelperina S, Kreuter J (2012) Transport of drugs across the blood-brain barrier by nanoparticles. J Control Release 161:264–273

122. Koffie RM, Farrar CT, Saidi LJ et al. (2011) Nanoparticles enhance brain delivery of blood-brain barrier-impermeable probes for in vivo optical and magnetic resonance imaging. Proc Natl Acad Sci USA 108:18837–18842

123. Schuster T, Muhlstein A, Yaghootfam C et al. (2017) Potential of surfactant-coated nanoparticles to improve brain delivery of arylsulfatase A. J Control Release 253:1–10

124. Loureiro JA, Gomes B, Coelho MA et al. (2014) Targeting nanoparticles across the blood-brain barrier with monoclonal antibodies. Nanomedicine (Lond) 9:709–722

125. Habgood MD, Begley DJ, Abbott NJ (2000) Determinants of passive drug entry into the central nervous system. Cell Mol Neurobiol 20:231–253

126. Jefferies WA, Brandon MR, Hunt SV et al. (1984) Transferrin receptor on endothelium of brain capillaries. Nature 312:162–163

127. Havrankova J, Roth J, Brownstein M (1978) Insulin receptors are widely distributed in the central nervous system of the rat. Nature 272:827–829

128. Johnsen KB, Bak M, Melander F et al. (2019) Modulating the antibody density changes the uptake and transport at the blood-brain barrier of both transferrin receptor-targeted gold nanoparticles and liposomal cargo. J Control Release 295:237–249

129. Cabezon I, Manich G, Martin-Venegas R et al. (2015) Trafficking of gold nanoparticles coated with the 8D3 anti-transferrin receptor antibody at the mouse blood-brain barrier. Mol Pharm 12:4137–4145

130. Clark AJ, Davis ME (2015) Increased brain uptake of targeted nanoparticles by adding an acid-cleavable linkage between transferrin and the nanoparticle core. Proc Natl Acad Sci USA 112:12486–12491

131. Wiley DT, Webster P, Gale A, Davis ME (2013) Transcytosis and brain uptake of transferrin-containing nanoparticles by tuning avidity to transferrin receptor. Proc Natl Acad Sci USA 110:8662–8667

132. Kim SS, Rait A, Kim E et al. (2015) Encapsulation of temozolomide in a tumor-targeting nanocomplex enhances anti-cancer efficacy and reduces toxicity in a mouse model of glioblastoma. Cancer Lett 369:250–258

133. Kolhar P, Anselmo AC, Gupta V et al. (2013) Using shape effects to target antibody-coated nanoparticles to lung and brain endothelium. Proc Natl Acad Sci USA 110:10753–10758

134. Ulbrich K, Hekmatara T, Herbert E, Kreuter J (2009) Transferrin- and transferrin-receptor-antibody-modified nanoparticles enable drug delivery across the blood-brain barrier (BBB). Eur J Pharm Biopharm 71:251–256

135. Ulbrich K, Knobloch T, Kreuter J (2011) Targeting the insulin receptor: nanoparticles for drug delivery across the blood-brain barrier (BBB). J Drug Target 19:125–132

136. Kwon EJ, Skalak M, Lo Bu R, Bhatia SN (2016) Neuron-targeted nanoparticle for siRNA delivery to traumatic brain injuries. ACS Nano 10:7926–7933

137. Bao Q, Hu P, Xu Y et al. (2018) Simultaneous blood-brain barrier crossing and protection for stroke treatment based on edaravone-loaded ceria nanoparticles. ACS Nano 12:6794–6805

138. Mann AP, Scodeller P, Hussain S et al. (2016) A peptide for targeted, systemic delivery of imaging and therapeutic compounds into acute brain injuries. Nat Commun 7:11980

139. Oyesiku NM, Evans CO, Houston S et al. (1999) Regional changes in the expression of neurotrophic factors and their receptors following acute traumatic brain injury in the adult rat brain. Brain Res 833:161–172

140. A controlled trial of recombinant methionyl human BDNF in ALS: The BDNF Study Group (Phase III) (1999). Neurology 52:1427–1433

141. Poduslo JF, Curran GL (1996) Permeability at the blood-brain and blood-nerve barriers of the neurotrophic factors: NGF, CNTF, NT-3, BDNF. Brain Res Mol Brain Res 36:280–286

142. Houlton J, Abumaria N, Hinkley SFR, Clarkson AN (2019) Therapeutic potential of neurotrophins for repair after brain injury: a helping hand from biomaterials. Front Neurosci 13:790

143. Clarkson AN, Parker K, Nilsson M et al. (2015) Combined ampakine and BDNF treatments enhance poststroke functional recovery in aged mice via AKT-CREB signaling. J Cereb Blood Flow Metab 35:1272–1279

144. Cook DJ, Nguyen C, Chun HN et al. (2017) Hydrogel-delivered brain-derived neurotrophic factor promotes tissue repair and recovery after stroke. J Cereb Blood Flow Metab 37:1030–1045

145. Fon D, Zhou K, Ercole F et al. (2014) Nanofibrous scaffolds releasing a small molecule BDNF-mimetic for the re-direction of endogenous neuroblast migration in the brain. Biomaterials 35:2692–2712

146. Santos D, Giudetti G, Micera S et al. (2016) Focal release of neurotrophic factors by biodegradable microspheres enhance motor and sensory axonal regeneration *in vitro* and *in vivo*. Brain Res 1636:93–106

147. Liu JJ, Wang CY, Wang JG, Ruan HJ, Fan CY (2011) Peripheral nerve regeneration using composite poly(lactic acid-caprolactone)/nerve growth factor conduits prepared by coaxial electrospinning. J Biomed Mater Res A 96:13–20

148. Wang H, Zhao Q, Zhao W et al. (2012) Repairing rat sciatic nerve injury by a nerve-growth-factor-loaded, chitosan-based nerve conduit. Biotechnol Appl Biochem 59:388–394

149. Fan J, Zhang H, He J et al. (2011) Neural regrowth induced by PLGA nerve conduits and neurotrophin-3 in rats with complete spinal cord transection. J Biomed Mater Res B Appl Biomater 97:271–277

150. Houweling DA, Lankhorst AJ, Gispen WH, Bar PR, Joosten EA (1998) Collagen containing neurotrophin-3 (NT-3) attracts regrowing injured corticospinal axons in the adult rat spinal cord and promotes partial functional recovery. Exp Neurol 153:49–59

151. Piantino J, Burdick JA, Goldberg D et al. (2006) An injectable, biodegradable hydrogel for trophic factor delivery enhances axonal rewiring and improves performance after spinal cord injury. Exp Neurol 201:359–367

152. Jinek M, Chylinski K, Fonfara I et al. (2012) A programmable dual-RNA-guided DNA endonuclease in adaptive bacterial immunity. Science 337:816–821

153. Mingozzi F, High KA (2013) Immune responses to AAV vectors: overcoming barriers to successful gene therapy. Blood 122:23–36

154. Ishida K, Gee P, Hotta A (2015) Minimizing off-target mutagenesis risks caused by programmable nucleases. Int J Mol Sci 16:24751–24771

155. Lee B, Lee K, Panda S (2018) Nanoparticle delivery of CRISPR into the brain rescues a mouse model of fragile X syndrome from exaggerated repetitive behaviours. Nat Biomed Eng 2:497–507

156. Kim J, Chhour P, Hsu J et al. (2017) Use of nanoparticle contrast agents for cell tracking with computed tomography. Bioconjug Chem 28:1581–1597

157. Mehta A, Ghaghada K, Mukundan S (2016) Molecular imaging of brain tumors using liposomal contrast agents and nanoparticles. Magn Reson Imaging Clin N Am 24:751–763

158. Rastinehad AR, Anastos H, Wajswol E et al. (2019) Gold nanoshell-localized photothermal ablation of prostate tumors in a clinical pilot device study. Proc Natl Acad Sci USA 116:18590–18596

159. Hainfeld JF, Smilowitz HM, O'Connor MJ et al. (2013) Gold nanoparticle imaging and radiotherapy of brain tumors in mice. Nanomedicine (Lond) 8:1601–1609

160. Cruz LJ, Que I, Aswendt M et al. (2016) Targeted nanoparticles for the non-invasive detection of traumatic brain injury by optical imaging and fluorine magnetic resonance imaging. Nano Research 9:1276–1289

161. Maier-Hauff K, Ulrich F, Nestler D et al. (2011) Efficacy and safety of intratumoral thermotherapy using magnetic iron-oxide nanoparticles combined with external beam radiotherapy on patients with recurrent glioblastoma multiforme. J Neurooncol 103:317–324

162. Han X, Xu K, Taratula O, Farsad K (2019) Applications of nanoparticles in biomedical imaging. Nanoscale 11:799–819

163. Starmans LW, Moonen RP, Aussems-Custers E et al. (2015) Evaluation of iron oxide nanoparticle micelles for magnetic particle imaging (MPI) of thrombosis. PLoS One 10:e0119257

164. Faucher L, Guay-Begin AA, Lagueux J et al. (2011) Ultra-small gadolinium oxide nanoparticles to image brain cancer cells *in vivo* with MRI. Contrast Media Mol Imaging 6:209–218

165. Jin AY, Tuor UI, Rushforth D et al. (2009) Magnetic resonance molecular imaging of post-stroke neuroinflammation with a P-selectin targeted iron oxide nanoparticle. Contrast Media Mol Imaging 4:305–311

166. van Kasteren SI, Campbell SJ, Serres S et al. (2009) Glyconanoparticles allow pre-symptomatic in vivo imaging of brain disease. Proc Natl Acad Sci USA 106:18–23

167. Frias JC, Ma Y, Williams KJ (2006) Properties of a versatile nanoparticle platform contrast agent to image and characterize atherosclerotic plaques by magnetic resonance imaging. Nano Lett 6:2220–2224

168. Kaijzel EL, van Beek ER, Stammes MA et al. (2017) Traumatic brain injury: preclinical imaging diagnostic(s) and therapeutic approaches. Curr Pharm Des 23:1909–1915

169. Leblond F, Davis SC, Valdes PA, Pogue BW (2010) Pre-clinical whole-body fluorescence imaging: Review of instruments, methods and applications. J Photochem Photobiol B 98:77–94

170. Wang LV, Hu S (2012) Photoacoustic tomography: *in vivo* imaging from organelles to organs. Science 335:1458–1462

171. Yao J, Wang LV (2014) Photoacoustic brain imaging: from microscopic to macroscopic scales. Neurophotonics 1:011003

172. Li M, Oh J, Xie X et al. (2008) Simultaneous molecular and hypoxia imaging of brain tumors *in vivo* using spectroscopic photoacoustic tomography. Proc IEEE 96:481–489

173. Kim S, Chen Y-S, Luke GP, Emelianov SY (2011) *In vivo* three-dimensional spectroscopic photoacoustic imaging for monitoring nanoparticle delivery. Biomed Opt Express 2:2540–2550

174. Lovell JF, Jin CS, Huynh E et al. (2011) Porphysome nanovesicles generated by porphyrin bilayers for use as multimodal biophotonic contrast agents. Nat Mater 10:324–332

175. Hannah A, Luke G, Wilson K, Homan K, Emelianov S (2014) Indocyanine green-loaded photoacoustic nanodroplets: dual contrast nanoconstructs for enhanced photoacoustic and ultrasound imaging. ACS Nano 8:250–259

176. Tang J, Xi L, Zhou J, Huang H, Zhang T, Carney PR, Jiang H (2015) Noninvasive high-speed photoacoustic tomography of cerebral hemodynamics in awake-moving rats. J Cerebr Blood F Met 35:1224–1232

177. Ray A, Wang X, Lee Y-EK et al. (2011) Targeted blue nanoparticles as photoacoustic contrast agent for brain tumor delineation. Nano Res 4:1163–1173

178. Li W, Brown PK, Wang LV, Xia Y (2011) Gold nanocages as contrast agents for photoacoustic imaging. Contrast Media Mol I 6:370–377

179. Yuan H, Wilson CM, Xia J et al. (2014) Plasmonics-enhanced and optically modulated delivery of gold nanostars into brain tumor. Nanoscale 6:4078–4082

180. Meola A, Rao J, Chaudhary N, Sharma M, Chang SD (2018) Gold nanoparticles for brain tumor imaging: a systematic review. Front Neurol 9:328

181. Hou R, Zhou D, Nie R et al. (2019) Brain CT and MRI medical image fusion using convolutional neural networks and a dual-channel spiking cortical model. Med Biol Eng Comput 57:887–900

182. Dao TT, Pouletaut P, Charleux F et al. (2015) Multimodal medical imaging (CT and dynamic MRI) data and computer-graphics multi-physical model for the estimation of patient specific lumbar spine muscle forces. Data Knowl Eng 96-97:3–18

183. Zhu J, Lu Y, Li Y et al. (2014) Synthesis of Au–Fe_3O_4 heterostructured nanoparticles for *in vivo* computed tomography and magnetic resonance dual model imaging. Nanoscale 6:199–202

184. Nebelung S, Brill N, Tingart M et al. (2016) Quantitative OCT and MRI biomarkers for the differentiation of cartilage degeneration. Skeletal Radiol 45:505–516

185. Wang X, Liu H, Chen D et al. (2013) Multifunctional Fe_3O_4@P(St/MAA)@Chitosan@Au core/shell nanoparticles for dual imaging and photothermal therapy. ACS App Mater Interfaces 5:4966–4971

186. Bogdanov AA, Dixon AJ, Gupta S et al. (2016) Synthesis and testing of modular dual-modality nanoparticles for magnetic resonance and multispectral photoacoustic imaging. Bioconjugate Chem 27:383–390

187. van den Berg MEL, Castellote JM, Mayordomo JI et al. (2017) Spinal cord injury due to tumour or metastasis in Aragón, Northeastern Spain (1991–2008): incidence, time trends, and neurological function. Biomed Res Int 2017:2478197

188. Types of Brain and Spinal Cord Tumors in Adults (2020) American Cancer Society. https://www.cancer.org/cancer/brain-spinal-cord-tumors-adults/about/types-of-brain-tumors.html. Cited 28 Sep 2021

189. Qian W, Qian M, Wang Y et al. (2018) Combination glioma therapy mediated by a dual-targeted delivery system constructed using OMCN-PEG-Pep22/DOX. Small 14:e1801905

190. Sun T, Patil R, Galstyan A et al. (2019) Blockade of a laminin-411-notch axis with CRISPR/Cas9 or a nanobioconjugate inhibits glioblastoma growth through tumor-microenvironment cross-talk. Cancer Res 79:1239–1251

191. Patil R, Ljubimov AV, Gangalum PR et al. (2015) MRI virtual biopsy and treatment of brain metastatic tumors with targeted nanobioconjugates: nanoclinic in the brain. ACS Nano 9:5594–5608

192. Pinel S, Thomas N, Boura C, Barberi-Heyob M (2019) Approaches to physical stimulation of metallic nanoparticles for glioblastoma treatment. Adv Drug Deliv Rev 138:344–357

193. ClinicalTrials.gov (2016) Radiosensitization of multiple brain metastases using AGuIX Gadolinium based nanoparticles (NANO-RAD). NIH—U.S. National Library of Medicine. https://clinicaltrials.gov/ct2/show/NCT02820454. Cited 28 Sep 2021

194. ClinicalTrials.gov (2009) Pilot study of AuroLase(tm) therapy in refractory and/or recurrent tumors of the head and neck. NIH—U.S. National Library of Medicine. https://clinicaltrials.gov/ct2/show/NCT00848042. Cited 28 Sep 2021

195. Yi G, Hong SH, Son J et al. (2018) Recent advances in nanoparticle carriers for photodynamic therapy. Quant Imaging Med Surg 8:433–443

196. Sun J, Kormakov S, Liu Y et al. (2018) Recent progress in metal-based nanoparticles mediated photodynamic therapy. Molecules 23:1704

197. Guzman R, Uchida N, Bliss TM et al. (2007) Long-term monitoring of transplanted human neural stem cells in developmental and pathological contexts with MRI. Proc Natl Acad Sci USA 104:10211–10216

198. Duan X, Lu L, Wang Y et al. (2017) The long-term fate of mesenchymal stem cells labeled with magnetic resonance imaging-visible polymersomes in cerebral ischemia. Int J Nanomedicine 12:6705–6719

199. Smith TT, Stephan SB, Moffett HF et al. (2017) *In situ* programming of leukaemia-specific T cells using synthetic DNA nanocarriers. Nat Nanotechnol 12:813–820

200. Sanz-Ortega L, Portilla Y, Perez-Yague S, Barber DF (2019) Magnetic targeting of adoptively transferred tumour-specific nanoparticle-loaded $CD8^+$ T cells does not improve their tumour infiltration in a mouse model of cancer but promotes the retention of these cells in tumour-draining lymph nodes. J Nanobiotechnol 17:87

201. Galanzha EI, Shashkov EV, Kelly T et al. (2009) *In vivo* magnetic enrichment and multiplex photoacoustic detection of circulating tumour cells. Nat Nanotechnol 4:855–860

202. Jin Y, Lee JU, Chung E et al. (2019) Magnetic control of axon navigation in reprogrammed neurons. Nano Lett 19:6517–6523

203. Pita-Thomas W, Steketee MB, Moysidis SN et al. (2015) Promoting filopodial elongation in neurons by membrane-bound magnetic nanoparticles. Nanomedicine 11:559–567

204. Steketee MB, Moysidis SN, Jin XL et al. (2011) Nanoparticle-mediated signaling endosome localization regulates growth cone motility and neurite growth. Proc Natl Acad Sci USA 108:19042–19047

205. Borisova T, Krisanova N, Borysov A et al. (2014) Manipulation of isolated brain nerve terminals by an external magnetic field using D-mannose-coated γ-Fe$_2$O$_3$ nano-sized particles and assessment of their effects on glutamate transport. Beilstein J Nanotechnol 5:778–788

206. Giannaccini M, Calatayud MP, Poggetti A et al. (2017) Magnetic nanoparticles for efficient delivery of growth factors: stimulation of peripheral nerve regeneration. Adv Healthc Mater 6:1601429
207. Zuidema JM, Provenza C, Caliendo T, Dutz S, Gilbert RJ (2015) Magnetic NGF-releasing PLLA/iron oxide nanoparticles direct extending neurites and preferentially guide neurites along aligned electrospun microfibers. ACS Chem Neurosci 6:1781–1788
208. Zhang XQ, Xu X, Bertrand N et al. (2012) Interactions of nanomaterials and biological systems: Implications to personalized nanomedicine. Adv Drug Deliv Rev 64:1363–1384
209. Gatoo MA, Naseem S, Arfat MY et al. (2014) Physicochemical properties of nanomaterials: implication in associated toxic manifestations. Biomed Res Int 2014:498420
210. Hu YL, Gao JQ (2010) Potential neurotoxicity of nanoparticles. Int J Pharm 394:115–121
211. Furtado D, Björnmalm M, Ayton S et al. (2018) Overcoming the blood-brain barrier: The role of nanomaterials in treating neurological diseases. Adv Mater 30:e1801362
212. Oberdörster G, Elder A, Rinderknecht A (2009) Nanoparticles and the brain: cause for concern? J Nanosci Nanotechnol 9:4996–5007
213. Xia T, Li N, Nel AE (2009) Potential health impact of nanoparticles. Annu Rev Public Health 30:137–150
214. Uttara B, Singh AV, Zamboni P, Mahajan RT (2009) Oxidative stress and neurodegenerative diseases: a review of upstream and downstream antioxidant therapeutic options. Curr Neuropharmacol 7:65–74
215. Gupta R, Xie H (2018) Nanoparticles in daily life: Applications, toxicity and regulations. J Environ Pathol Toxicol Oncol 37:209–230
216. Maiti D, Tong X, Mou X, Yang K (2018) Carbon-based nanomaterials for biomedical applications: A recent study. Front Pharmacol 9:1401
217. Onoda A, Takeda K, Umezawa M (2017) Pretreatment with N-acetyl cysteine suppresses chronic reactive astrogliosis following maternal nanoparticle exposure during gestational period. Nanotoxicology 11:1012–1025
218. Samiei F, Shirazi FH, Naserzadeh P et al. (2020) Toxicity of multi-wall carbon nanotubes inhalation on the brain of rats. Environ Sci Pollut Res Int 27:12096–12111
219. Ayton S, Lei P (2014) Nigral iron elevation is an invariable feature of Parkinson's disease and is a sufficient cause of neurodegeneration. Biomed Res Int 2014:581256
220. Núñez MT, Urrutia P, Mena N et al. (2012) Iron toxicity in neurodegeneration. Biometals 25:761–776
221. Valdiglesias V, Fernandez-Bertolez N, Kilic G et al. (2016) Are iron oxide nanoparticles safe? Current knowledge and future perspectives. J Trace Elem Med Biol 38:53–63
222. Gremo F, Sogos V, Ennas MG et al. (1997) Features and functions of human microglia cells. Adv Exp Med Biol 429:79–97
223. Combs CK, Karlo JC, Kao SC, Landreth GE (2001) Beta-Amyloid stimulation of microglia and monocytes results in TNFalpha-dependent expression of inducible nitric oxide synthase and neuronal apoptosis. J Neurosci 21:1179–1188
224. Wu J, Sun J, Xue Y (2010) Involvement of JNK and P53 activation in G2/M cell cycle arrest and apoptosis induced by titanium dioxide nanoparticles in neuron cells. Toxicol Lett 199:269–276
225. Wu J, Wang C, Sun J, Xue Y (2011) Neurotoxicity of silica nanoparticles: brain localization and dopaminergic neurons damage pathways. ACS Nano 5:4476–4489
226. Choi J, Zheng Q, Katz HE, Guilarte TR (2010) Silica-based nanoparticle uptake and cellular response by primary microglia. Environ Health Perspect 118:589–595
227. Feng X, Chen A, Zhang Y et al. (2015) Central nervous system toxicity of metallic nanoparticles. Int J Nanomedicine 10:4321–4340
228. Márquez-Ramírez SG, Delgado-Buenrostro NL, Chirino YI et al. (2012) Titanium dioxide nanoparticles inhibit proliferation and induce morphological changes and apoptosis in glial cells. Toxicology 302:146–156
229. Xue Y, Wu J, Sun J (2012) Four types of inorganic nanoparticles stimulate the inflammatory reaction in brain microglia and damage neurons in vitro. Toxicol Lett 214:91–98

230. Shimizu M, Tainaka H, Oba T et al. (2009) Maternal exposure to nanoparticulate titanium dioxide during the prenatal period alters gene expression related to brain development in the mouse. Part Fibre Toxicol 6:20

231. Lee SH, Jun BH (2019) Silver nanoparticles: synthesis and application for nanomedicine. Int J Mol Sci 20:865

232. Sun C, Yin N, Wen R et al. (2016) Silver nanoparticles induced neurotoxicity through oxidative stress in rat cerebral astrocytes is distinct from the effects of silver ions. Neurotoxicology 52:210–221

233. Alkilany AM, Murphy CJ (2010) Toxicity and cellular uptake of gold nanoparticles: what we have learned so far? J Nanopart Res 12:2313–2333

234. Turner M, Golovko VB, Vaughan OP et al. (2008) Selective oxidation with dioxygen by gold nanoparticle catalysts derived from 55-atom clusters. Nature 454:981–983

235. El-Drieny E, Sarhan NI, Bayomy NA et al. (2015) Histological and immunohistochemical study of the effect of gold nanoparticles on the brain of adult male albino rat. J Microsc Ultrastruct 3:181–190

236. Kreyling WG, Abdelmonem AM, Ali Z et al. (2015) *In vivo* integrity of polymer-coated gold nanoparticles. Nat Nanotechnol 10:619–623

237. Lam PL, Wong WY, Bian Z, Chui CH, Gambari R (2017) Recent advances in green nanoparticulate systems for drug delivery: efficient delivery and safety concern. Nanomedicine (Lond) 12:357–385

238. Mignani S, El Kazzouli S, Bousmina M, Majoral JP (2013) Expand classical drug administration ways by emerging routes using dendrimer drug delivery systems: a concise overview. Adv Drug Deliv Rev 65:1316–1330

Chapter 8
Therapeutic Approaches for Stroke: A Biomaterials Perspective

Artur Filipe Rodrigues, Catarina Rebelo, Tiago Reis, João André Sousa, Sónia L. C. Pinho, João Sargento-Freitas, João Peça, and Lino Ferreira

Abstract Stroke is a leading cause of death worldwide and poses significant societal and healthcare challenges due to functional impairment of the brain. In order to fully restore brain function, innovative approaches have aimed to regenerate the injured tissue and to restore neuronal circuitry. In the last 5 years, stem cells have been consistently explored in clinical trials for tissue regeneration. Recent technological progress regarding the use of stem cell-derived extracellular vesicles has also shown promise toward the administration of cell-based therapies exploiting paracrine signaling. In addition, neuromodulation using different stimulation modalities has become increasingly investigated in the clinic as a non-invasive strategy to promote functional recovery. This approach contrasts with invasive strategies using devices capable of delivering electrical pulses in deep regions of the brain, which nonetheless are well-established in the clinic for the treatment of other neurological disorders. This chapter reviews the latest approaches covering brain tissue regeneration and neuromodulation, and discusses their limitations for clinical translation. Preclinical investigations on the use of light for neuromodulation in optogenetics have sparked the development of biocompatible interfaces capable of coupling optical stimulation with electrical recording. These biointerfaces require novel materials whose physicochemical properties are discussed herein.

A. F. Rodrigues · S. L. C. Pinho
Center for Neuroscience and Cell Biology (CNC), University of Coimbra, Coimbra, Portugal
e-mail: afcdrodrigues@cnc.uc.pt; slpinho@uc-biotech.pt

C. Rebelo · T. Reis · J. Peça · L. Ferreira (✉)
Center for Neuroscience and Cell Biology (CNC), University of Coimbra, Coimbra, Portugal

Faculty of Medicine, University of Coimbra, Coimbra, Portugal
e-mail: tj.reis@campus.fct.unl.pt; jpeca@cnc.uc.pt; lino@uc-biotech.pt

J. A. Sousa · J. Sargento-Freitas
Faculty of Medicine, University of Coimbra, Coimbra, Portugal

Department of Neurology, Centro Hospitalar e Universitário de Coimbra, Coimbra, Portugal
e-mail: 7589@chuc.min-saude.pt

Keywords Biointerfaces · Cell-based therapies · Electrical stimulation · Neuromodulation · Optogenetics · Stem cells · Stroke · Tissue regeneration

8.1   Introduction

Stroke is the second leading cause of death worldwide and it is characterized by neurological impairment caused by vascular failure, which deprives focal areas of the central nervous system (CNS) from oxygen and nutrients supplied via the bloodstream (Fig. 8.1) [1]. Stroke encompasses clinical events, primarily occurring in arteries, which are triggered by different vascular pathologies: ischemic stroke, intracerebral hemorrhage, subarachnoid hemorrhage, and cerebral venous thrombosis [2]. On the one hand, ischemic stroke is the clinical consequence of local obstructions in the brain vasculature by blood clots, resulting in extensive cell death in the ischemic area [1]. On the other hand, hemorrhagic stroke results from a rupture of a weakened intracranial blood vessel, which can be caused by high blood pressure, amyloid angiopathy, coagulopathies, or a structural blood vessel abnormality (e.g. aneurysm, arteriovenous malformation, neoplasm) [2]. These etiological features underlying hemorrhagic stroke require immediate action to control blood pressure and, in certain cases, the administration of procoagulant agents and/or surgery to drain intracranial blood. In contrast, the most effective strategy to treat ischemic stroke is simply removing the blockage to the blood flow, either by intravenously administered drugs or endovascular mechanical therapy [2, 3]. Compared to hemorrhagic stroke, the variety of treatments for ischemic stroke has increased the chances of survival by 5-fold, saving every year the lives of around 80 million people worldwide [1]. However, current clinical practice has not evolved in the management of long-term associated morbidities [4]. It has primarily focused on the patients' behavioral changes to prevent relapses by adopting correct occupational habits such as a poor diet, physical inactivity, and smoking. These have been associated with metabolic and cardiovascular risk factors including high blood pressure and cholesterol levels in the blood, as well as cardiac arrhythmia and diabetes [1]. In addition, focused physical therapy has enabled functional rehabilitation of muscle movement and mobility, albeit with limited recovery, especially from other common impairments such as speech, language, vision, swallowing, and cognition [5].

Although these efforts have reconfigured neuronal networks disrupted by extensive brain damage, they are insufficient to fully restore function. In this context, biomaterials have assisted the development of advanced therapies such as electrical stimulation and cell-based therapies, which have been employed to remodel neural circuitry and to trigger regeneration of the affected brain tissue. The present chapter describes the existing state-of-the-art for the treatment of stroke and some of the most recent innovations in cell-based therapies and neuromodulation using light and electricity, whose combination is anticipated to be of clinical relevance in the near future. Stroke therapies have mostly relied on non-invasive strategies such

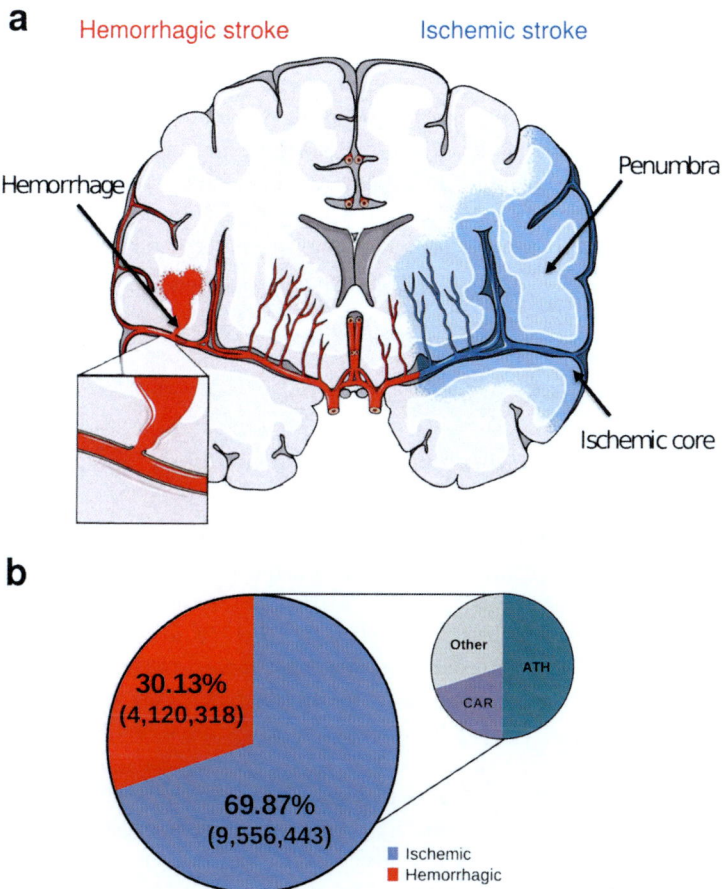

Fig. 8.1 Stroke etiology and prevalence. (**a**) Schematic representation of the main stroke subtypes, which can be classified by the deprivation of brain regions from access to oxygen and nutrients due to either the disruption (hemorrhagic) or occlusion (ischemic) of blood vessels. Adapted from images from Servier Medical Art by Servier (http://smart.servier.com), licensed under a Creative Commons Attribution 3.0 Unported License. (**b**) Although the overall number of stroke events has decreased in recent years, more than 13 million events were registered in 2016. Hemorrhagic stroke was less frequent than ischemic stroke, and it can be characterized by its onset in the brain or the subarachnoid space [1]. Contrarily, ischemic stroke is most frequently triggered by the rupture of atherosclerotic plaques from major vessels (i.e. large-vessel atherosclerosis, ATH) [6, 7]. Another frequent subtype of ischemic stroke is cardioembolism (CAR), which consists of the release of blood clots or atherosclerotic plaques accumulating in the cardiac tissue. Other ischemic stroke subtypes include small vessel occlusion triggered in patients suffering from hypertension or diabetes and rare events caused by non-atherosclerotic pathologies or other unknown factors [6, 7]

as transcranial stimulation of the brain and the administration of medicines to minimize tissue damage. Invasive neuromodulation techniques such as deep brain stimulation are highly effective in remodeling neural circuitry; albeit still generate long-term complications resulting from poor device biocompatibility. We propose the development of novel devices with biodegradable materials and minimally invasive implantation strategies to expand the therapeutic possibilities for stroke. The use of biomaterials to modulate cell activity will be discussed, with particular emphasis on material properties leading to improved biocompatibility and electrical conductivity.

8.2   Therapeutic Approaches to Stroke

8.2.1   Stroke Epidemiology and Pathophysiology

A variety of etiological mechanisms may trigger an ischemic stroke. The TOAST study has classified ischemic stroke based on the following causes: large artery atherosclerosis, cardioembolism, small vessel occlusion, stroke of other determined etiology, and stroke of undetermined etiology [6, 7]. Knowledge of these mechanisms for each patient is crucial to adjust secondary prevention with tailored therapies. Clinically, there are some noticeable symptoms associated with stroke, ranging from a minor central facial palsy to an acute coma. Other symptoms include numbness in one side of the body, difficulty understanding other people, difficulty in seeing with one or both eyes, gait problems and discoordination, dizziness/vertigo, and severe headache. These symptoms correspond to a cerebral loss of function of sudden onset, whose severity depends on the anatomy of the occluded/ruptured artery and collateral systems, as well as the patient's age and gender, and the presence of comorbidities [8]. The common triad of face drooping, arm/leg weakness, and speech difficulties (FAST acronym) should warrant an immediate call for help through pre-hospital emergency systems, as response time is critical at this stage [9]. Indeed, determining etiology and location of the infarct and rapidly restoring an adequate systemic blood pressure and irrigation will dictate the final infarct size and subsequent neurological consequences [8, 10].

Histologically, stroke is characterized by an ischemic core surrounded by a "penumbra" region, which can be monitored using non-invasive imaging techniques such as computerized tomography (CT) or magnetic resonance imaging (MRI) [2, 7]. Although imaging tools are a valuable asset to identify anatomical regions that are damaged during and after stroke, the quality of patient recovery requires specific functional predictors to guide rehabilitation. Clinical management of stroke has relied on biomarkers for the molecular processes taking place in the brain, including inflammation, hemostasis, and cell death [11]. At the ischemic core, where blood flow is most severely restricted, excitotoxic and necrotic cell death occurs within minutes due to oxygen and glucose deprivation, which causes glutamate

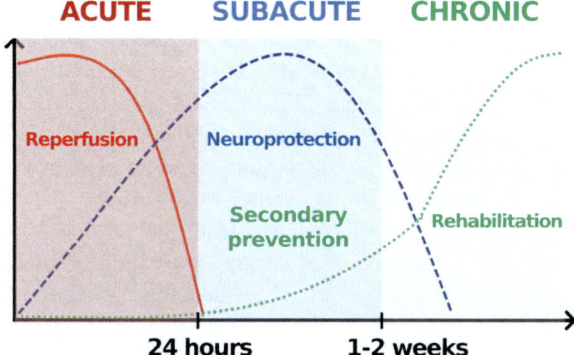

Fig. 8.2 Ischemic stroke management over time. During an ischemic stroke event, the initial priority is to rapidly irrigate the brain tissue deprived from blood circulation. In the following days, oxidative stress and extensive cell death are mitigated by the administration of neuroprotective agents. Continuous monitoring of brain activity is required to prevent secondary stroke events. Finally, long-term rehabilitation and physiotherapy aims to restore brain functions

release and mitochondrial dysfunction [12]. Activation of apoptosis, necrosis, and autophagy pathways disrupt the blood–brain barrier (BBB) and trigger peripheral immune responses to the lesion site, which further enhance oxidative degradation of several biomolecules, such as proteins, lipids, and DNA. As a result, cell death is progressive toward the penumbra, where collateral blood flow can buffer the effects of tissue damage at the ischemic core [13]. Although elevated serum cytokine levels and increased production of inflammatory mediators in circulating and splenic immune cells can be detected within hours after ischemia [14], there are currently no specific biomarkers to detect brain damage [11].

8.2.2 Clinical Standard of Care

The management of an ischemic stroke is multiphasic and time-bound (Fig. 8.2). First, an acute/early stage prioritizes the reperfusion of the occluded artery, followed by a subacute stage where monitoring, prevention of stroke complications, preservation of the surviving brain, and etiologic investigation take place. Finally, a chronic stage focuses on rehabilitation and prevention of secondary stroke events.

Current treatment options in the acute phase, although with a relevant improvement on the clinical outcome, have a limited time window to be applied. Stroke patients may be subjected either to pharmacological treatment for the dissolution of blood clots in ischemic strokes and/or to mechanical removal of the clot by endovascular procedures [15]. Tissue plasminogen activator (tPA) is the only thrombolytic drug that has been clinically approved by both the Food and Drug Administration of the USA (FDA) and the European Medicines Agency (EMA).

Patients treated with tPA are at least 30% more likely to have minimal or no disability 3 months after stroke [16]. However, treatment time is crucial for this outcome. No significant improvements were observed when tPA was administered more than 4.5 h after symptoms onset [17, 18]. Systemic delivery of tPA promotes the conversion of plasminogen to plasmin, which will bind and degrade fibrin, dissolving blood clots. Efficacy of tPA treatment can be extended up to 24 h after the development of symptoms by mechanically destroying the blood clot [19]. Thrombectomy is a catheter-based, image-guided intervention for the mechanical removal of blood clots in large vessels through aspiration or stent retrieval. This procedure showed remarkable improvement in the recovery of neurological function of patients suffering from large-vessel occlusion [20]. Nonetheless, patient selection and timely reperfusion are crucial for a successful outcome. Only 13%–20% of total acute ischemic stroke patients are eligible for endovascular therapy [21], due to factors such as patient's age, stroke severity, and anatomical location of the occlusion, as well as the history of previous disability/dependence episodes [22].

The aforementioned pharmacological and mechanical therapies rely on the re-implementation of blood flow to stop the onset of tissue damage. In contrast, adjuvant neuroprotective treatments attempt to minimize the signaling pathways that are subsequently activated after loss of blood flow and lead to neuronal death [23]. Currently, there are no approved pharmacological treatments with neuroprotective effects [15]. Nevertheless, several agents have been studied and are under development, particularly now that restoring blood flow to the occluded artery has become clinically established [24]. The aimed neuroprotective strategies are focused on addressing excitotoxicity, i.e. cell death associated with an excess of excitatory neurotransmitters [25], immune and inflammatory responses [26], and apoptosis [27]. Among these, statins are a main group of neuroprotective agents that act inhibiting hydroxylmethylglutaryl coenzyme A reductase, which cause a reduction in low-density lipoprotein (LDL) cholesterol levels. In addition to this anti-thrombotic effect, statins seem to have other roles in the treatment of the pathophysiology of ischemic stroke [28], which have been investigated in clinical trials [29, 30].

Altogether, clinical management of stroke requires comprehensive hospital units with multidisciplinary teams dedicated to mitigate permanent neurological disabilities which, if unrecovered, pose a huge burden to society [31, 32]. However, this strategy has not been fully successful. Recently, there is a shift toward inno-vative neurorestorative treatments focused on restoring brain tissue and improving neurological function after damage. They aim to solve some of the aforementioned caveats, including the short time window for therapy and the inclusion of patients that were otherwise excluded from a therapeutic solution.

8.3   Advanced Therapies for Stroke

8.3.1   Cell-Based Therapies

Due to the limitations of conventional therapies and innovative adjuvant approaches, regenerative medicine has emerged with the aim of restoring brain function in a post-acute stage of stroke. The generation of neurons in some parts of the adult mammalian brain (e.g. the subgranular zone of the hippocampal dentate gyrus and the subventricular zone [SVZ] located outside of the lateral ventricles) provides a possible therapeutic solution for restoring neural function. However, this is still a debated topic following recent evidence with apparently contradicting outcomes [33, 34]. In fact, endogenous repair mechanisms including neurogenesis, synaptogenesis, glial cell activation, and angiogenesis are triggered after ischemic stroke [35]. Nevertheless, if any novel neurons are generated, they are not enough to repopulate the injured site. In addition, angiogenesis is compromised in older patients [34], which poses additional barriers to the restoration of lost neural circuitries [36]. Cell-based therapies are therefore positioned to potentiate endogenous mechanisms and overcome pathophysiological boundaries set by ischemic stroke. Two conceptually different approaches for regenerative therapy after stroke involve cell transplantation and cell recruitment (Fig. 8.3a).

8.3.1.1   Cell Transplantation

It implies the use of stem/progenitor cells that can be originated from the patient itself (autologous) or from donors that are genetically similar (allogenic) or identical (syngeneic). These cells can be derived from either fetal tissues (e.g. umbilical cord and placenta) or adult tissues (e.g. bone marrow, adipose tissue, olfactory mucosa, and dental pulp) and have been tested over the last 10 years for the treatment of ischemic stroke in the clinic [40]. The most advanced technology consists of the extraction of multipotent adult progenitor cells from the bone marrow of healthy donors (e.g. MultiStem® from Athersys). An exploratory Phase II clinical trial with MultiStem® pointed a favorable clinical outcome for patients that received a single dose of the product 24–48 h after the occurrence of the stroke [41]. The MASTERS-2 Phase III trial to employ MultiStem® as an "off-shelf" product for stroke treatment is now underway [42]. With such a variety of cells according to their source and tissue origin, a main challenge toward clinically relevant cell therapies is to generate high amounts of the optimal cell type. Neural stem cells (NSCs) have the capacity to differentiate into neurons, astrocytes, and oligodendrocytes, what makes them good candidates for effective transplantation and attenuation of the cell loss associated with ischemic stroke. Mesenchymal stem cells (MSCs) have been also investigated to arrest stroke-associated cell death [43]. Compared to NSCs, MSCs can be readily isolated from non-invasive tissue sources such as dental tissue and amplified ex vivo for autologous transplantation. Dental pulp tissue

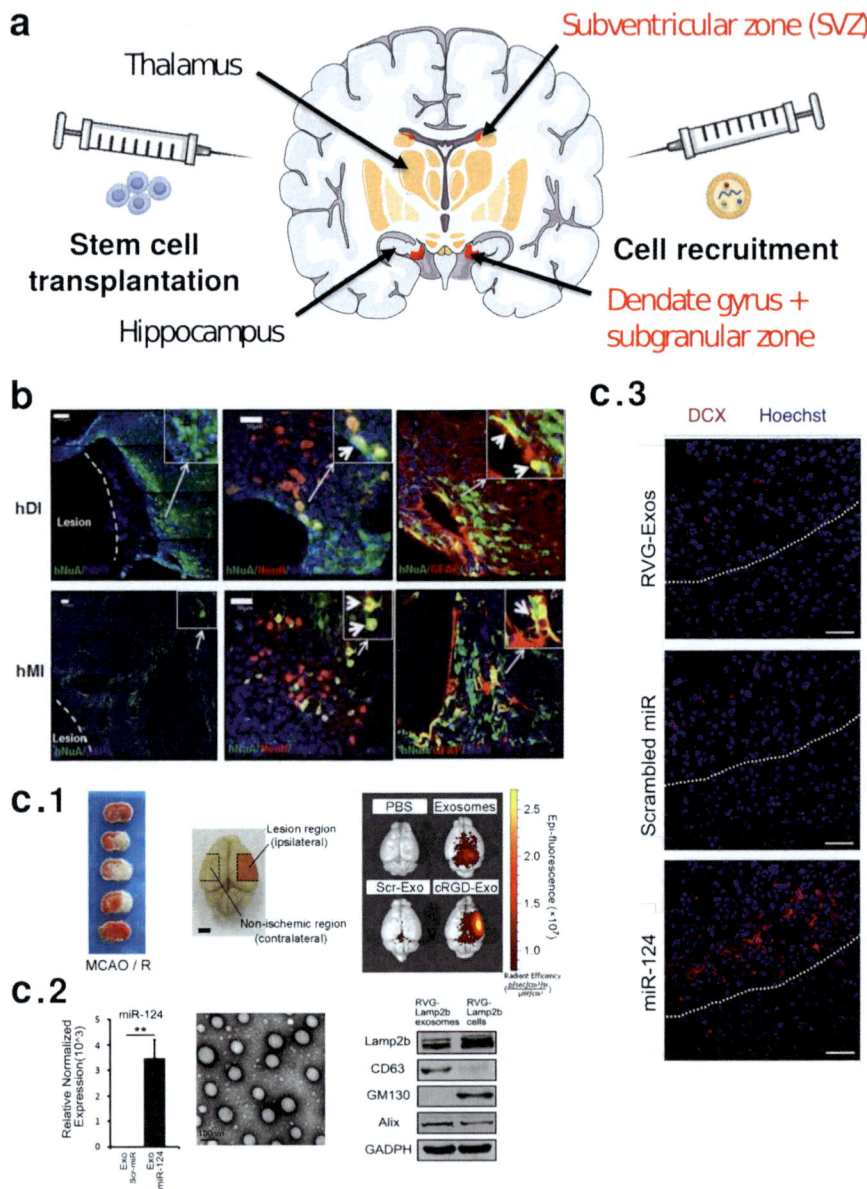

Fig. 8.3 Remarkable strategies for brain tissue regeneration. (**a**) Schematic representation of a coronal section of the brain, highlighting the putative reservoirs of NSCs capable of generating new neurons (red). These include the subventricular zone (SVZ), along the lateral wall of the lateral ventricles, and the subgranular zone of the dentate gyrus in the hippocampus. Because the adult brain is not capable of completely restore function after tissue damage, therapeutic approaches to promote neurogenesis consist of stem cell transplantation and the delivery of biomolecules to activate endogenous NSCs. Adapted from Servier Medical Art by Servier (http://smart.servier.com), which is licensed under a Creative Commons Attribution 3.0 Unported License. (**b**) Human dental stem cells (hDI) revealed superior performance than bone marrow-derived stem cells (hMI)

offers very interesting prospects for neurogenesis because it is derived from the ectoderm/neural crest and endogenously mark for several neuronal markers [44]. In addition, dental pulp stem cells were demonstrated to differentiate into functionally active neurons and secrete neurotrophic factors, thus revealing superior therapeutic potential for brain regeneration after stroke than other stem cell sources (Fig. 8.3b) [37]. Clinical investigation of the beneficial effects of intravenously administered dental pulp stem cells is now underway in a Phase I clinical trial [45].

Other cell types of interest to improve neuronal cell function include immune cells, hematopoietic stem cells, and endothelial progenitor cells (EPCs). EPCs have the potential to reduce inflammation and apoptosis, to promote angiogenesis, and even to promote endogenous repair mechanisms. EPCs can be derived from the bone marrow and are classically defined by their surface expression of antigen CD34 [46]. Their presence at the ischemic core is associated with improved clinical outcome after stroke [47], due to their capability of remodeling brain vasculature and promoting angiogenesis [48], which peaks at the subacute phase [49, 50]. These promising results have supported the transplantation of CD34$^+$ cells for the treatment of ischemic stroke. Their clinical efficacy is currently under investigation in an ongoing interventional Phase IIa trial [51].

8.3.1.2 Cell Modulation Strategies

Although cell transplantation is a promising strategy for the generation of new neural cells and the replacement of lost neuronal circuitries with appropriate synaptic integration in the host tissue [52–54], there is still no definitive evidence with respect to clinical outcome improvements [40, 55]. This could be due to inefficient cell transplantation, which is still limited by their homing to the injured area [56] and cell survival on the damaged tissue microenvironment [57]. Numerous solutions have been tested to improve engraftment efficiency, from preconditioning or genetically modifying transplanted cells to adopting biomaterials (e.g. scaffolds) in order to facilitate their integration in the brain tissue. Recent approaches have coupled the manipulation of stem cells with electrical stimulation, which led to enhanced neurogenesis and angiogenesis [58]. Furthermore, NSCs from the own

Fig. 8.3 (continued) in promoting neurogenesis in rat brains 28 days after middle cerebral artery occlusion. This was demonstrated by immunohistochemical analysis of proliferating neurons (NeuN$^+$) and astrocytes (GFAP$^+$), which stained positive for human nucleus (hNuA). Scale bars = 100 μm. Adapted with permission from SAGE Publishing [37]. (c.1) EVs secreted by bone marrow-derived MSCs can be functionalized with brain-targeting peptides for local delivery of bioactive molecules. Reprinted from [38], Copyright (2018), with permission from Elsevier. (c.2) EVs are nano-sized vehicles which are characterized by the enriched expression of surface markers (e.g. CD63, Alix) and can be loaded with bioactive molecules by electroporation. (c.3) Delivery of microRNA-124 by EVs functionalized with targeting peptide RVG enhanced neurogenesis after ischemic stroke as demonstrated by the expression of the neuronal marker doublecortin (DCX) at the infarct site 7 days after administration. Reprinted from [39], Copyright (2017), with permission from Elsevier

patient can be modulated to enhance neurogenesis [59]. We have demonstrated that polymeric nanoparticles (NPs) could mediate delivery of bioactive molecules to the SVZ in order to control differentiation of NSCs and EPCs, as well as to promote cell survival and normalize inflammatory responses occurring during ischemia [60–62]. NP-based formulations are attractive systems for cell modulation due to their efficacy, biocompatibility, and chemical versatility. They can be rendered compatible with imaging techniques, such as MRI [63, 64], or responsive to external stimuli (e.g. light) to confer spatiotemporal control over drug release to the brain [65, 66].

Besides cell replacement in the damaged brain, stem cell-mediated regenerative processes after stroke have been attributed to a paracrine effect characterized by the release of trophic factors and genetic modulators that activate brain remodeling pathways [67]. These biomolecules were found to be enriched in extracellular vesicles (EVs), which are nano-sized mediators playing key roles in intercellular communication [68]. EVs provide a cell-free option to modulate neural repair and overcome some of the limitations inherent to stem cell transplantation, including their scarcity and immunogenicity, which not only affects cell survival and motility after transplantation but can also cause significant adverse effects. Therapeutic EVs can be produced by MSCs and their content can be modulated for the delivery of proteins, lipids, and nucleic acids to enhance endogenous repair mechanisms (Fig. 8.3c). For instance, we and others have identified a panel of microRNAs associated with good prognosis after ischemic stroke, which affected migration of $CD34^+$ cells and their angiogenic activity [48, 49]. Further investigation is warranted to understand the effects of cell source and culture conditions on EVs content and, therefore, in their therapeutic potential.

8.3.2   Brain Electrical Stimulation

In addition to replacing damaged tissue with new cells, neurological functions can be restored after stroke by restructuring and rewiring functional networks [69, 70]. These restructuring processes are mainly due to the sprouting of spared axons, which innervate the affected regions, and create new neuronal circuitry [71, 72]. However, the brain alone does not have enough capacity to regenerate and reprogram neuronal circuits to the same complexity as that prior to stroke. Several strategies have been employed to maximize the chances of restoring sensory and motor functions by reestablishing neuronal connections [73, 74]. For instance, electrical stimulation of specific regions of the cortex has been explored to reorganize neural circuitry and restore brain functions after stroke (Fig. 8.4) [75]. Compared to pharmacological therapies, which can indiscriminately affect all neurons in the brain, this strategy allows for fewer adverse effects and much lower treatment associated costs [76]. Nevertheless, this has been employed mainly in patients with significant neurological impairment. Considering the extensive tissue damage in these patients, neuromodulation has been primarily performed using minimally invasive techniques.

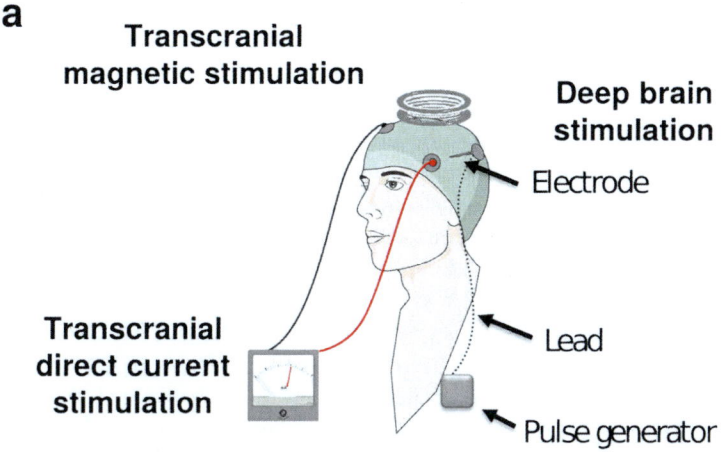

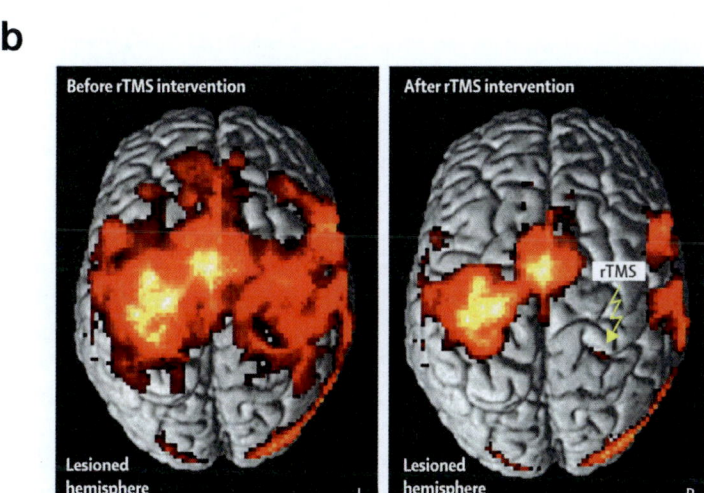

Fig. 8.4 Neuromodulation strategies for the management of stroke. (**a**) Schematic representation of stimulation modalities to modulate brain activity. Non-invasive modalities such as tDCS and transcranial magnetic stimulation (TMS) have been more frequently employed in the clinic. Adapted from Servier Medical Art by Servier (http://smart.servier.com), which is licensed under a Creative Commons Attribution 3.0 Unported License. (**b**) Functional MRI revealed that repetitive TMS of the contralesional primary motor cortex at 1 Hz inhibited excessive neural activity, which was associated with significant functional improvements. Reprinted from The Lancet [77], Copyright (2014), with permission from Elsevier

8.3.2.1 Transcranial Stimulation

Non-invasive modalities such as transcranial direct current stimulation (tDCS) and functional electrical stimulation (FES) consist in the application of electrodes on the skin surfacing the target region of interest. Typically in tDCS, one electrode targets the primary motor cortex, whereas the other acts over the contralateral supraorbital region. Based on the choice of anodal or cathodal electrodes [78], tDCS can induce either long-term potentiation or depression of neuronal activity, respectively, by modulating sodium- and calcium-dependent channels, as well as the NMDA receptor activity [79–81]. Although tDCS was shown to have an effect on upper limb functions in stroke patients, this occurred mostly during follow-up treatments, raising doubts about its long-term clinical efficacy [82].

On its turn, FES elicited moderate improvement in limb function by promoting muscle movement and mobility [83]. Due to the dissipation of the delivered current through the skull, high voltages are required to penetrate the brain tissue with enough power to activate neurons [84]. However, as high voltages were reported to cause patient discomfort, they were replaced by magnetic fields which have greater penetration depth [84]. Fast-oscillating magnetic fields along a copper coil external to the skull generate a strong electric current that can be directed to the motor cortex [84]. Specifically, transcranial magnetic stimulation (TMS) has been applied in the chronic setting of stroke in a strategy for interhemispheric inhibition [85–87]. It consists of exciting the ipsilesional primary motor cortex with high frequencies (>5 Hz) [88–90], whereas the contralesional primary motor cortex is inhibited using low frequencies (<1 Hz) [79, 91]. Other parameters such as stimulation time, coil shape, and magnetic field strength have been optimized to regulate cortical activity [79, 92, 93]. Despite some promising results particularly in the management of discrete neuropsychiatric conditions [94], magnetic stimulation of the brain and peripheral nerves is still at an early stage, and thus it has little clinical evidence of functional improvement in stroke patients [95, 96]. It is still unclear which protocol is more effective for improving motor function after stroke, given the lack of randomized controlled trials and small sample sizes [96]. New protocols have emerged, including the application of intermittent or continuous bursts of even higher frequencies than conventional TMS, thus requiring lower intensities [97, 98]. Such a variety of stimulation protocols warrants careful design of clinical trials to validate their safety and efficacy after stroke.

8.3.2.2 Deep Brain Stimulation

Although more invasive, modalities such as deep brain stimulation (DBS) are clinically well-established in movement disorders such as Parkinson's disease, and enable the stimulation of target regions with significant reproducibility [76]. DBS addresses the aforementioned issues of transcranial stimulation by implanting electrodes in regions adjacent to the target site [99]. Medical devices performing electrical stimulation have been tested in the clinic since the 1950s [100] and are successfully employed in the management of several neurological disorders where pharmacological options alone are inefficient, such as epilepsy, dementia, Alzheimer's disease and Parkinson's disease [99–101]. Currently, DBS is approved for the treatment of refractory Parkinson's disease, essential tremor, dystonia, obsessive–compulsive disorders, and drug-resistant partial onset epilepsy [102]. In the context of stroke, two objectives may arise from the use of DBS: the symptomatic treatment of extrapyramidal signs, following the same paradigm as in parkinsonian disorders, and the more conceptual goal of recovering brain function. First clinical evidence compiling several trials with small cohorts suggests that DBS could enhance motor status in stroke patients, particularly from disorders such as tremors, dyskinesia, and dystonia [103]. Such signs and symptoms represent post-stroke maladaptive responses where DBS could potentially have a role. In all these conditions, external electrical fields are thought to activate voltage-sensitive ion channels in neurons, which in turn generate chemical or electrical depolarization at their membranes, with subsequent release of neurotransmitters. As a result, irregular firing patterns in brain regions can be precisely modulated according to stimulus parameters such as signal amplitude, frequency, and duration [101].

Nevertheless, electrical stimulation performed by clinically approved DBS devices is experienced by all local cells, not only the targeted neurons. Other cell types including glia, fibroblasts, endothelium, and immune cells can also respond to these electrical cues, with significant effects in their phenotypes [104]. This could have an impact on the overall process of restoring brain function after stroke. Interestingly, transmembrane voltage for each cell type was associated with their differentiation state, with stem and proliferative cells being less polarized than terminally differentiated cells [104]. Hence, electrical stimulation could force membrane depolarization in neurons and glial cells that populate the infarcted area after stroke and promote tissue regeneration. *Post mortem* analysis showed that chronic stimulation (0.5–6 years) of the subthalamic nucleus enhanced neurogenesis in the neighboring SVZ in patients suffering from Parkinson's disease [105]. These findings encourage the investigation of potential in situ brain tissue regeneration following electrical stimulation. Yet no clinical trials to date have specifically demonstrated such effect, which could be attributed to the advanced disease progression by the time patients enroll in these studies [106].

8.3.2.3 Limitations of Deep Brain Electrical Stimulation

Indiscriminate stimulation of brain regions through conventional electrical stimulation devices might result in significant adverse effects, as reported in approximately 50–60% of patients and, in most cases, more than once [107–109]. Some of the most common causes of failure were improper electrode localization, inefficient device programming, infections, and hemorrhages resulting from surgical implantation [110]. Electrode positioning can be corrected with the guidance of imaging techniques (e.g. MRI, CT), while correct device programming overcomes issues such as overstimulation of undesired cells with high frequencies, which may impair physiological neuronal communication [111]. The recent development of closed-loop devices that adjust their stimulation parameters according to electro-physiological information recorded in real time paves the way for multifunctional neural interfaces, with further improvements expected in the following years [112]. The remaining caveats related to the implantation of DBS devices include their poor long-term stability and need for multiple surgeries to replace the electrodes.

Conventional electrodes are typically made of metals such as gold and iridium [113]. Metallic conductors are utilized because of their capability to readily mediate charge transfer between electrons at their interface with ions from the surrounding tissue (Fig. 8.5a). Most metals conduct electricity based on local reduction and oxidation reactions at the electrode surface, in a process known as Faradaic charge conduction. Repeated redox reactions at the metallic surface generate a hydrated oxide film that dramatically increases the amount of electric current that can be transferred to the adjacent tissue [113]. Although this electrochemical process is mostly reversible, changes in the electrolyte composition at the interface with the tissue can limit the rate of Faradaic reactions that can be performed without irre-versibly modifying the material. Otherwise, not only the electrode can be degraded but also induce oxidative stress to the surrounding tissue. Conversely, capacitive charge conduction is a more desirable feature for implanted electrodes, since it involves solely the redistribution of charges at the electrode–electrolyte interface, thus avoiding redox reactions. However, capacitive materials such as titanium nitride suffer from limited charge injection capacity [113]. Pseudocapacitive materials such as platinum and its alloys with iridium have become then clinically adopted because they combine both Faradaic and capacitive conduction, hence increasing charge injection while minimizing redox effects [113]. For further details on electroactive materials with large charge capacity, readers are referred to Chap. 5 in this book.

Alongside charge transfer processes, the mechanical properties of the implanted materials are of upmost importance. Despite considerable efforts in the design of sterile, non-toxic materials with long-term chemical and electrical stability, they tend to trigger foreign body response because of their rigidity ($>1\,GPa$) compared to the soft brain tissue ($<10\,kPa$) (Fig. 8.5b) [118]. Mechanical mismatch of the implant promotes adverse biomechanical interactions leading to the formation of glial scars at the electrode interface as soon as few weeks after surgical implantation [118]. Ultimately, the efficacy of electrical stimulation and recording is dampened by the increased distance between the electrode and the target cells, as well as the impedance derived from the scar tissue [119]. Although device architecture can be engineered to minimize biological impact by decreasing local strain imposed by the electrodes [120], there is a clinical need for biocompatible electrodes that can be seamlessly integrated in the brain microenvironment. Electrodes can be incorporated in soft polymer mesh electronics (Fig. 8.5c), which facilitate their implantation by direct injection into the target brain region [115]. Besides being minimally invasive, mesh electronics are mechanically compliant to the brain tissue and, thus, more biocompatible, showing in vivo stability of up to 1 year without gliosis.

Additional challenges for neural interfaces include targeted stimulation of specific sites without affecting other physiological functions. These devices should

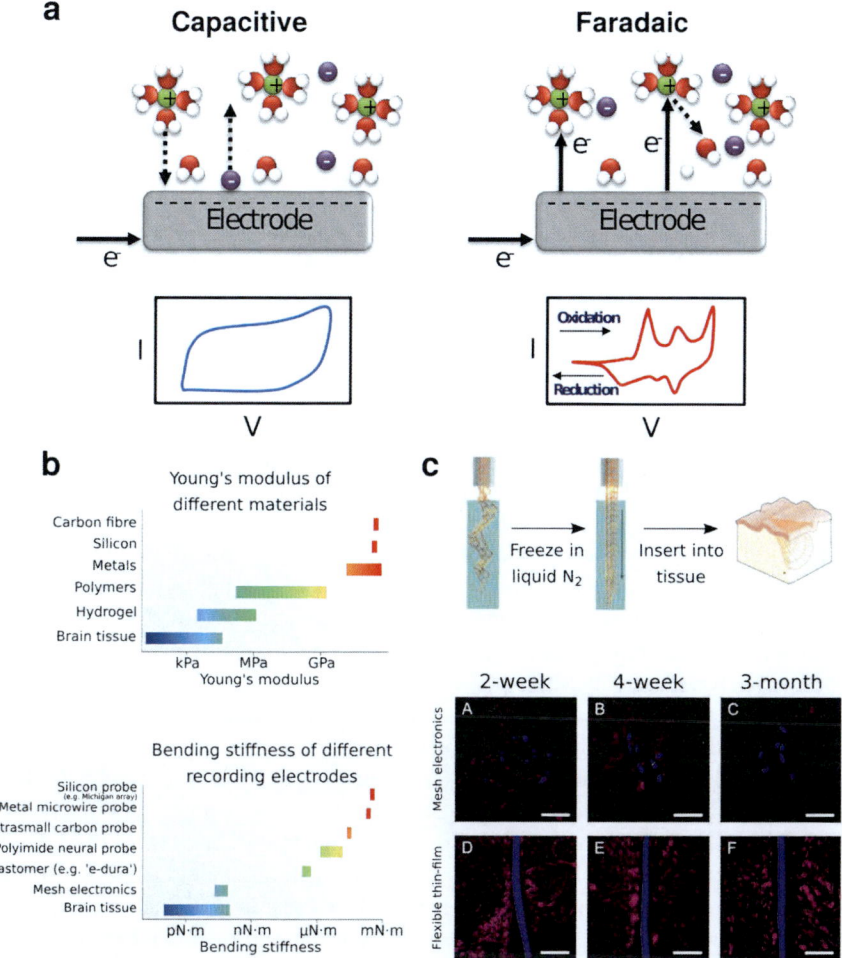

Fig. 8.5 Material properties determine long-term device biocompatibility and performance. (**a**) Electrical stimulation performed by electrodes depends on their electronic properties. Upon injection of electric current, capacitive materials such as titanium nitride, carbon nanotubes, and graphene generate a double layer at the electrode–electrolyte interface, attracting adsorbed water molecules and ionic species to the electrode surface [113, 114]. Because this process solely involves charge redistribution, the amount of charge injected from the electrode is limited by its surface. Although they enable greater amount of charge injected to the electrolyte, iridium oxide and PEDOT mediate Faradaic processes, which consist of the ejection of electrons from the electrode, leading to changes in the electrolyte composition and pH adjacent to the electrode [113]. Platinum and its alloys are attractive for brain stimulation because they combine capacitive and Faradaic processes, which result in higher charge injection with limited electrode degradation. Although these pseudocapacitive materials generate double layer charging, Faradaic processes may occur when specifically adsorbed ions react with the electrode surface [113]. (**b**) Typically used materials for implanted electrodes such as silicon, carbon, and metals are very rigid compared to brain tissues, presenting extremely high Young's moduli and bending stiffness values. Adapted by permission from Springer Nature Customer Service Centre GmbH: Springer Nature, Nature Reviews Neuroscience [115]. Copyright© 2019. (**c**) Mechanically compliant mesh electronics

be also capable of recording their physiological environment in order to coordinate neural stimulation parameters [119].

8.3.3 Optogenetic Neuromodulation

Exploring the intrinsic electrical properties of neurons, electrical stimulation has remained one of the main strategies to restore functional activity. However, because of its invasiveness, alternative approaches to DBS are being developed. One of the most promising is optogenetics, which combines light and genetic techniques to control and/or monitor cellular activity [121]. Although light has long been known to alter the behavior of neurons [122], this effect was only exploited in 2005, following their genetic modification with light-sensitive opsins [123]. Channelrhodopsins (ChR) are rapidly gated light-sensitive cation channels, commonly expressed in algae [124], and have provided unprecedented control over neuronal activity in well-defined neuronal populations with temporal precision. Upon light exposure, neuronal depolarization can be employed to investigate the functions of specific neurological circuitries and the mechanisms underlying neurological disorders [125, 126]. Even though optogenetics has been used mainly as a tool for neuroscience research in animals, therapeutic applications of this technology are under investigation [127–129].

Optogenetic tools have been applied in preclinical models of stroke (Fig. 8.6). In combination with voltage-sensitive dyes, the plasticity of the somatosensory cortex could be monitored after stroke, helping not only to understand the functional impact of the infarction but also to map potential regions of interest for stimulation [132]. Recovery of sensorimotor functions could be achieved after optogenetic stimulation of unaffected regions surrounding the infarcted cortex, such as corticospinal and thalamocortical neurons [130, 133]. In particular, stimulation of the ipsilesional primary motor cortex could contribute to functional recovery after stroke [129]. Repeated stimulation significantly improved neurovascular coupling and enhanced neuronal plasticity in the contralesional cortex. The cerebellum was also demonstrated to be a powerful target for brain stimulation due to the widespread activation of multiple motor and sensory regions via neuronal projections to the thalamus [134, 135]. All these studies have reported that optogenetic stimulation promoted axon growth and subsequent neuronal projections to the damaged site to

Fig. 8.5 (continued) are attractive for brain implantation owing to their long-term biocompatibility and minimal inflammatory response. Immunohistochemical staining for Iba-1 (magenta) demonstrated that mesh electronics can be implanted in mice brains for several months and seamlessly integrate in the brain tissue with minimal glial response. Implanted probes were pseudo-colored blue. Scale bars = 100 μm. (c-top) Adapted by permission from Springer Nature Customer Service Centre GmbH: Springer Nature, Nature Reviews Materials [116]. Copyright© 2017. (c-bottom) Reprinted from [117], with permission from the National Academy of Sciences

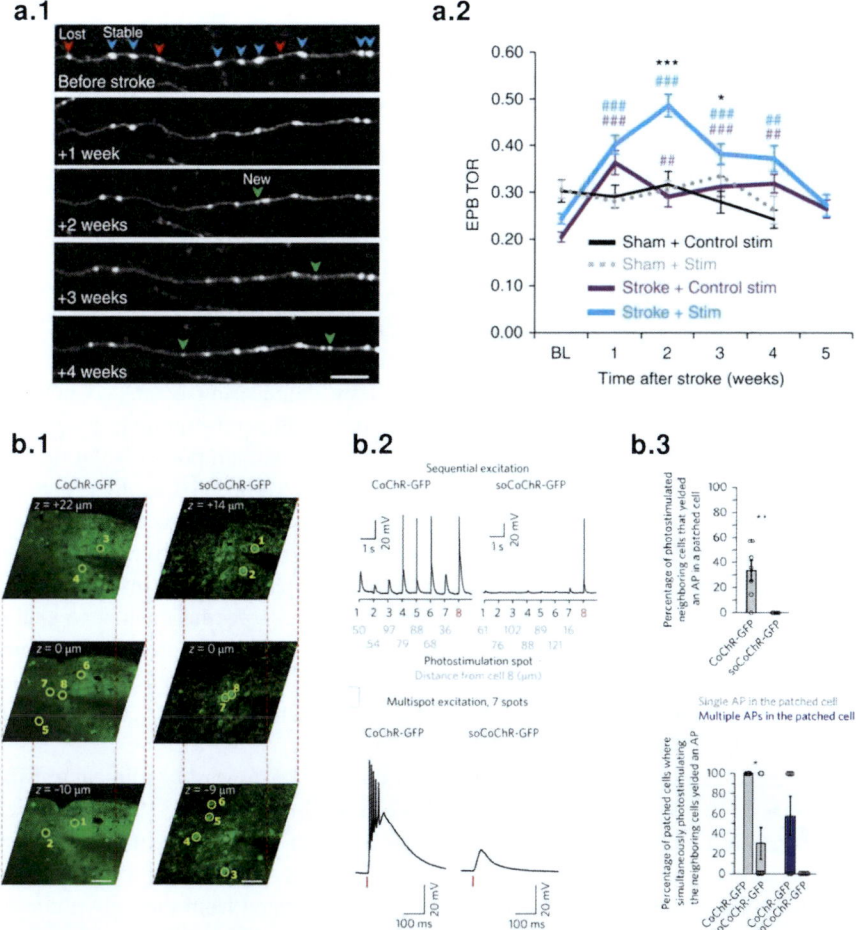

Fig. 8.6 Optogenetic stimulation for the treatment of stroke. (**a**) Optogenetic stimulation of ChR2-expressing thalamocortical neurons for up to 4 weeks after ischemic stroke significantly contributed to the formation of synaptic boutons, which play an important role in learning and memory processes. Reprinted by permission from Springer Nature Customer Service Centre GmbH: Springer Nature, Nature Communications [130]. Copyright© 2017. (b.1) Soma-targeted opsins (soCoChR) are selectively expressed in the cell body of neurons. (b.2) Precise activation of soCoChR neurons by two-photon microscopy ($\lambda = 1030$ nm, 100 μW/μm^2) without affecting neighboring cells. (b.3) Engineered opsins enabled unprecedented precision over the stimulation of single cells, yielding well-defined action potentials in a given patched cell with minimal detection of action potentials from neighboring cells. Adapted by permission from Springer Nature Customer Service Centre GmbH: Springer Nature, Nature Neuroscience [131]. Copyright© 2017

remodel neural circuitry. A recent avenue of research resides in the possibility of enhancing neurogenesis in the SVZ. Considering that the striatum has neuronal projections to the physically adjacent SVZ, optogenetic stimulation of striatum glutamatergic neurons enhanced regeneration and functional recovery after ischemic stroke by evoking membrane currents and calcium influx in proliferating SVZ neuroblasts [136].

These promising results are encouraged by technological advances to enhance control over neuronal stimulation. While channelrhodopsins enable precisely timed depolarization of neurons, halorhodopsins derived from archaeal species can be stimulated with light of the same wavelength to hyperpolarize neurons [137]. The combination of these two rhodopsins can be used to accurately and bidirectionally control neuronal activity and cells native spiking patterns. Furthermore, spatiotemporal resolution could be enhanced by engineering opsins to potently respond to short light pulses (<1 ms), enabling single-cell stimulation by two-photon microscopy [131]. Other strategies to achieve spatiotemporal resolution over optogenetics include conditional expression of opsins using cell-specific promoters [138], which can be specifically activated using gene editing tools such as the Cre-*loxP* technology [139–141]. Because some cell-specific promoters have a weak transcriptional activity resulting on reduced levels of opsins in the cell membrane, Cre recombinase can be expressed in a cell-specific manner to enable expression of rhodopsins under the control of stronger ubiquitous promoters. Thus, optogenetic stimulation is controlled spatiotemporally by modulating the activity of Cre recombinase in specific cells, through either chemical [142, 143] or light-inducible [144, 145] Cre-*loxP* recombination systems.

Nevertheless, optogenetics faces considerable hurdles toward its clinical translation. One of them is the requirement of using either blue or green light as a trigger. Since visible light poorly penetrates biological tissues, invasive light sources such as fiber optics and light-emitting diodes have been applied in preclinical models, which may damage local tissues due to the heat dissipated from the light emission point [146]. Recently, a step-function opsin was engineered to respond to blue light with enhanced sensitivity and slower kinetics, which enabled transcranial activation owing to neuron depolarization for longer periods of time. Prolonged light accumulation compensates for its dissipation across biological tissues, allowing for transcranial stimulation in deeper regions of the brain down to 5 mm [147].

Considering the minimal absorbance of hemoglobin and water in this region (650–900 nm), the use of near-infrared (NIR) light is an attractive alternative due to its minimal scattering in biological tissues. NIR light not only penetrates deeper than visible light (up to 2 cm), but can also be less attenuated by the human skull (approximately, 0.5–5% of emitted light) [148]. For instance, lanthanide-doped up-conversion nanoparticles (UCNPs) have enabled deep tissue activation of rhodopsins by emitting visible light after exposure to NIR radiation [149–152]. These nanoparticles have promising optical properties including low autofluorescence background and minimal photobleaching and heat-mediated photodamage. Hence, UCNPs enable safer and minimally invasive stimulation compared to the use of NIR radiation alone [153] or combined with plasmonic nanoparticles such as gold nanorods to activate heat-sensitive proteins [154]. Moreover, UCNPs can act as remote actuators for transcranial NIR activation of neuronal depolarization [149, 151, 155], enabling control over animal behavior in optogenetics studies. Finally, their chemical composition can be tuned to modulate light emission in order to selectively activate different channelrhodopsins and enhance the control over specific neural circuits [156]. These strategies open new opportunities to simultaneously control cell activity with spatiotemporal resolution and monitor neural circuits over time to improve recovery. However, the need for long-term expression of light-sensitive proteins, which is typically achieved by lentiviral vectors [128], carries numerous ethical and safety concerns regarding the possible genomic integration of undesired gene products after transfection, as well as potential adverse immune responses.

8.3.4 Coupling Optical and Electrical Stimulation of the Brain

Safety concerns related to the clinical use of optogenetics have prompted the investigation of numerous strategies to circumvent the need for genetic modification, while maintaining the capacity of specifically stimulating neurons with unprecedented resolution. This could be achieved by using photoactive nanomaterials and surfaces that generate an electric field when exposed to light, thus resulting in localized neuronal stimulation. This would avoid the need of implantable energy sources commonly used in DBS and prolong device lifetime. Moreover, device implantation would be desirably less invasive, with minimal foreign body response compromising long-term performance. However, this approach has not been investigated in preclinical stroke models yet because there are important biocompatibility considerations to minimize potential adverse effects in patients suffering from severe brain trauma. The section below explores the use of innovative polymers and nanomaterials, and the potential integration of light-responsive materials in such devices.

8.3.4.1 Novel Polymeric Materials for DBS

A main avenue of research consists of the design of minimally invasive devices using biodegradable materials (Fig. 8.7). These devices are based on biocompatible polymers, such as silk fibroin [159] and poly(lactic-co-glycolic acid) (PLGA) [160], and have been already developed for wireless electronic stimulation of peripheral nerves. This technology operates in a similar fashion to cochlear implants, where an external source of radiofrequency signals generates magnetic coupling with an antenna at the implanted device, which transduces that signal to electric current at the interfacing electrode. Although its application may be limited by the necessary power input to cross deeper regions such as those stimulated by DBS devices, the concept of bioresorbable devices is attractive for rehabilitation regimes in stroke because it avoids an additional surgical procedure to remove them. For instance,

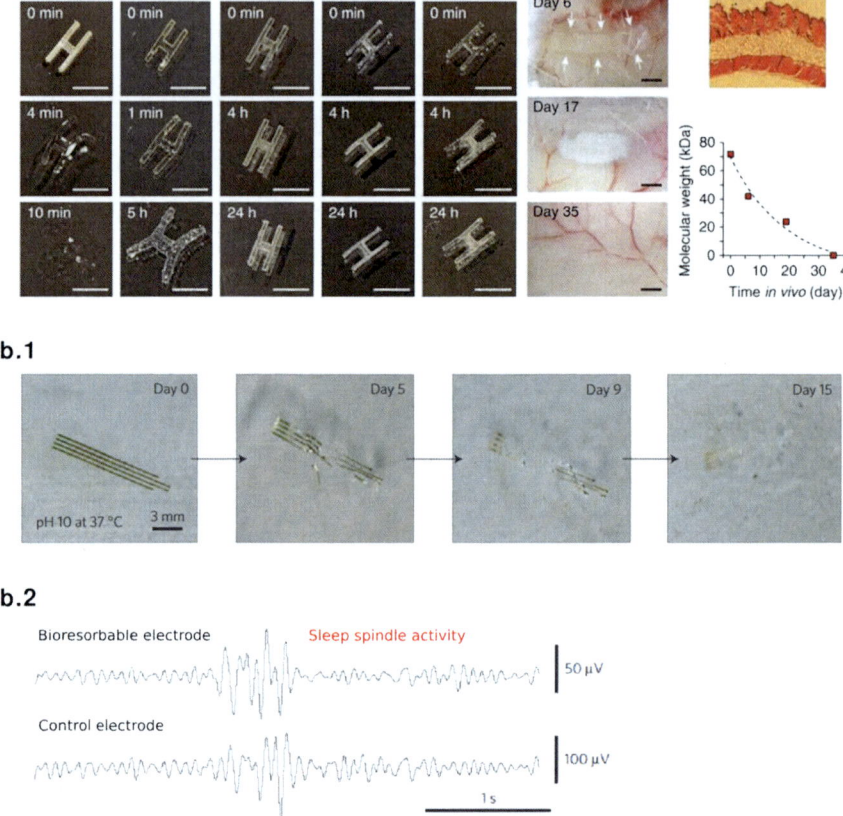

Fig. 8.7 Biodegradable electrodes enable transient monitoring and stimulation of the brain. (**a.1**) Biodegradability of silicon-based electrodes was tuned by adjusting the composition of PLGA films (50:50), in order to maintain their structural properties in phosphate buffer saline for several days, but were completely degraded within 35 days after subcutaneous implantation in a mouse model. (**a.2**) No signs of inflammatory response to the implant were observed. Reprinted by permission from Springer Nature Customer Service Centre GmbH: Springer Nature, Nature Communications [157]. Copyright© 2016. (**b.1**) Dissolution profile in aqueous buffer solution (pH 10) at 37°C and (**b.2**) electrophysiological recording of cortical activity in rat brains during sleep and drug-induced epilepsy, compared to commercial stainless steel microwire electrodes. Silicon-based electrodes exhibited high signal-to-noise ratio. Adapted by permission from Springer Nature Customer Service Centre GmbH: Springer Nature, Nature Materials [158]. Copyright© 2016

silicon-based electrodes deposited on PLGA films recorded electrophysiological information from the rat cortex with comparable performance to clinically used electrodes [158], as well as intracranial pressure and temperature [161]. Other biocompatible polymer substrates and device operation modalities are currently under investigation to ensure long-term safety and improved electrical stimulation over more conventional methods [162].

Alternatively to biodegradable materials, a variety of biopersistent materials are well-established in the medical device industry. Device miniaturization could minimize their biological impact in the CNS. However, this comes at the expense of greater impedance, which is highly undesired in neural interfaces due to increased noise in recording electrodes and decreased amount of current that can be injected in stimulating electrodes [113, 163]. Impedance can be also detrimental for electrode longevity and biocompatibility because of local generation of heat from stimulating electrodes and potential toxic by-products from electrochemical reactions [113]. Finally, platinum is sensitive to various imaging techniques, producing artifacts in CT and MRI and interfering with optogenetics tools due to its lack of transparency [163]. Transparent materials that are not comprised of heavy elements and have low magnetic susceptibility are therefore preferred.

Indium tin oxide (ITO) is a transparent and electrically conductive material that is well-known for its application in touchscreens and solar cells. Despite its attractive features, ITO is expensive and brittle, which limits the available area of the electrode for recording and stimulation [164]. Alternatively, ITO could be deposited on flexible substrates such as parylene, poly(dimethylsiloxane) (PDMS), polymethylmethacrylate (PMMA), polyimide, and SU-8 epoxy [120]. However, ITO deposition requires temperatures that are higher than the glass transition temperature of most flexible polymer substrates [165]. Moreover, ITO has reduced optical transmittance toward the ultraviolet (UV)/blue and IR regions, maybe unsuitable for optogenetics. Although less conductive than ITO, flexible polymers such as poly(3,4-ethylenedioxythiophene) (PEDOT) surpass these challenges (Fig. 8.8a) [166]. PEDOT is a pseudocapacitive polymer stabilized in aqueous formulations by poly(styrenesulfonate) (PSS), which is also important in charge transfer processes resulting in the oxidation of PEDOT [113]. Despite its high electrical conductivity and low impedance [166], PEDOT:PSS lacks long-term stability in physiological milieu and delaminates from its substrate at higher charge densities [113], thus precluding its application in high-frequency recording and stimulation (Fig. 8.8b).

8.3.4.2   Novel Nanomaterials for DBS

Aiming device miniaturization, nanomaterials have been increasingly applied either as an electrode coating for already existing devices or as electrodes themselves (Fig. 8.8b–c) [163, 170]. Owing to the network comprised by π electrons resulting from the sp^2 hybridization of carbon atoms, carbon nanomaterials such as carbon nanotubes (CNTs) and graphene have emerged as promising candidates for neural interfaces due to their high capacitive charge conductivity and physicochemical

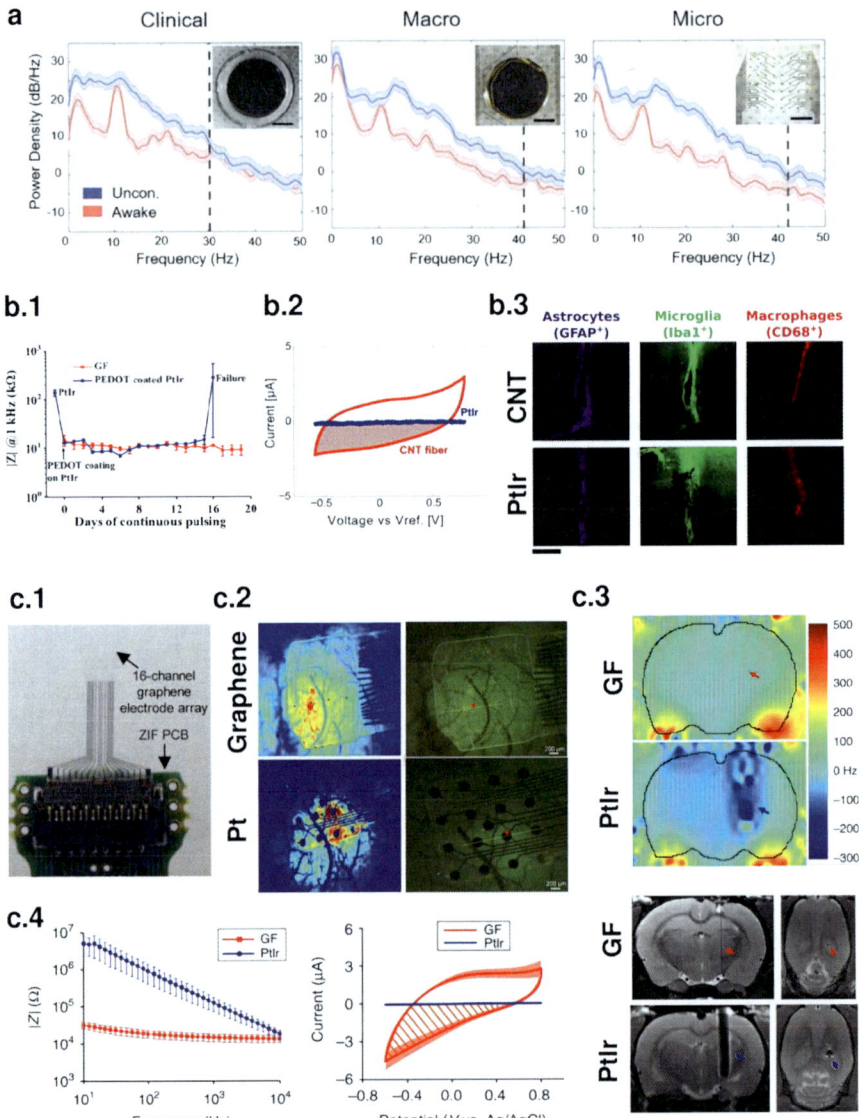

Fig. 8.8 Optically compatible materials for brain stimulation. (**a**) PEDOT:PSS electrodes showed comparable electrocorticography differences to clinically used platinum electrodes in recording brain activity of awake and unconscious rats. PEDOT:PSS maintained its sensitivity irrespective of electrode size, thus enabling device miniaturization. Reproduced from [166], with permission from John Wiley and Sons. © 2017 WILEY-VCH Verlag GmbH & Co. KGaA, Weinheim. (**b.1**) PEDOT:PSS-coated platinum–iridium (PtIr) electrodes show poor stability under prolonged continuous overpulsing at 1 kHz, demonstrated by the increased impedance comparable to uncoated PtIr electrodes. (**b.2**) CNT fibers mediated capacitive charge conduction and showed greater stability, (**b.3**) but their rigidity triggered significant glial response 6 weeks after implantation. (**c.1**) Transparent graphene-based electrodes enabled multimodal imaging to monitor brain activity

stability [163]. For instance, microelectrodes containing vertically aligned CNTs enabled highly sensitive electrochemical measurements and precise stimulation of brain regions at the nanotube tip [168, 171, 172], with CNT coatings enhancing the electrode stability [172, 173]. In addition, the well-defined electronic energy levels of single-walled CNTs (also known as Van Hove singularities) could guide the design of electrodes with minimal light-induced artifacts during optogenetic stimulation and record electrophysiological activity with high fidelity [174]. However, biomedical research involving CNTs has become somewhat controversial [175]. For instance, a type of long multi-walled CNT fibers with high aspect ratio (MWCNT-7) has been classified as "potentially carcinogenic to humans" based on extensive preclinical evidence of tumor formation due to excessive fibrotic and inflammatory responses [176].

Sharing similar electronic features with CNTs, graphene has emerged as a strong candidate for the development of neural interfaces [114]. Despite its potentially slow degradation profile [177], graphene is more flexible and biocompatible than CNTs, evidenced by the lack of significant fibrosis in multiple tissues after different administration routes [178, 179]. In fact, graphene substrates were shown to improve neural cell growth and differentiation by potentiating electric circuits [180–182]. Moreover, the application of graphene as surface coatings not only protected metal electrodes from corrosive electrochemical reactions at their surface, but also shielded them from electromagnetic interference during MRI, hence minimizing image artifacts [183]. Such compatibility with functional MRI has facilitated the mechanistic study of the therapeutic effects of DBS in Parkinsonian rats using graphene-based fiber electrodes (Fig. 8.8c) [167].

Altogether, these properties enabled graphene to be employed in flexible interfaces for multimodal imaging, which couple recording neural activity with high sensitivity and spatiotemporal resolution. For instance, graphene-based transistor arrays designed for electrocorticography were demonstrated to map electrical activity in the brain with greater spatial resolution and lower electronic noise than clinically used platinum and gold [184, 185]. Furthermore, a neural interface comprised of graphene-based sensing and stimulating electrodes was shown to regulate thalamocortical circuits and effectively correct abnormal epileptic activity using high-frequency discharges, after epidural implantation [186]. Graphene-based electrode arrays have been also developed to couple optogenetics stimulation with electrophysiological recording [165, 187]. Despite superior performance compared

---

Fig. 8.8 (continued) and (**c.2**) minimal artifacts in fluorescence imaging compared to clinically used platinum-based electrodes. (**c.3**) Graphene fiber electrodes insulated with Parylene C enabled brain stimulation of the subthalamic nucleus of rat brains with minimal interference in MRI. (**c.4**) Graphene exhibits lower electrical impedance than PtIr and greater charge injection by capacitive charge conduction, thus demonstrating superior performance for brain stimulation. (b1,c3,c4) Adapted by permission from Springer Nature Customer Service Centre GmbH: Springer Nature, Nature Communications [167]. Copyright© 2020. (b2,b3) Adapted with permission from [168]. Copyright (2015) American Chemical Society. (c1,c2) Adapted with permission from [169]. Copyright (2018) American Chemical Society

to platinum, graphene electrodes could suffer from artifacts derived from photoelectric effects upon exposure to blue light. As these artifacts were mostly limited to the immediate vicinity of the irradiated electrode, this phenomenon was attributed to the photovoltaic effect, which is characterized by the generation of electric current upon light exposure. Also, similarly to what is commonly observed in metals, light-induced artifacts depended on incident laser power and exposure time. The observed light-induced artifacts could compromise the use of graphene electrodes in combination with optogenetics tools. Nevertheless, this intrinsic capability of generating electricity upon light exposure could offer a promising alternative to optogenetics by avoiding the need of genetic modifications. To this regard, Savchenko et al. discovered that graphene substrates could elicit cell contraction upon light stimulation [188]. Consistent with the aforementioned photoelectric effect, light stimulation elicited capacitive charge injection. In these studies, cellular activity was manipulated by adjusting light intensity rather than wavelength.

Alternatively, silicon nanowires (SiNWs) have been also recently explored toward the development of photoresponsive electrodes mediating optoelectronic stimulation of cardiomyocytes and neurons [189–191]. SiNWs convert light into electricity via photothermal and photoelectrochemical reactions catalyzed by atomic gold used to nucleate and generate these nanostructures. In addition, conductive polymers have been employed in the preclinical development of retinal implants and could provide a platform for optoelectronic stimulation [192]. Further investigation on their photosensitivity, as well as their long-term biocompatibility and stability, is warranted to determine their clinical applicability.

8.4   Conclusions and Future Perspectives

Recent improvements in critical care of acute ischemic stroke have saved the lives of millions of patients worldwide. However, most survivors experience noticeable deficits in neurological function, which could affect independence in their daily lives. Novel therapies and devices have been developed with the aim of resolving or attenuating these disabilities.

Stem cell transplantation has been the most investigated strategy to date for restoring brain functions. However, key factors determining the success of this strategy remain unknown. First, the influence of donor cell type and tissue origin for the transplant needs to be considered to ensure their integration in the injured brain site. Furthermore, the patient clinical history (e.g. age, sex, presence of comorbidities, and recent surgical procedures such as recanalization), delivery method for the treatment (e.g. intravenous, intra-arterial, and stereotaxic), and timeline may also play important roles in choosing the appropriate regime. Transplanted stem cells are more effective when delivered at early stages to modulate tissue regeneration and reintegration in the neuronal circuitry. However, the exacerbated immune response to traumatic injuries may limit their efficacy. In this sense, clinical evidence shows limited efficacy of stem cells in improving neuronal function after stroke. This could be explained by late interventions performed at subacute and chronic stages after stroke, when neuronal circuitry has been already reestablished [40, 70]. Further investigation is required to evaluate whether immune and angiogenic responses dominating the subacute stage could have a beneficial impact on neurogenesis and synaptogenesis [193, 194]. Considering the high cost of cell transplantation, the delivery of EVs arises as an attractive cell-free option to mimic some of the beneficial effects of stem cells. However, this therapeutic strategy requires further development and testing [68].

Medical devices for brain stimulation are expected to undergo significant technological development in the following years, following the clinical acceptance of different materials from the conventionally used metals as electrodes. Silicon- and graphene-based nanomaterials rank among the most promising candidates for bioelectronics, owing to their biocompatibility. However, current fabrication processes are laborious and involve high temperatures which are not conducive to their application in flexible polymer substrates. Cost-effective procedures such as inkjet printing should yield electrically conductive nanomaterials which can be formulated to facilitate their incorporation in soft interfaces, thus making them more accessible [195, 196]. Nonetheless, the effects of long-term exposure to these nanomaterials require extensive assessment of device biocompatibility along its life cycle, including the careful characterization of dissolution and/or degradation by-products. Covalent functionalization and chemical doping strategies will provide added control over nanomaterial biocompatibility and biodegradability for biomedical applications [197–200].

Acknowledgments This work was supported by the ERA Chair (ERA@UC, ref: 669088) and Twinning (RESETageing, ref: 952266) projects through the European Union's Horizon 2020 research and innovation programme. The authors would also like to thank the financial support from the European Regional Development Fund (ERDF), through QREN-COMPETE funding (Projects "NeuroAtlantic", Ref: EAPA_791/2018, and "2IQBioNeuro", Ref: 0624_2IQBIONEURO_6_E), which is co-funded by Program Interreg Atlantic Space, and through the COMPETE 2020—Operational Programme for Competitiveness and Internationalization, as well as Portuguese national funds via FCT—Fundação para a Ciência e a Tecnologia (under projects PTDC/NAN-MAT/28060/2017 and UID/NEU/04539/2019) and FEDER (CENTRO-01-0145-FEDER-028060). AFR acknowledges funding from the EU Horizon 2020 programme under grant agreement No. 101003413. CR is grateful for the PhD fellowship sponsored by FCT under the MIT Portugal Program (SFRH/BD/52337/2013).

References

1. Johnson CO, Nguyen M, Roth GA et al (2019) Global, regional, and national burden of stroke, 1990–2016: A systematic analysis for the Global Burden of Disease Study 2016. The Lancet Neurology 18:439–458
2. Sacco RL, Kasner SE, Broderick JP et al (2013) An updated definition of stroke for the 21st century: A statement for healthcare professionals. Stroke 44:2064–2089
3. Ramee SR, White CJ (2014) Acute stroke intervention. Curr Probl Cardiol 39:59–76

4. Bosetti F, Koenig JI, Ayata C et al (2017) Translational stroke research: Vision and opportunities. Stroke, 48:2632–2637
5. Langhorne P, Bernhardt J, Kwakkel G (2011) Stroke rehabilitation. Lancet 377:1693–1702
6. Adams Jr HP, Bendixen BH, Kappelle LJ et al (1993) Classification of subtype of acute ischemic stroke: Definitions for use in a multicenter clinical trial. Stroke 24:35–41
7. Hankey GJ (2017) Stroke. Lancet 389:641–654
8. Sommer CJ (2017) Ischemic stroke: experimental models and reality. Acta Neuropathol 133:245–261
9. Harbison J, Hossain O, Jenkinson D et al (2003) Diagnostic accuracy of stroke referrals from primary care, emergency room physicians, and ambulance staff using the face arm speech test. Stroke 34:71–6
10. Martins AI, Sargento-Freitas J, Jesus-Ribeiro J et al (2018) Blood pressure variability in acute ischemic stroke: the role of early recanalization. Eur Neurol 80:63–67
11. Branco JP, Oliveira S, Sargento-Freitas J et al (2018) S100beta protein as a predictor of poststroke functional outcome: A prospective study. J Stroke Cerebrovasc Dis 27:1890–1896
12. Rodrigo R, Fernandez-Gajardo R, Gutierrez R et al (2013) Oxidative stress and pathophysiology of ischemic stroke: novel therapeutic opportunities. CNS Neurol Disord Drug Targets 12:698–714
13. Yoo AJ, Hu R, Hakimelahi R et al (2012) CT angiography source images acquired with a fast-acquisition protocol overestimate infarct core on diffusion weighted images in acute ischemic stroke. J Neuroimaging 22:329–35
14. Offner H, Subramanian S, Parker SM et al (2006) Experimental stroke induces massive, rapid activation of the peripheral immune system. J Cereb Blood Flow Metab 26:654–65
15. Powers WJ, Rabinstein AA, Ackerson T et al (2019) Guidelines for the early management of patients with acute ischemic stroke: 2019 update to the 2018 guidelines for the early management of acute ischemic stroke. Stroke 50:e344–e418
16. National Institute of Neurological Disorders and Stroke rt-PA Stroke Study Group (1995) Tissue plasminogen activator for acute ischemic stroke. N Engl J Med 333:1581–7
17. Hacke W, Kaste M, Bluhmki E et al (2008) Thrombolysis with alteplase 3 to 4.5 hours after acute ischemic stroke. N Engl J Med 359:1317–29
18. Lees KR, Bluhmki E, Von Kummer R et al (2010) Time to treatment with intravenous alteplase and outcome in stroke: an updated pooled analysis of ECASS, ATLANTIS, NINDS, and EPITHET trials. Lancet 375:1695–703
19. Nogueira RG, Jadhav AP, Haussen DC (2018) Thrombectomy 6 to 24 hours after stroke with a mismatch between deficit and infarct. N Engl J Med 378:11–21
20. Goyal M, Menon BK, Van Zwam WH et al (2016) Endovascular thrombectomy after large-vessel ischaemic stroke: a meta-analysis of individual patient data from five randomised trials. Lancet 387:1723–31
21. Mokin M, Pendurthi A, Ljubimov V et al (2018) ASPECTS, large vessel occlusion, and time of symptom onset: estimation of eligibility for endovascular therapy. Neurosurgery 83:122–127
22. Mokin M, Ansari SA, McTaggart RA et al (2019) Indications for thrombectomy in acute ischemic stroke from emergent large vessel occlusion (ELVO): report of the SNIS Standards and Guidelines Committee. J Neurointerv Surg 11:215–220
23. Karsy M, Brock A, Guan J et al (2017) Neuroprotective strategies and the underlying molecular basis of cerebrovascular stroke. Neurosurg Focus 42:E3
24. Savitz SI, Baron JC, Yenari MA et al (2017) Reconsidering neuroprotection in the reperfusion era. Stroke 48:3413–3419
25. da Silva-Candal A, Pérez-Díaz A, Santamaria M et al (2018) Clinical validation of blood/brain glutamate grabbing in acute ischemic stroke. Ann Neurol 84:260–273
26. Malhotra K, Chang JJ, Khunger A et al (2018) Minocycline for acute stroke treatment: a systematic review and meta-analysis of randomized clinical trials. J Neurol 265:1871–1879
27. Miyaji Y, Yoshimura S, Sakai N et al (2015) Effect of edaravone on favorable outcome in patients with acute cerebral large vessel occlusion: subanalysis of RESCUE-Japan Registry. Neurol Med Chir (Tokyo) 55:241–7

28. Zhao J, Zhang X, Dong L et al (2014) The many roles of statins in ischemic stroke. Curr Neuropharmacol 12:564–74
29. Tuttolomondo A (2014) Atorvastatin in acute stroke treatment [Online]. University of Palermo, Italy: ClinicalTrials.gov Identifier: NCT02225834. https://clinicaltrials.gov/ct2/show/NCT02225834. Cited 29 Sep 2021
30. Lou M (2015) The safety and efficacy study of high dose atorvastatin after thrombolytic treatment in acute ischemic stroke (SEATIS) [Online]. Second Affiliated Hospital, School of Medicine, Zhejiang University: ClinicalTrials.gov Identifier: NCT02452502. https://clinicaltrials.gov/ct2/show/NCT02452502. Cited 29 Sep 2021
31. Rudd AG, Hoffman A, Irwin P et al (2005) Stroke unit care and outcome: results from the 2001 National Sentinel Audit of Stroke (England, Wales, and Northern Ireland). Stroke 36:103–106
32. Stroke Unit Trialists C (2013) Organised inpatient (stroke unit) care for stroke. Cochrane Database Syst Rev CD000197
33. Sorrells SF, Paredes MF, Cebrian-Silla A et al (2018) Human hippocampal neurogenesis drops sharply in children to undetectable levels in adults. Nature 555:377–381
34. Boldrini M, Fulmore CA, Tartt AN et al (2018) Human hippocampal neurogenesis persists throughout aging. Cell Stem Cell 22:589e5–599e5
35. Kokaia Z, Lindvall O (2003) Neurogenesis after ischaemic brain insults. Curr Opin Neurobiol 13:127–32
36. Dillen Y, Kemps H, Gervois P et al (2020) Adult neurogenesis in the subventricular zone and its regulation after ischemic stroke: implications for therapeutic approaches. Transl Stroke Res 11:60–79
37. Song M, Lee JH, Bae J, Bu Y, Kim EC (2017) Human dental pulp stem cells are more effective than human bone marrow-derived mesenchymal stem cells in cerebral ischemic injury. Cell Transplant 26:1001–1016
38. Tian T, Zhang HX, He CP et al (2018) Surface functionalized exosomes as targeted drug delivery vehicles for cerebral ischemia therapy. Biomaterials 150:137–149
39. Yang J, Zhang X, Chen X et al (2017) Exosome mediated delivery of miR-124 promotes neurogenesis after ischemia. Mol Ther Nucleic Acids 7:278–287
40. Kalladka D, Sinden J, Pollock K et al (2016) Human neural stem cells in patients with chronic ischaemic stroke (PISCES): a phase 1, first-in-man study. Lancet 388:787–96
41. Mays R, Deans R (2016) Adult adherent cell therapy for ischemic stroke: clinical results and development experience using MultiStem. Transfusion 56:6S–8S
42. Mays RW (2018) MultiStem® administration for stroke treatment and enhanced recovery study (MASTERS-2) [Online]. Athersys, Inc., United States: ClinicalTrials.gov Identifier: NCT03545607. https://clinicaltrials.gov/ct2/show/NCT03545607. Cited 29 Sep 2021
43. Yasuhara T, Kawauchi S, Kin K et al (2020) Cell therapy for central nervous system disorders: Current obstacles to progress. CNS Neurosci Ther 26:595–602
44. Martens W, Bronckaers A, Politis C et al (2013) Dental stem cells and their promising role in neural regeneration: an update. Clin Oral Investig 17:1969–83
45. Nagpal A, Kremer KL, Hamilton-Bruce MA et al (2016) TOOTH (The Open study Of dental pulp stem cell Therapy in Humans): Study protocol for evaluating safety and feasibility of autologous human adult dental pulp stem cell therapy in patients with chronic disability after stroke. Int J Stroke 11:575–85
46. Banerjee S, Bentley P, Hamady M et al (2014) Intra-arterial immunoselected CD34+ stem cells for acute ischemic stroke. Stem Cells Transl Med 3:1322–30
47. Bogoslovsky T, Chaudhry A, Latour L et al (2010) Endothelial progenitor cells correlate with lesion volume and growth in acute stroke. Neurology 75:2059–62
48. Sargento-Freitas J, Aday S, Nunes C et al (2018) Endothelial progenitor cells enhance blood-brain barrier permeability in subacute stroke. Neurology 90:e127–e134
49. Esquiva G, Grayston A, Rosell A (2018) Revascularization and endothelial progenitor cells in stroke. Am J Physiol Cell Physiol 315:C664–C674

50. Sargento-Freitas J, Aday S, Nunes C et al (2018) Endothelial progenitor cells influence acute and subacute stroke hemodynamics. J Neurol Sci 385:119–125
51. Sargento-Freitas J, Pereira A, Gomes A et al (2018) STROKE34 study protocol: a randomized controlled phase IIa trial of intra-arterial CD34+ cells in acute ischemic stroke. Front Neurol 9:302
52. Tsupykov O, Kyryk V, Smozhanik E et al (2014) Long-term fate of grafted hippocampal neural progenitor cells following ischemic injury. J Neurosci Res 92:964–74
53. Muñetón-Gómez VC, Doncel-Pérez E, Fernandez AP et al (2012) Neural differentiation of transplanted neural stem cells in a rat model of striatal lacunar infarction: light and electron microscopic observations. Front Cell Neurosci 6:30
54. Tornero D, Tsupykov O, Granmo M et al (2017) Synaptic inputs from stroke-injured brain to grafted human stem cell-derived neurons activated by sensory stimuli. Brain 140:692–706
55. Qiao LY, Huang FJ, Zhao M et al (2014) A two-year follow-up study of cotransplantation with neural stem/progenitor cells and mesenchymal stromal cells in ischemic stroke patients. Cell Transplant 23 Suppl 1:S65–S72
56. Acosta SA, Tajiri N, Hoover J et al (2015) Intravenous bone marrow stem cell grafts preferentially migrate to spleen and abrogate chronic inflammation in stroke. Stroke 46:2616–27
57. Korshunova I, Rhein S, García-González D et al (2020) Genetic modification increases the survival and the neuroregenerative properties of transplanted neural stem cells. JCI Insight 5:e126268
58. Huang Y, Li Y, Chen J et al (2015) Electrical stimulation elicits neural stem cells activation: new perspectives in CNS repair. Front Hum Neurosci 9:586
59. Santos T, Boto C, Saraiva CM, Bernardino L, Ferreira L (2016) Nanomedicine approaches to modulate neural stem cells in brain repair. Trends Biotechnol 34:437–439
60. Aday S, Zoldan J, Besnier M et al (2017) Synthetic microparticles conjugated with VEGF165 improve the survival of endothelial progenitor cells via microRNA-17 inhibition. Nat Commun 8:747
61. Santos T, Ferreira R, Maia J et al (2012) Polymeric nanoparticles to control the differentiation of neural stem cells in the subventricular zone of the brain. ACS Nano 6:10463–10474
62. Machado-Pereira M, Santos T, Ferreira L et al (2018) Intravenous administration of retinoic acid-loaded polymeric nanoparticles prevents ischemic injury in the immature brain. Neurosci Lett 673:116–121
63. Aday S, Paiva J, Sousa S et al (2014) Inflammatory modulation of stem cells by Magnetic Resonance Imaging (MRI)-detectable nanoparticles. RSC Adv 4:31706–31709
64. Saraiva C, Paiva J, Santos T, Ferreira L, Bernardino L (2016) MicroRNA-124 loaded nanoparticles enhance brain repair in Parkinson's disease. J Control Release 235:291–305
65. Santos T, Ferreira R, Quartin E et al (2017) Blue light potentiates neurogenesis induced by retinoic acid-loaded responsive nanoparticles. Acta Biomater 59:293–302
66. Praça C, Rai A, Santos T et al (2018) A nanoformulation for the preferential accumulation in adult neurogenic niches. J Control Release 284:57–72
67. Leong WK, Henshall TL, Arthur A et al (2012) Human adult dental pulp stem cells enhance poststroke functional recovery through non-neural replacement mechanisms. Stem Cells Transl Med 1:177–87
68. Zhang ZG, Buller B, Chopp M (2019) Exosomes—beyond stem cells for restorative therapy in stroke and neurological injury. Nat Rev Neurol 15:193–203
69. Murphy TH, Corbett D (2009) Plasticity during stroke recovery: from synapse to behaviour. Nat Rev Neurosci 10:861–872
70. Steinberg GK, Kondziolka D, Wechsler LR et al (2016) Clinical outcomes of transplanted modified bone marrow-derived mesenchymal stem cells in stroke: a phase 1/2a study. Stroke 47:1817–1824
71. Clarkson AN, López-Valdes HE, Overmann JJ et al (2013) Multimodal examination of structural and functional remapping in the mouse photothrombotic stroke model. J Cereb Blood Flow Metab 33:716–723

72. Hilton BJ, Anenberg E, Harrison TC et al (2016) Re-establishment of cortical motor output maps and spontaneous functional recovery via spared dorsolaterally projecting corticospinal neurons after dorsal column spinal cord injury in adult mice. J Neurosci 36:4080–4092
73. Zeiler SR, Krakauer JW (2013) The interaction between training and plasticity in the poststroke brain. Curr Opin Neurol 26:609–16
74. Dancause N, Nudo RJ (2011) Shaping plasticity to enhance recovery after injury. Prog Brain Res 192:273–295
75. Grefkes C, Fink GR (2011) Reorganization of cerebral networks after stroke: new insights from neuroimaging with connectivity approaches. Brain 134:1264–1276
76. Lozano AM (2017) Waving hello to noninvasive deep-brain stimulation. N Engl J Med 377:1096–1098
77. Grefkes C, Fink GR (2014) Connectivity-based approaches in stroke and recovery of function. Lancet Neurol 13:206–216
78. Nitsche MA, Nitsche MS, Klein CC et al (2003) Level of action of cathodal DC polarisation induced inhibition of the human motor cortex. Clin Neurophysiol 114:600–604
79. Hummel FC, Cohen LG (2006) Non-invasive brain stimulation: a new strategy to improve neurorehabilitation after stroke? Lancet Neurol 5:708–712
80. Paulus W (2003) Transcranial direct current stimulation (tDCS). Suppl Clin Neurophysiol 56:249–54
81. Wassermann EM, Grafman J (2005) Recharging cognition with DC brain polarization. Trends Cogn Sci 9:503–505
82. Elsner B, Kugler J, Pohl M, Mehrholz J (2016) Transcranial direct current stimulation (tDCS) for improving activities of daily living, and physical and cognitive functioning, in people after stroke. Cochrane Database Syst Rev 3:CD009645
83. Howlett OA, Lannin NA, Ada L, McKinstry C (2015) Functional electrical stimulation improves activity after stroke: a systematic review with meta-analysis. Arch Phys Med Rehabil 96:934–943
84. Rossini PM, Burke D, Chen R, Cohen LG et al (2015) Non-invasive electrical and magnetic stimulation of the brain, spinal cord, roots and peripheral nerves: Basic principles and procedures for routine clinical and research application. An updated report from an I.F.C.N. Committee. Clin Neurophysiol 126:1071–1107
85. Nowak DA, Grefkes C, Ameli M, Fink GR (2009) Interhemispheric competition after stroke: brain stimulation to enhance recovery of function of the affected hand. Neurorehabil Neural Repair 23:641–656
86. Murase N, Duque J, Mazzocchio R, Cohen LG (2004) Influence of interhemispheric interactions on motor function in chronic stroke. Ann Neurol 55:400–409
87. Duque J, Murase N, Celnik P et al (2007) Intermanual differences in movement-related interhemispheric inhibition. J Cogn Neurosci 19:204–213
88. Pascual-Leone A, Valls-Sole J, Wassermann EM, Hallett M (1994) Responses to rapid-rate transcranial magnetic stimulation of the human motor cortex. Brain 117(Pt 4):847–858
89. Muellbacher W, Ziemann U, Boroojerdi B, Hallett M (2000) Effects of low-frequency transcranial magnetic stimulation on motor excitability and basic motor behavior. Clin Neurophysiol 111:1002–1007
90. Kim YH, You SH, Ko MH et al (2006) Repetitive transcranial magnetic stimulation-induced corticomotor excitability and associated motor skill acquisition in chronic stroke. Stroke 37:1471–1476
91. Takeuchi N, Chuma T, Matsuo Y, Watanabe I, Ikoma K (2005) Repetitive transcranial magnetic stimulation of contralesional primary motor cortex improves hand function after stroke. Stroke 36:2681–2686
92. Maeda F, Keenan JP, Tormos JM et al (2000) Interindividual variability of the modulatory effects of repetitive transcranial magnetic stimulation on cortical excitability. Exp Brain Res 133:425–430
93. Hallett M (2000) Transcranial magnetic stimulation and the human brain. Nature 406:147–150

94. Deng ZD, Luber B, Balderston NL et al (2020) Device-based modulation of neurocircuits as a therapeutic for psychiatric disorders. Annu Rev Pharmacol Toxicol 60:591–614

95. Sakai K, Yasufuku Y, Kamo T et al (2019) Repetitive peripheral magnetic stimulation for impairment and disability in people after stroke. Cochrane Database Syst Rev 11:CD011968

96. Hao Z, Wang D, Zeng Y, Liu M (2013) Repetitive transcranial magnetic stimulation for improving function after stroke. Cochrane Database Syst Rev 5:CD008862

97. Huang YZ, Edwards MJ, Rounis E et al (2005) Theta burst stimulation of the human motor cortex. Neuron 45:201–206

98. Cardenas-Morales L, Nowak DA, Kammer T et al (2010) Mechanisms and applications of theta-burst rTMS on the human motor cortex. Brain Topogr 22:294–306

99. Perlmutter JS, Mink JW (2006) Deep brain stimulation. Annu Rev Neurosci 29:229–257

100. Miocinovic S, Somayajula S, Chitnis S, Vitek JL (2013) History, applications, and mechanisms of deep brain stimulation. JAMA Neurol 70:163–171

101. Thompson DM, Koppes AN, Hardy JG, Schidt CE (2014) Electrical stimuli in the central nervous system microenvironment. Annu Rev Biomed Eng 16:397–430

102. Lee DJ, Lozano CS, Dallapiazza RF, Lozano AM (2019) Current and future directions of deep brain stimulation for neurological and psychiatric disorders. J Neurosurg 131:333–342

103. Elias GJB, Namasivayam AA, Lozano AM (2018) Deep brain stimulation for stroke: Current uses and future directions. Brain Stimul 11:3–28

104. Levin M (2012) Molecular bioelectricity in developmental biology: new tools and recent discoveries: control of cell behavior and pattern formation by transmembrane potential gradients. Bioessays 34:205–217

105. Vedam-Mai V, Gardner B, Okun MS et al (2014) Increased precursor cell proliferation after deep brain stimulation for Parkinson's disease: a human study. PLoS One 9:e88770

106. Jakobs M, Fomenko A, Lozano AM, Kiening KL (2019) Cellular, molecular, and clinical mechanisms of action of deep brain stimulation-a systematic review on established indications and outlook on future developments. EMBO Mol Med 11:e9575

107. Buhmann C, Huckhagel T, Engel K et al (2017) Adverse events in deep brain stimulation: A retrospective long-term analysis of neurological, psychiatric and other occurrences. PLoS One 12:e0178984

108. Burdick AP, Fernandez HH, Okun MS (2010) Relationship between higher rates of adverse events in deep brain stimulation using standardized prospective recording and patient outcomes. Neurosurg Focus 29:E4

109. Hariz MI, Rehncrona S, Quinn NP et al (2008) Multicenter study on deep brain stimulation in Parkinson's disease: an independent assessment of reported adverse events at 4 years. Mov Disord 23:416–421

110. Okun MS (2012) Deep-brain stimulation for Parkinson's disease. N Engl J Med 367:1529–1538

111. Meidahl AC, Tinkhauser G, Herz DM et al (2017) Adaptive deep brain stimulation for movement disorders: the long road to clinical therapy. Mov Disord 32:810–819

112. Hell F, Palleis C, Mehrkens JH et al (2019) Deep brain stimulation programming 2.0: Future perspectives for target identification and adaptive closed loop stimulation. Front Neurol 10:314

113. Cogan SF (2008) Neural stimulation and recording electrodes. Annu Rev Biomed Eng 10:275–309

114. Kostarelos K, Vincent M, Hebert C, Garrido JA (2017) Graphene in the design and engineering of next-generation neural interfaces. Adv Mater 29:1700909

115. Hong G, Lieber CM (2019) Novel electrode technologies for neural recordings. Nat Rev Neurosci 20:330–345

116. Feiner R, Dvir T (2018) Tissue–electronics interfaces: from implantable devices to engineered tissues. Nat Rev Mater 3:17076

117. Zhou T, Hong G, Fu TM et al (2017) Syringe-injectable mesh electronics integrate seamlessly with minimal chronic immune response in the brain. Proc Natl Acad Sci USA 114:5894–5899

118. Lacour SP, Courtine G, Guck J (2016) Materials and technologies for soft implantable neuroprostheses. Nat Rev Mater 1:16063

119. Frank JA, Antonini MJ, Anikeeva P (2019) Next-generation interfaces for studying neural function. Nat Biotechnol 37:1013–1023
120. Wellman SM, Eles JR, Ludwig KA et al (2018) A materials roadmap to functional neural interface design. Adv Funct Mater 28:170129
121. Yizhar O, Fenno LE, Davidson TJ et al (2011) Optogenetics in neural systems. Neuron 71:9–34
122. Fork RL (1971) Laser stimulation of nerve cells in Aplysia. Science 171:907–908
123. Boyden ES, Zhang F, Bamberg E et al (2005) Millisecond-timescale, genetically targeted optical control of neural activity. Nat Neurosci 8:1263–1268
124. Nagel G, Ollig D, Fuhrmann M et al (2002) Channelrhodopsin-1: a light-gated proton channel in green algae. Science 296:2395–2398
125. Proville RD, Spolidoro M, Guyon N et al (2014) Cerebellum involvement in cortical sensorimotor circuits for the control of voluntary movements. Nat Neurosci 17:1233–1239
126. Tye KM, Deisseroth K (2012) Optogenetic investigation of neural circuits underlying brain disease in animal models. Nat Rev Neurosci 13:251–266
127. Kondabolu K, Kowalski MM, Roberts EA, Han X (2015) Optogenetics and deep brain stimulation neurotechnologies. Handb Exp Pharmacol 228:441–450
128. Delbeke J, Hoffman L, Mols K et al (2017) And then there was light: Perspectives of optogenetics for deep brain stimulation and neuromodulation. Front Neurosci 11:663
129. Cheng MY, Wang EH, Woodson WJ et al (2014) Optogenetic neuronal stimulation promotes functional recovery after stroke. Proc Natl Acad Sci USA 111:12913–12918
130. Tennant KA, Taylor SL, White ER, Brown CE (2017) Optogenetic rewiring of thalamocortical circuits to restore function in the stroke injured brain. Nat Commun 8:15879
131. Shemesh OA, Tanese D, Zampini V et al (2017) Temporally precise single-cell-resolution optogenetics. Nat Neurosci 20:1796–1806
132. Lim DH, Ledue JM, Mohajerani MH, Murphy TH (2014) Optogenetic mapping after stroke reveals network-wide scaling of functional connections and heterogeneous recovery of the peri-infarct. J Neurosci 34:16455–16466
133. Wahl AS, Buchler U, Brandli A et al (2017) Optogenetically stimulating intact rat corticospinal tract post-stroke restores motor control through regionalized functional circuit formation. Nat Commun 8:1187
134. Shah AM, Ishizaka S, Cheng MY et al (2017) Optogenetic neuronal stimulation of the lateral cerebellar nucleus promotes persistent functional recovery after stroke. Sci Rep 7:46612
135. Machado AG, Baker KB, Schuster D et al (2009) Chronic electrical stimulation of the contralesional lateral cerebellar nucleus enhances recovery of motor function after cerebral ischemia in rats. Brain Res 1280:107–116
136. Song M, Yu SP, Mohamad O et al (2017b) Optogenetic stimulation of glutamatergic neuronal activity in the striatum enhances neurogenesis in the subventricular zone of normal and stroke mice. Neurobiol Dis 98:9–24
137. Zhang F, Wang LP, Brauner M et al (2007) Multimodal fast optical interrogation of neural circuitry. Nature 446:633–639
138. Han X (2012) In vivo application of optogenetics for neural circuit analysis. ACS Chem Neurosci 3:577–584
139. Atasoy D, Aponte Y, Su H et al (2008) A FLEX switch targets Channelrhodopsin-2 to multiple cell types for imaging and long-range circuit mapping. J Neurosci 28:7025–7030
140. Kuhlman SJ, Huang ZJ (2008) High-resolution labeling and functional manipulation of specific neuron types in mouse brain by Cre-activated viral gene expression. PLoS One 3:e2005
141. Madisen L, Mao T, Koch H et al (2012) A toolbox of Cre-dependent optogenetic transgenic mice for light-induced activation and silencing. Nat Neurosci 15:793–802
142. Taniguchi H, He M, Wu P et al (2011) A resource of Cre driver lines for genetic targeting of GABAergic neurons in cerebral cortex. Neuron 71:995–1013
143. Mansuy IM, Winder DG, Moallem TM et al (1998) Inducible and reversible gene expression with the rtTA system for the study of memory. Neuron 21:257–265

144. Kawano F, Okazaki R, Yazawa M, Sato M (2016) A photoactivatable Cre-loxP recombination system for optogenetic genome engineering. Nat Chem Biol 12:1059–1064
145. Zhang W, Lohman AW, Zhuravlova Y (2017) Optogenetic control with a photocleavable protein, PhoCl. Nat Methods 14:391–394
146. Gonçalves SB, Ribeiro JF, Silva AF et al (2017) Design and manufacturing challenges of optogenetic neural interfaces: a review. J Neural Eng 14:041001
147. Gong X, Mendoza-Halliday D, Ting JT et al (2020) An ultra-sensitive step-function opsin for minimally invasive optogenetic stimulation in mice and macaques. Neuron 107:38–51
148. Salehpour F, Cassano P, Rouhi N et al (2019) Penetration profiles of visible and near-infrared lasers and light-emitting diode light through the head tissues in animal and human species: A review of literature. Photobiomodul Photomed Laser Surg 37:581–595
149. Hososhima S, Yuasa H, Ishizuka T et al (2015) Near-infrared (NIR) up-conversion optogenetics. Sci Rep 5:16533
150. Bansal A, Liu H, Jayakumar MK et al (2016) Quasi-continuous wave near-infrared excitation of upconversion nanoparticles for optogenetic manipulation of C. elegans. Small 12:1732–43
151. Wang Y, Lin X, Chen X et al (2017) Tetherless near-infrared control of brain activity in behaving animals using fully implantable upconversion microdevices. Biomaterials 142:136–148
152. Lin X, Chen X, Zhang W et al (2018) Core-shell-shell upconversion nanoparticles with enhanced emission for wireless optogenetic inhibition. Nano Lett 18:948–956
153. Shapiro MG, Homma K, Villarreal S et al (2012) Infrared light excites cells by changing their electrical capacitance. Nat Commun 3:736
154. Yong J, Needham K, Brown WG et al (2014) Gold-nanorod-assisted near-infrared stimulation of primary auditory neurons. Adv Healthc Mater 3:1862–1868
155. Chen S, Weitemier AZ, Zeng X et al (2018) Near-infrared deep brain stimulation via upconversion nanoparticle-mediated optogenetics. Science 359:679–684
156. Lin X, Wang Y, Chen X et al (2017) Multiplexed optogenetic stimulation of neurons with spectrum-selective upconversion nanoparticles. Adv Healthc Mater 6:1700446
157. Nizamoglu S, Gather MC, Humar M et al (2016) Bioabsorbable polymer optical waveguides for deep-tissue photomedicine. Nat Commun 7:10374
158. Yu KJ, Kuzum D, Hwang SW et al (2016) Bioresorbable silicon electronics for transient spatiotemporal mapping of electrical activity from the cerebral cortex. Nat Mater 15:782–791
159. Tao H, Hwang SW, Marelli B et al (2014) Silk-based resorbable electronic devices for remotely controlled therapy and in vivo infection abatement. Proc Natl Acad Sci USA 111:17385–17389
160. Koo J, Macewan MR, Kang SK et al (2018) Wireless bioresorbable electronic system enables sustained nonpharmacological neuroregenerative therapy. Nat Med 24:1830–1836
161. Kang SK, Murphy RK, Hwang SW et al (2016) Bioresorbable silicon electronic sensors for the brain. Nature 530:71–76
162. Yu X, Shou W, Mahajan BK et al (2018) Materials, processes, and facile manufacturing for bioresorbable electronics: a review. Adv Mater 30:e1707624
163. Won SM, Song E, Zhao J et al (2018) Recent advances in materials, devices, and systems for neural interfaces. Adv Mater 30:e1800534
164. Ledochowitsch P, Yazdan-Shahmorad A, Bouchard KE (2015) Strategies for optical control and simultaneous electrical readout of extended cortical circuits. J Neurosci Methods 256:220–231
165. Park DW, Schendel AA, Mikael S et al (2014) Graphene-based carbon-layered electrode array technology for neural imaging and optogenetic applications. Nat Commun 5:5258
166. Ganji M, Kaestner E, Hermiz J et al (2017) Development and translation of PEDOT:PSS microelectrodes for intraoperative monitoring. Adv Funct Mater 28:1700232
167. Zhao S, Li G, Tong C et al (2020) Full activation pattern mapping by simultaneous deep brain stimulation and fMRI with graphene fiber electrodes. Nat Commun 11:1788
168. Vitale F, Summerson SR, Aazhang B et al (2015) Neural stimulation and recording with bidirectional, soft carbon nanotube fiber microelectrodes. ACS Nano 9:4465–4474

169. Park DW, Ness JP, Brodnick SK et al (2018). Electrical neural stimulation and simultaneous in vivo monitoring with transparent graphene electrode arrays implanted in GCaMP6f mice. ACS Nano 12:148–157

170. Wang M, Mi G, Shi D et al (2017) Nanotechnology and nanomaterials for improving neural interfaces. Adv Funct Mater 28:1700905

171. Wang K, Fishman HA, Dai H, Harris JS (2006) Neural stimulation with a carbon nanotube microelectrode array. Nano Lett 6:2043–2048

172. Keefer EW, Botterman BR, Romero MI, Rossi AF, Gross GW (2008) Carbon nanotube coating improves neuronal recordings. Nat Nanotechnol 3:434–439

173. Luo X, Weaver CL, Zhou DD et al (2011) Highly stable carbon nanotube doped poly(3,4-ethylenedioxythiophene) for chronic neural stimulation. Biomaterials 32:5551–5557

174. Zhang J, Liu X, Xu E et al (2018) Stretchable transparent electrode arrays for simultaneous electrical and optical interrogation of neural circuits in vivo. Nano Lett 18:2903–2911

175. Fadeel B, Kostarelos K (2020) Grouping all carbon nanotubes into a single substance category is scientifically unjustified. Nat Nanotechnol 15:164

176. Grosse Y, Loomis D, Guyton KZ et al (2014) Carcinogenicity of fluoro-edenite, silicon carbide fibres and whiskers, and carbon nanotubes. Lancet Oncol 15:1427–1428

177. Kurapati R, Mukherjee SP, Martin C et al (2018) Degradation of single-layer and few-layer graphene by neutrophil myeloperoxidase. Angew Chem Int Ed Engl 57:11722–11727

178. Rodrigues AF, Newman L, Jasim DA et al (2018) Immunological impact of graphene oxide sheets in the abdominal cavity is governed by surface reactivity. Arch Toxicol 92:3359–3379

179. Schinwald A, Murphy FA, Jones A, MacNee W, Donaldson K (2012) Graphene-based nanoplatelets: a new risk to the respiratory system as a consequence of their unusual aerodynamic properties. ACS Nano 6:736–46

180. Fabbro A, Scaini D, Leon V et al (2016) Graphene-based interfaces do not alter target nerve cells. ACS Nano 10:615–623

181. Pampaloni NP, Lottner M, Giugliano M (2018) Single-layer graphene modulates neuronal communication and augments membrane ion currents. Nat Nanotechnol 13:755–764

182. Zhang Z, Klausen LH, Chen M, Dong M (2018) Electroactive scaffolds for neurogenesis and myogenesis: graphene-based nanomaterials. Small 14:e1801983

183. Zhao S, Liu X, Xu Z et al (2016) Graphene encapsulated copper microwires as highly MRI compatible neural electrodes. Nano Lett 16:7731–7738

184. Hébert C, Masvida-Codina E, Suarez-Pérez A et al (2017) Flexible graphene solution-gated field-effect transistors: Efficient transducers for micro-electrocorticography. Adv Funct Mater 28:1703976

185. Masvidal-Codina E, Illa X, Dasilva M et al (2019) High-resolution mapping of infraslow cortical brain activity enabled by graphene microtransistors. Nat Mater 18:280–288

186. Park SW, Kim J, Kang M et al (2018) Epidural electrotherapy for epilepsy. Small 14:e1801732

187. Kuzum D, Takano H, Shim E et al (2014) Transparent and flexible low noise graphene electrodes for simultaneous electrophysiology and neuroimaging. Nat Commun 5:5259

188. Savchenko A, Cherkas V, Liu C et al (2018) Graphene biointerfaces for optical stimulation of cells. Sci Adv 4:eaat0351

189. Jiang Y, Li X, Liu B et al (2018) Rational design of silicon structures for optically controlled multiscale biointerfaces. Nat Biomed Eng 2:508–521

190. Parameswaran R, Carvalho-de-Souza JL, Jiang Y et al (2018) Photoelectrochemical modulation of neuronal activity with free-standing coaxial silicon nanowires. Nat Nanotechnol 13:260–266

191. Parameswaran R, Koehler K, Rotenberg MY et al (2019) Optical stimulation of cardiac cells with a polymer-supported silicon nanowire matrix. Proc Natl Acad Sci USA 116:413–421

192. Ghezzi D, Antognazza MR, Maccarone R et al (2013) A polymer optoelectronic interface restores light sensitivity in blind rat retinas. Nat Photonics 7:400–406

193. Honmou O, Houkin K, Matsunaga T et al (2011) Intravenous administration of auto serum-expanded autologous mesenchymal stem cells in stroke. Brain 134:1790–807

194. Kenmuir CL, Wechsler LR (2017) Update on cell therapy for stroke. Stroke Vasc Neurol 2:59–64

195. Yuk H, Lu B, Zhao X (2019) Hydrogel bioelectronics. Chem Soc Rev 48:1642–1667
196. Kamyshny A, Magdassi S (2019) Conductive nanomaterials for 2D and 3D printed flexible electronics. Chem Soc Rev 48:1712–1740
197. Ali-Boucetta H, Nunes A, Sainz R et al (2013) Asbestos-like pathogenicity of long carbon nanotubes alleviated by chemical functionalization. Angew Chem Int Ed Engl 52:2274–2278
198. Sureshbabu AR, Kurapati R, Russier J et al (2015) Degradation-by-design: Surface modification with functional substrates that enhance the enzymatic degradation of carbon nanotubes. Biomaterials 72:20–28
199. Bussy C, Hadad C, Prato M et al (2016) Intracellular degradation of chemically functionalized carbon nanotubes using a long-term primary microglial culture model. Nanoscale 8:590–601
200. Kang SK, Koo J, Lee YK, Rogers JA (2018) Advanced materials and devices for bioresorbable electronics. Acc Chem Res 51:988–998

Chapter 9
Orthotic and Robotic Substitution Devices for Central Nervous System Rehabilitation and Beyond

Raquel Madroñero-Mariscal, Ana de los Reyes Guzmán, Joana Mestre Veiga, Alejandro Babin Contreras, Ángel Gil-Agudo, and Elisa López-Dolado

Abstract During the last decade, the scientific community has enormously advanced in the development of innovative wearable devices, orthoses, and robotic exoskeletons that are revolutionizing rehabilitation in patients who suffer any kind of neurological, neuromuscular, or orthopedical disorder. In this context, different materials and manufacturing design processes are being under investigation with the aim of enhancing the recovery outcomes and the level of functional independence of the patients. In this chapter, current progress in materials applied to the development of orthotics, robotics, and other wearable devices is reviewed, as well as their manufacturing design processes, with a major attention in the rehabilitation of

R. Madroñero-Mariscal
Fundación del Lesionado Medular, Madrid, Spain

Hospital Universitario Infanta Leonor, SERMAS, Madrid, Spain

Laboratory of Interfaces for Neural Repair, Hospital Nacional de Parapléjicos, SESCAM, Toledo, Spain
e-mail: medico.reh@medular.org

A. R. Guzmán
Biomechanics and Technical Aids Laboratory, Hospital Nacional de Parapléjicos, SESCAM, Toledo, Spain
e-mail: adlos@sescam.jccm.es

J. M. Veiga · A. B. Contreras
Center for Prosthetics and Orthotics PRIM, Madrid, Spain
e-mail: j.mestre@prim.es; a.babin@prim.es

A. Gil-Agudo (✉)
Biomechanics and Technical Aids Laboratory, Hospital Nacional de Parapléjicos, SESCAM, Toledo, Spain

Rehabilitation Department, Hospital Nacional de Parapléjicos, SESCAM, Toledo, Spain

Department of Medicine and Medical Specialities, School of Medicine, Universidad de Alcalá, Madrid, Spain
e-mail: amgila@sescam.jccm.es

central nervous system disorders. Specifically, orthoses and exoskeleton devices will be classified according to their manufacturing materials. We will later discuss how material and manufacturing choices affect the features of the resulting devices, indicating both associated advantages and disadvantages, and then condition their clinical applicability along the rehabilitation process and patient recovery. Final considerations and future research directions will be proposed in an attempt to improve the usability of these devices, both during the clinical practice and, even more importantly, in the patients' daily life.

Keywords Dynamic orthosis · Exoskeleton · Manufacturing · Material · Rigid orthosis · Robotic device · Soft orthosis · Wearable device

9.1 Introduction: An Overview of Current Technologies

During the last decade, the scientific community has achieved numerous advances in the development of innovative wearable devices, orthoses, and robotic exoskeletons. This progress is revolutionizing the current way of rehabilitation treatment for patients suffering any kind of neurological, neuromuscular, or orthopedical disorder. In this sense, different materials and manufacturing design processes are being studied and proposed with the aim of enhancing the recovery outcomes and the level of functional independence of these patients. Orthotics refers to devices that provide assistance and support in the context of neurological or orthopedical pathologies, contributing to the stabilization, immobilization, and correct posture and helping to maintain functions and prevent or reduce pain [1]. From a classic perspective, there are two main types of orthoses. On the one hand, rigid orthoses have been made of stiff materials that prevent the voluntary or involuntary motion of the joints. For instance, this type of orthoses has been used in children with dystonia disorders. In this particular case, they are thought to assist the promotion of neuroplasticity processes in the brain by resetting the cortical map to the previous normal topography. This removal of all motor and sensory input to the limbs would help to reorganize control resources in order to accomplish different upper limb motor tasks [2]. On the other hand, soft orthoses are fabricated from materials

E. López-Dolado (✉)
Laboratory of Interfaces for Neural Repair, Hospital Nacional de Parapléjicos, SESCAM, Toledo, Spain

Rehabilitation Department, Hospital Nacional de Parapléjicos, SESCAM, Toledo, Spain

Department of Medicine and Medical Specialities, School of Medicine, Universidad de Alcalá, Madrid, Spain

Research Unit of "Design and development of biomaterials for neural regeneration", Hospital Nacional de Parapléjicos, Joint Research Unit with CSIC, Toledo, Spain
e-mail: elopez@sescam.jccm.es

that allow a more compliant and adaptable performance. Nonetheless, they may not provide the required forces, thus becoming ineffective for patients with a high loss of strength. The main benefits of this type of orthoses are their tolerance for the occurrence of some voluntary movements done by the patient and the hold of some proprioceptive information. When residual motor function is preserved, this type of orthoses can be also simultaneously combined with the use of a tactile, visual, haptic, and auditory bio-feedback system supply, which is thought to enhance the rehabilitation treatment outcomes providing a more accurate way of skilled movements while training with immersive technologies in real time [2]. Depending on the residual muscle strength of the affected limb and the aim of the rehabilitation treatment, it is important to select the most appropriate orthoses within these two types, which are essentially different in their component materials. These materials would provide adaptable response capacity to different exposure loads. Interestingly, Garavagia et al. have described that different forces applied to a muscle promote different myoelectrical responses from the muscle [2]. Specifically, inertial loads prolong the duration of agonist and antagonist muscle contractions, while elastic loads need some static torsional forces in order to maintain the position. External viscous loads reduce the contraction of antagonists, with the particularity of promoting velocity changes that do not cause a change in movement time. Based on this, the electromyogram (EMG) signals of agonist muscles are not affected by the type of load, but those of the antagonist muscles increase with inertial loads and decrease with viscous and elastic ones [2]. As the response to a load is mainly determined by the orthosis component materials, their selection will deeply depend on the focused aim of the rehabilitation treatment. Alternatively, a new trend in the development of orthoses includes dynamic ones. For instance, Garavaglia et al. have developed a dynamic wearable orthosis made of a pseudoelastic Nitinol alloy as a metallic material with non-linear mechanical features [3]. This orthosis provides a dynamic force that promotes changes in the way of controlling different strategies that are easily learned by dystonic patients. Dynamic elastomeric fabric orthoses (DEFOs) are another example of dynamic orthoses with a customized design, useful to treat diseases which can evolve with movement disorders and lack of postural control [4].

Wearable devices are not only being used to create orthoses, but also platforms called e-textiles, which act as a tool for monitoring the rehabilitation process during conventional treatments [5]. For example, wearable monitoring devices for upper limbs integrating electrocardiogram (ECG) and EMG sensors help to analyze parameters such as heart rate frequency and EMG signals. Thanks to this, these devices provide valuable information on patient's alertness, tiredness, and muscle activation in order to establish the best training period for them, contributing to optimize rehabilitation in real time [5, 6].

Embracing more recent and advanced technological progresses, sophisticated and automated electromechanical robot devices have been developed in order to provide a more efficient rehabilitation by increasing the accuracy and intensity of the exercise, offering many repetitions of oriented tasks in a suitable environment and providing bio-feedback from the patient performance in real time. These robots differ from the orthoses mentioned above, as they increase the performance of an unable-bodied wearer by using an external source of energy that will be transformed into motion [1]. They can be classified in: (A) distal end-effector devices [7], which use a single distal point of contact to guide the movement of the entire limb and provide combined movements, being difficult to isolate a pure single movement; and (B) exoskeletons, being their structure located in parallel to the affected limbs, which makes possible to provide a direct control of each segment of the limb by using different motors, also called actuators, fitted at the anatomical axis of the different joints of the patient [8, 9].

Exoskeletons are used for both upper and lower limbs rehabilitation training. Focusing on lower limbs, stationary robots are considered the most stable and safest, as they are usually fixed to either the ground or the wall and are formed by large and powerful actuators and controllers that involve the different leg segments of the patient, who is usually suspended from an overhead guide rail while walking on a treadmill. Within this type of exoskeletons, some of the most commonly employed are Lokomat, Walk-Trainer, LOPES, and ReoAmbulator [10–13]. Moreover, lower limb exoskeletons could be also ambulatory, which are made of a different kind of electrically powered actuators that get adapted to the legs. These ambulatory devices are able to produce joint motion to obtain an automatic overground gait without the need of a partial weight support or a treadmill. For selected spinal cord injured patients (SCIPs), this makes possible to regain the ability of walking short distances outdoors. As these patients usually suffer from balance disorders and a lack of body trunk strength, they typically need technical aids such as crutches to avoid fallings. One of the most popular ambulatory exoskeletons is ReWalk, but we could also cite Indego, Ekso, and HAL [14, 15].

As with orthotics, the search for combining the best patient adaptability, safety, efficiency, and comfort has conduced to the development of different lower limb wearable exoskeletons depending on their actuation, structure, and interface attachment components, thus offering different solutions. Based on their mechanical components, which condition their intrinsic compliance, exoskeletons can be classified into: (1) exoskeletons with compliant actuators (elastic, variably stiff, and pneumatic); (2) exoskeletons with soft structure and rigid actuators; and (3) exoskeletons with soft structure and compliant actuators. Compliant actuators are nowadays under active investigation in an attempt to obtain more lightweight, efficient, and personalized options. Choosing the best structure will dramatically influence patient comfort and the actuation principles of the resulting robotic device. Additionally, it is important to consider the physical interface between the patient and the exoskeleton, as it is involved in the comfort, adaptation, and efficiency of the device. In general terms, soft exoskeleton-based technologies need further research and validation to become a practical solution for these assistive applications when compared to rigid exoskeleton-based technologies. This is due to the lack of a rigid structure responsible for leading soft tissue deformations, which patients are typically not able to control because of their neuromuscular inability to support their weight, derived from their disease-driven lack of motor strength [16]. Nevertheless, soft robots are currently a promising research field, as they provide a more comfortable and natural appearance option for rehabilitation. For instance, Chen et al. have recently developed a soft glove that acts as a soft wearable exoskeleton for patients suffering from dementia and Parkinson's disease [17]. This glove assists an efficient passive rehabilitation treatment aimed to recover hand range of motion and strength by repeating flexion and extension exercises in the hand and fingers [17].

Based on the state-of-the-art of orthotic and robotic devices, we herein review the current materials trends applied in the design and fabrication of orthotics, robotics, and other wearable devices, as well as their manufacturing design processes with a main focus on the rehabilitation of central nervous system (CNS) disorders. We also discuss how materials and manufacturing choices might affect the features of the different devices, conditioning the rehabilitation process and because of that, patient recovery.

9.2   Orthotic Devices

Orthotics includes those devices which provide assistance and support in the context of neurological or orthopedical limb pathologies. Although biomechanical principles leading the manufacturing design processes of orthotic and prosthetic devices have not suffered variations in their fundamental concepts, a remarkable progress on available materials has emerged to improve their fabrication design.

9.2.1   Types of Orthoses Depending On Their Material Structure

From the materials perspective, orthoses can be divided into three main groups, according to the characteristics of their components and resulting mechanical performance: rigid, soft, and other dynamic types. Table 9.1 summarizes the manufacturing materials, main goals, and exemplary clinical applications of these main orthoses.

Rigid Orthoses

Traditionally, rigid orthoses have been made of stiff materials that provide them the capacity of preventing voluntary and/or involuntary joints motions. As mentioned above, they aim to contribute to the stabilization, immobilization, and correct posture of the limb, which can help to maintain its function and prevent or reduce pain [1]. Fracture immobilizations and maintenance of the functional posture of spastic muscles, which tend to adopt contracted, rigid, and painful postures

Table 9.1 Summary of the manufacturing materials, main goals, and exemplary clinical applications of the main types of orthoses according to their material mechanical properties

Orthosis type	Manufacturing materials	Main goals	Exemplary clinical applications
Rigid orthoses	– Three-ply materials with polyurethane foams of different densities – Textile layers to cover and protect those previous foams – Thermoplastic materials – High- or low-density polyethylene materials – Metals (aluminum, steel) – Nickel–titanium shape memory alloys	– To prevent voluntary or involuntary joints motion – To stabilize – To immobilize – To correct limb posture – To prevent or reduce pain	– Fracture immobilizations – Maintenance of functional postures of spastic muscles to avoid rigid and painful contractions
Soft orthoses	Elastic textiles are made of synthetic lineal macromolecules of stiff and soft alternated segments linked by polyurethane bonds – Polycotton Base fabric (made of polyamide, elastane, and cotton) – Powernet fabric (made of nylon and polyurethane filaments) allows to achieve textile reinforcements which are positioned aligned with the muscle chains of the patient body	– To provide a more compliant and adaptable performance – To allow some voluntary movements – To provide some proprioceptive information – To enhance reached movements while training with immersive technologies in real time	Dynamic elastic soft orthoses – dynamic lycra soft orthoses (for neuromuscular disorders and children with cerebral palsy) – dynamic elastic tissue orthoses called DEFO (for diseases with movement alterations and lack of postural control)
Other dynamic orthoses	They combine soft and rigid structures made of a mixture of the materials mentioned above	Orthopedical rehabilitation treatments in which different tissues are involved needing diverse immobilization periods	Hand surgery immobilization involving osteosynthesis in a carpal fracture

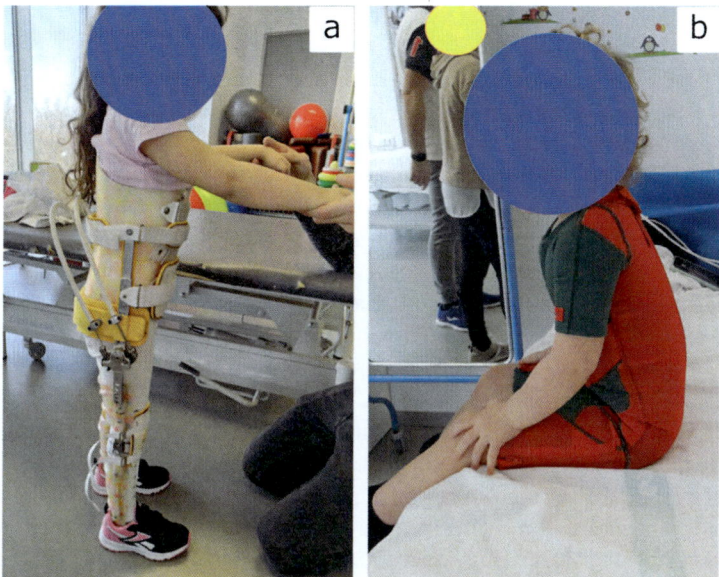

Fig. 9.1 Exemplary images of the main two types of orthoses: Rigid orthosis (**a**) and soft orthosis (**b**)

impairing some activities of daily living (ADLs) such as bathing and clothing, are two clear examples for which rigid orthoses are remarkably useful.

Current trends in the design of rigid orthoses include the use of materials that lead to the thinnest walls in the device, while providing an adequate compression and stabilization of the involved limb tissues. For instance, their structure might mostly consist of three-ply materials composed of polyurethane foams of different densities [18]. Then, these foams are covered and protected by textile layers that are usually fabricated with materials of high softness and breathability to avoid frictions and potential discomfort in patients. Finally, for a greater immobilization, thermoplastic materials such as high- or low-density polyethylene materials, and metallic elements such as aluminum or steel are added to attain the stiff support that characterizes rigid orthoses (Fig. 9.1a). Shape memory alloys made of nickel–titanium have been also proposed to build orthotic, prosthetic, and assistive rehabilitation devices due to their biocompatibility, great ductility, and corrosion resistance. However, these materials show difficulties when acting as an actuator for prosthetics because of their control complexity and temperature needs for a correct function. Even when used for orthotic purposes, they provide a very stiff profile. In spite of these impairments, these shape memory alloys show promise for the fabrication of other assistive devices due to their lightweight, silent operation, and high force properties [19, 20].

Soft Orthoses

Soft orthoses are fabricated from materials able to achieve a more compliant and adaptable performance, although not capable of providing the required forces and then becoming ineffective for patients with a high loss of strength. Dynamic, elastic soft orthoses are usually custom-made and especially useful in neuromuscular diseases with potential for functional recovery. In this context, they promote proximal stability and position of the trunk and limbs, allow for better static and dynamic muscle tone, and decrease involuntary movements. They also facilitate proprioceptive perception, which ultimately drives to greater functionality and independence and increased patient motor skills [21]. Examples in which this type of orthoses are useful include dynamic lycra soft orthoses for neuromuscular disorders and cerebral palsy in children [22] and DEFOs, customized designs to treat diseases which can evolve with movement disorders and lack of postural control [4].

As a general rule, materials used to fabricate this type of orthoses are typically the same than those for rigid ones, but the outer layer of thermoplastics, polyethylene, or metallic materials is missing in order to avoid their characteristic stiffness. It is important to note that there is an increasingly high tendency to substitute the traditional sewing of the different textile parts by introducing new technologies that allow the fabrication of orthoses without couture, providing greater comfort during their use. In the particular case of DEFOs, these are totally customized textile clothing fabricated with elastomers, which are designed to be in direct contact with the skin. They can be made following different designs to cover concrete anatomical regions of the body, or even the whole body, depending on patient needs (Fig. 9.1b) [23]. Two main textiles are being used for DEFOs: Polycotton Base, which is a base fabric made of polyamide, elastane, and cotton, and Powernet, a fabric composed of nylon and polyurethane filaments that allow textile reinforcements aligned with the muscle chains of the patient body [24]. Elastic textiles are made of synthetic linear macromolecules with stiff and soft alternated segments bonded together by polyurethane links. The structural composition given by these materials provides particular mechanical properties, such as biaxial elasticity, that are necessary to enforce biomechanical functions [25].

Dynamic Orthoses

In addition to the soft dynamic orthoses of elastic tissues mentioned above, there are other different dynamic orthoses that are used in the clinical practice, especially for orthopedical rehabilitation treatments. By combining a soft and a rigid structure made of a mixture of the materials mentioned above, these orthoses are useful when dissimilar tissues requiring different immobilization periods are involved. Such is the case of a hand surgery involving osteosynthesis of a carpal fracture, where the wrist may need a longer time of immobilization than the metacarpophalangeal (MCP) or interphalangeal (IP) joints of the fingers. This disparity in the immobilization periods intends to avoid posterior complications of rigidness in finger

Fig. 9.2 Dynamic orthosis that combines a soft and a rigid structure. Courtesy of Kathy Brou and Walter Bureck from the Reconstructive Orthopedic Center at Houston, Texas, USA

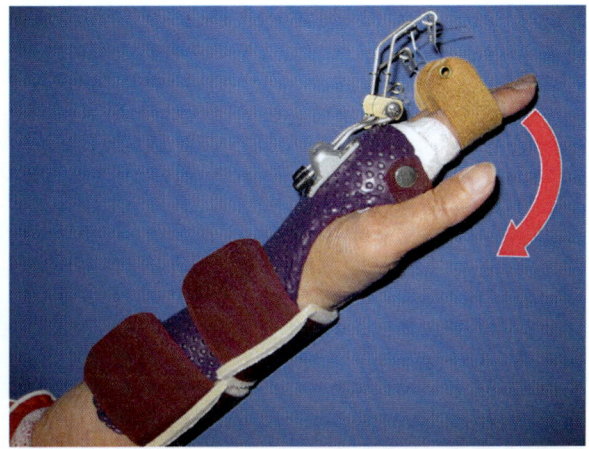

joints. Based on this, the structure of the orthosis for the wrist should be rigid, while the MCP and IP joints should be supported by a softer material that allows the mobilization of the fingers from the beginning of the rehabilitation treatment (Fig. 9.2).

9.2.2   Manufacture Design Process of Orthotic Devices

The need for designing perfectly customized orthotic devices to satisfy patient's necessities for the best comfort, esthetical appearance, and function has prompted the emergence of 3D-printed manufacture design methods. These methods aim to achieve exact precision in the way materials are used in the construction of orthopaedic devices. With the aid of new technologies, it is possible to create digital computerized images by scanning an anatomical part of the body. The images are used afterwards to create the shape of the orthosis device or other functional devices for rehabilitation purposes such as technical aids and prostheses (Fig. 9.3). Besides, this virtual method for obtaining the patient's shape is faster than traditional manufacturing processes, which have geometry limitations that involve the need of subsequent steps for the acquisition of multiple geometries. On the contrary, 3D-printing manufacturing technologies attain final geometries of great freedom and complexity in a unique process [26], being also more pleasant for patients as plaster molds are not needed [27, 28]. One example of the current use of this application is the assistive technology of wrist-driven orthoses (WDOs), used by SCIPs to improve their hand function. Importantly, it is thought that this more personalized design approach can increase patients comfort and then decrease the abandonment of the orthosis in their daily use. Besides, this new technological way of manufacturing orthotics is believed to contribute for a more efficient way of building orthotic devices for orthopedical technicians. Moreover, it reduces their

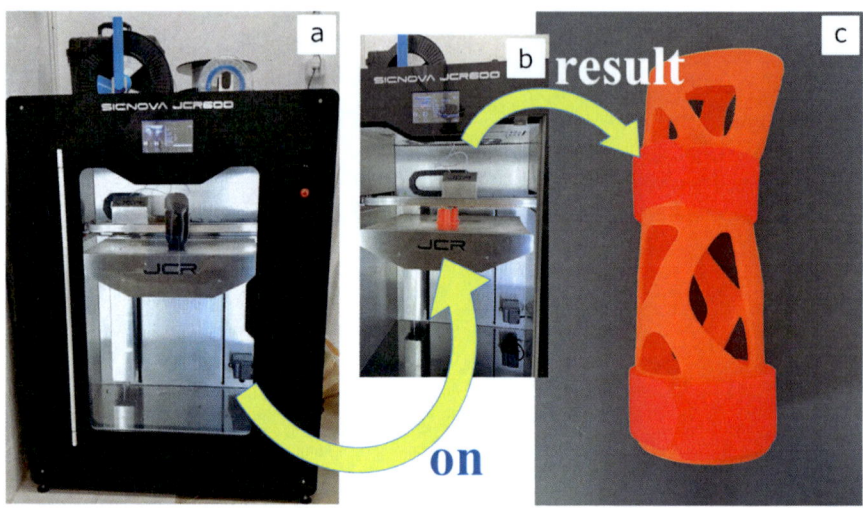

Fig. 9.3 New manufacturing methods based on 3D-printing machines. 3D-printing machine turned off (**a**), 3D-printing manufacture process on going (**b**) and final result of a 3D-printed orthotic device (**c**)

fabrication time, as well as their cost. An example of the utility of these 3D-printed orthotic devices has been published by Portnova et al. [29]. The authors reported functional improvements in SCIPs wearing this type of orthotics, in comparison with traditional ones. Furthermore, production time was reduced 1.5 h, as well as the cost compared to traditional manufacturing methods [29]. Other authors, although claiming that these 3D-printing manufacture processes may be more expensive than traditional ones, still support their use as cost-effective based on the better functional outcomes obtained with a 3D-printed myoelectric hand orthosis in SCIPs [30]. Conversely, Parker et al. found no significant differences in pressure reduction when comparing traditional and digital methods to create a customized orthotic device for the feet of patients at risk of diabetic plantar ulceration, being the digital ones more expensive and requiring more orthopedical technician time to produce it [31].

While 3D-printed solutions seem to be promising for upper limb applications, they may not be as useful for lower limb dysfunction because of the difficulty of building orthotic devices able to support the biomechanical forces caused by the patient weight during walking. Traditional materials considered for this purpose were plastic polymers such as acrylonitrile–butadiene–styrene, polylactic acid, and acrylonitrile–styrene–acrylate, which are too stiff, so they were used for immobilization devices acting as a cast [28]. More recently, the appearance of new materials that combine polyamide with carbon fiber offers a new range of opportunities because of their capacity to support strong forces while providing sufficient consistency and flexibility to obtain acceptable functional outcomes in rehabilitation [26]. This new manufacturing method is not only used for fabricating orthotic devices, but also in the design of active wearable exoskeletons [28].

Encouraged by the worldwide pandemic due to the Coronavirus disease 2019 (Covid-19), some authors have suggested the incorporation of copper nanoparticles to polymers used in the manufacturing design process of medical devices to achieve a protective surface against viruses transmission and infection. The virus SARS-COV-2 has a viability of 4 h in copper surfaces, while surviving 3 days in steel and polypropylene. When virus gets in contact with copper, water diffusion forms ions of copper with reactive oxygen species that are able to damage the virus membrane and later its RNA, thus inactivating the virus [32]. Other microbial agents can be also inactivated by copper. According to this, copper nanocomposites may be of great interest for their incorporation in the manufacturing design process of orthotic devices to protect users from microbial transmission and infection. Indeed, copper ions have been widely explored in the functionalization of materials for biomedical applications due to their beneficial biological effects including osteogenic, angiogenic, and antibacterial properties. Nonetheless, further studies are needed to clearly elucidate the specific actions of these ions on different cells and tissues and the safe range of concentrations to be used in copper-doped implants and devices [33–36].

9.3   Other Wearable Devices and Their Applications

Some material platforms are being used to fabricate other wearable devices, apart from orthotics, also important as rehabilitation tools. Such is the case of e-textiles which enable to monitor the rehabilitation progress during conventional treatments [5]. These advanced textiles provide a solution to hide the technology during wearing, thus improving patient comfort, while offering high levels of functionality. Some examples include wearable monitoring devices for upper limbs, integrating ECG and EMG sensors that help to analyze parameters such as heart rate frequency and EMG signals [5, 6]. This additional information permits to know, more accurately, the degree of alertness, tiredness and muscle activation, fitness, fatigue, endurance level, and gesture in the patients. Consequently, the best training period for each particular patient can be decided, thus contributing to optimize rehabilitation in real time. In a different approach, Zhao et al. developed a physiological monitoring system for upper limb rehabilitation based on a soft robotic rehabilitation glove carrying sensors for ECG and EMG detection. Thanks to the assistance provided by this device, the fatigue status of the patient could be easily identified based on ECG and EMG signals, then decreasing the pressure of the glove to protect muscles from injuries. Contrarily, when the patient is in an excited status, this pressure is increased to optimize training intensity. This manner of optimizing rehabilitation according to the real time status of the patient improves the rehabilitation effects, neuroplasticity, and patient adherence to the treatment [6].

Another example of our close clinical practice is the use of bracelets as wearable devices with sensors to monitor various corporal parameters during rehabilitation by using the Leap Motion Controller. This device is used by free interaction,

Fig. 9.4 Exemplary hand rehabilitation applying virtual reality provided by the Leap Motion Controller device accompanied by a wrist bracelet informing about the status of the patient (e.g. heart rate frequency, pulse oximetry, and muscle activation)

supposing methodological approaches that do not require additional instrumentation with wires and other technologies. The Leap Motion Controller, commonly used for rehabilitation after neurological diseases, is a small, physical, and upward-facing desktop device connected to the computer via USB. It can also be mounted on a virtual reality headset. The device observes a nearly hemispherical area at a distance of about 1 m using three infrared sensors and two cameras. These LED sensors emit patternless infrared light and cameras generate almost 200 frames per second of reflected data. The acquired data is sent via a USB cable to the host computer. The overall calculated average accuracy of the controller is 0.7 mm. Additionally, the Leap Motion Controller allows interaction with some virtual reality applications by executing gestures and hand movements (Fig. 9.4) [37]. With the aid of a bracelet with sensors informing on muscle contraction during the practiced exercises and heart rate frequency or pulse oximetry (i.e. dictating the patient status during exercise), the treatment can be easily optimized.

Other medical fields could also benefit from the use of these e-textiles wearable devices. For instance, textile-based moisture activated batteries and sensors can be used to detect incontinence disorders; cardiovascular diseases can be monitored by using heart rate frequency sensors in wearable devices; pressure sensing sheets could complement clothing in SCIPs to detect the risk of bedsores occurrence; textile-based glucose sensors can monitor glucose levels in the diabetic population; and wearable inert sensors detecting balance could alert about the risk of falling, especially in fragile elderly patients. In any case, before incorporation in the clinical practice, further research to improve sensor integration into textiles, reliable interconnections, regulation, and validation needs to be implemented [5].

Alternatively, silicone is being widely used for the manufacture of multiple orthotic and prosthetic devices, including splints for realigning upper and lower limbs and electrode patches for conducting electrical impulses from neurostimulation devices to muscles innervated by damaged nerves, as well as other orthopedical devices such as podiatric silicones (Fig. 9.5a). Silicone is a synthetic polymer formed by chains of alternate silicon and oxygen atoms, with organic groups attached to silicon. With high purity, its chemical inertness, waterproofing, resistance to oxidation, and stability at both high and low temperatures make silicones the preferred material to manufacture orthotic and prosthetic devices [38]. Within silicones, those vulcanized are the most commonly used. They can be prepared by using two different methods: (1) Vulcanization at low temperatures (i.e. room-temperature-vulcanizing silicones, RTV-silicones) and (2) Vulcanization at high temperatures (i.e. high-temperature-vulcanizing silicones, HTV-silicones). RTV-silicones are liquids of low molecular weight that can be melt and mold in desired shapes, getting hardened at room temperature. RTV-silicone-based elastomers can be, in their turn, fabricated by polycondensation or polyaddition, resulting in polymers with different properties elected depending on the molding technique and the user requirements. Alternatively, HTV-silicones are prepared

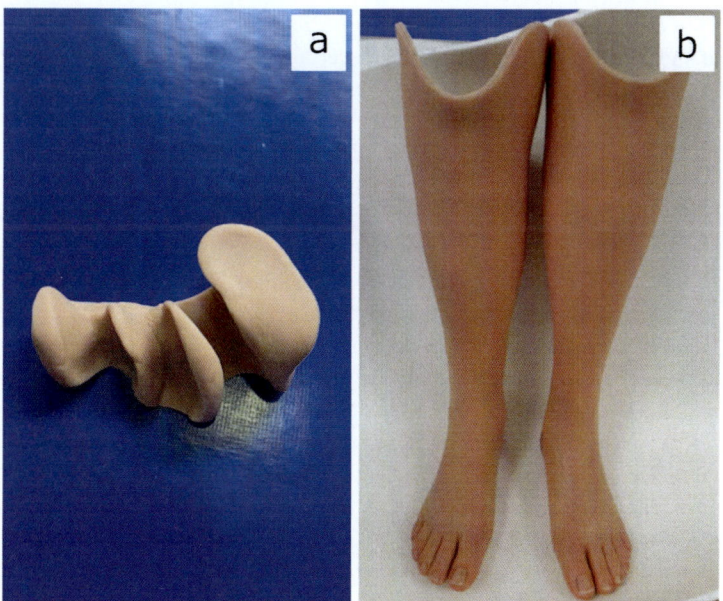

Fig. 9.5 Examples of different types of silicones: (**a**) RTV-silicone for podiatric purposes and (**b**) HTV-silicone hyperrealistic prosthetics for lower limb amputation

from a mixture of HTV-silicone, an inhibitor, a catalyzer, and a pigment, made with the aid of a manual or electronic grille. These silicones are presented as a paste with almost rigid consistency from which different degrees of stiffness could be attained. HTV-silicones have multiple applications for orthotic and prosthetic devices. Indeed, the physical properties of these silicones make them the preferred choice for the manufacture of prosthetic sockets for lower limb amputees. Traditional rigid sockets for these patients used to produce complaints when supporting weight during walking, which resulted in gait disturbances, increased energy consumption, and even psychological alterations that led to the abandonment of the use of the prosthetic device. Contrarily, HTV-silicone sockets allow not only for a comfortable support during walking, but also for better suspension, adhesion, and proprioception of the prosthesis while walking. These silicones are commonly used to manufacture hyperrealistic prosthetics for partial upper and lower limb amputations, with esthetically perfect outcomes (Fig. 9.5b).

Besides, silicone rubbers are useful materials to build soft robots, as they provide a continuous deformation and a large range of motion that rigid materials are not able to attain [39]. Moreover, recent studies support nitride-based materials (e.g. aluminum nitride and silicon nitride) as promising elements to build sensors for robotic devices, as they offer the required flexibility to adapt to deformation and the desired stiffness to avoid breaks. Furthermore, the piezoelectric properties of aluminum nitride confer remarkable electromechanical features when deposited in thin layers, thus making this material the preferred choice to build flexible tactile sensors. On their turn, the use of silicon nitride layers sustains piezo-resistive mechanisms for detecting and measuring mechanical deformations caused by fluid flow and shear forces, thus mimicking, to certain extent, the proprioception system of the biological lateral line organ of fishes. This artificial lateral line system can be a promising tool to test and improve other sensors features such as impulse, sensing ability and feedback control in water environments [40].

9.4   Robotic Devices

In the recent decades, sophisticated and automated electromechanical robotic devices are becoming a solid alternative to provide a more efficient rehabilitation by increasing the accuracy and intensity of the treatment. Different wearable exoskeletons depending on their actuation, structure, and interface attachment components have been developed in an attempt to offer the best patient adaptability, safety, efficiency, and comfort. For instance, AMADEO is an example of an end-effector and rigid stationary robotic device for the rehabilitation of the upper limb. It is a hand and finger robot which does not follow the contours of the human hand, but directly connects to the fingers. Table 9.2 summarizes the main types of exoskeletons and their characteristics. In what follows, we will provide an overview of some of the most remarkable soft and rigid exoskeletons used for traumatic brain and spinal cord injury patients in the clinical practice nowadays. Further details

Table 9.2 Main types of exoskeletons and their characteristics including actuators, interface attachment components, main goals, disadvantages, and examples of clinical applications

Exoskeleton type	Actuators	Interface attachment components	Main goals	Disadvantages	Exemplary clinical applications
Rigid	Compliant actuators – SEAs – VSAs – Pneumatic actuators	– Braces – Cuffs – Straps – Other orthopedic components	To provide an enhanced interaction with security and strong support, protecting from bumps, with consumption of energy, and providing backdrivability during rehabilitation	– Oversized and heavy devices – Motion restriction – Not natural performance of motion – Not natural looking – Discomfort	– Armeo® Spring for upper limb rehabilitation – Exo-H2 and Exo-H3 for lower limb rehabilitation
Soft	Composed of motors with multiple gearbox mechanisms located at the back or waist level	– Braces – Straps	To assist the patient to perform different tasks achieving a more natural performance of their joints movements	Requirement of acceptable balance, body trunk force, and residual limb muscle strength preserved to perform desired tasks	– Carbonhand for upper limb rehabilitation after stroke, multiple sclerosis, or Parkinson's disease – ReStore^{TM} for lower limb rehabilitation after stroke

of those clinically used for the rehabilitation of traumatic brain injury patients are discussed in Chap. 10.

9.4.1 Rigid Exoskeletons

Exoskeletons with compliant actuators usually have rigid structures located in parallel with patient limbs. This rigid structure is heavy, especially for bilateral exoskeletons. There are three types of compliant actuator systems: series elastic actuators (SEAs), variable stiffness actuators (VSAs), and pneumatic actuators. SEAs are the most commonly used actuators. They have elastic elements that do not vary their elasticity and are located in series with the motor. They confer an enhanced interaction, security, consumption of energy, protection from bumps, and backdrivability. Alternatively, VSAs are a novel category derived from SEAs that imitate the human joints stiffness in order to adapt to different ground surfaces and decrease the consumption of energy by modulating the degree of compliance provided. Both, SEAs and VSAs need to function with a defined spring stiffness that will affect the stored and released energy during the gait cycle. Particularly, high stiffness will increase the impedance, while lower ones will reduce the bandwidth. Finally, pneumatic actuators are those which produce forces through compressed air. These actuators have the advantage of low weight and cost and a great backdrivability and power. However, they involve a difficult mechanical design and control due to their high hysteresis and non-linear force-contraction characteristics. Besides, these exoskeletons usually need braces, cuffs, straps, and orthopedical components as interface attachment elements to adapt the exoskeleton structure to the patient [16]. In what follows, we will describe the main features of some of the most relevant rigid exoskeletons used in the clinical practice: Armeo(R) Spring, Exo-H2 and Exo-H3.

Armeo® Spring

Armeo® Spring is a rigid exoskeleton used for upper limb rehabilitation. Its solid structure supports arm weight, so the patient can focus on performing a higher number of movements (i.e. reaching and grasping) to attain specific therapeutic goals by using any remaining motor functions. During these repetitive training tasks, patients are developing their own movements, thus leading to faster results and improving long-term outcomes. This exoskeleton has ergonomic and adjustable braces that cover the whole arm and counterbalance its weight. Importantly, every active movement and joint angles performed during training can be recorded by sensors. Additionally, Armeo® Spring allows simultaneous arm and hand training in an extensive 3D workspace with six degrees of freedom. This enables patients to practice movements with an immerse therapy of virtual reality, thus helping their

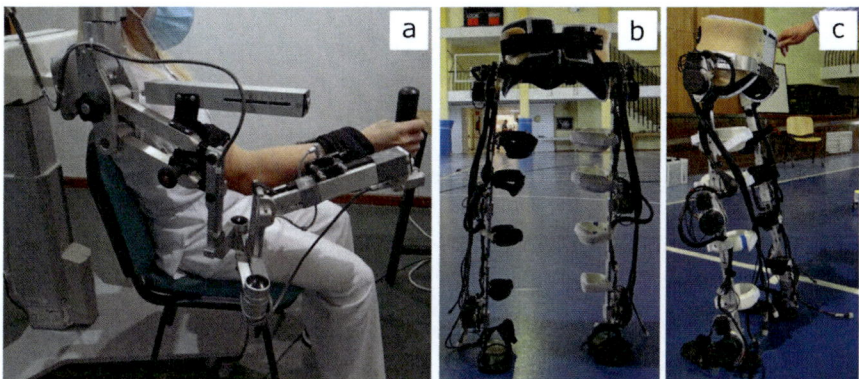

Fig. 9.6 Exemplary rigid robotic exoskeletons: (**a**) Armeo® Spring, used for upper limb rehabilitation; (**b**) Exo-H2 (front view); and (**c**) Exo-H2 (lateral view), used for lower limb rehabilitation

progression in ADLs. Armeo® Spring supports 1D (i.e. joint-specific), 2D and 3D movements (Fig. 9.6a).

Exo-H2 and Exo-H3

Exo-H2 is a rigid exoskeleton for gait rehabilitation training able to provide hips, knees, and ankles assistance. It is energetically supplied by rechargeable batteries and has sensors to control the robotic device while carrying out the elected task (i.e. to walk, to stand up, and to sit down) (Fig. 9.6b, c) [41]. The main difference between this exoskeleton and other previous models is its open architecture design, which allows the physiotherapist or scientist to easily adjust the control parameters of the system (i.e. joint angle, interaction torque, and ground reaction force measurement) for the best performance based on the patient needs. Moreover, the interaction information between the patient and the therapist can be recorded and analyzed. This recorded interaction data can be later used to develop robotic controllers that emulate the physical interaction that physiotherapists provide. With robotic controllers that emulate physiotherapists, patients could receive continuous care at remote locations without direct contact with them. The Exo-H2 exoskeleton emerged as the result of many years of research carried out by the Bioengineering Group at the Consejo Superior de Investigaciones Científicas (CSIC), who is the current proprietary of its know-how rights and has conceded an exclusive license to Technaid S.L. for the design, manufacturing, and commercial exploitation of the system [42].

Exo-H3 is the latest version of the lower limb robotic exoskeleton made by Technaid. This new version incorporates several improvements that make it more versatile, robust, and reliable for research. It has improved security mechanisms such as a power shut-off button and mechanical stops and a wider area for hips attachment. Exo-H3 also provides wider mobility ranges for a more human-like walking performance. It incorporates better electronic integration for an improved form factor, and its connectors have been redesigned for faster transmission. With the aid of six actuated joints in the sagittal plane, Exo-H3 imitates the natural process of human walking. Its main current use is being research for improving neurorehabilitation of patients who have lost their ability to walk after stroke. Its particular design for research provides the implementation of own algorithms and different robotic control strategies. Besides, it allows a great adaptability to different sizes, which is very useful for research purposes. Exo-H3 could be further equipped with an on-board computer as an interface between the internal main controller and a robot operating system network. Finally, the device has an Android app as interface to operate the basic functions of the exoskeleton, including gait speed, motor assistance, and stand up and sit-down commands [43].

9.4.2 Soft Exoskeletons

Soft exoskeletons use soft exosuits, composed of non-rigid components such as neoprene and other flexible textiles, that enable patients who have suffered, for instance, stroke, multiple sclerosis, and Parkinson's disease to improve their gait training. The actuators of this type of exoskeletons are usually integrated by motors with multiple gearbox mechanisms and located at the back or waist levels. This composition allows some torques to the patient joints. Their attachment components, such as braces and straps, are considered a part of the textiles and compliant structure [16].

Traditionally, exoskeletons have been built using rigid structures located in parallel with the different anatomical body segments of the patient in order to provide a strong support that protect patients from falling and physical stress (e.g. helping SCIPs to restore their ability to walk). These rigid exoskeletons are heavy devices and avoid the natural performance of joints movements. Encouraged by these limitations, researchers are developing new soft exoskeletons that, instead of containing stiff materials in their structure, are composed of designed lighter fabrics. This fact allows the patient to develop motion without any restriction, avoiding joint misalignment troubles and looking for a more natural performance of their joints movements. These lighter structures are also able to generate forces that help the patient to perform different tasks, achieved by mechanical actuators that are embedded in the suit [44]. The main disadvantage of these soft structures may be that the patients need to preserve an acceptable balance and body truck force to provide stability, and residual involved limb muscles strength to develop the different tasks with the aid of these robots. If the patients do not preserve these features, the forces provided by this exoskeleton may not be sufficient to perform the desired tasks, as it may happen in SCIPs with an ASIA Impairment Scale (AIS) classification grade A or B and complete hemiplegic patients after suffering brain

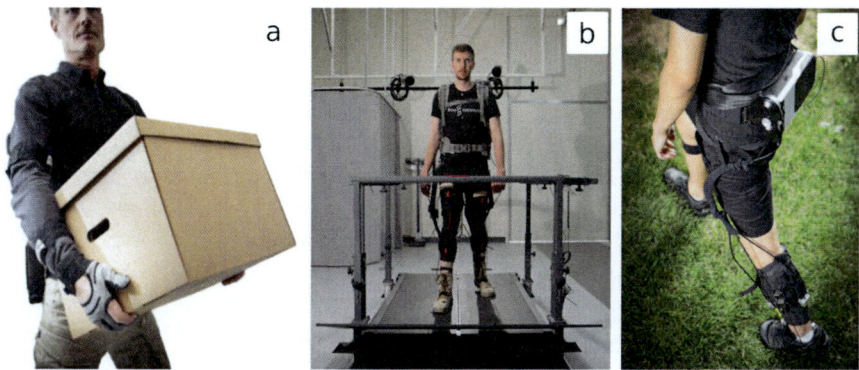

Fig. 9.7 Exemplary soft robotic exoskeletons: (**a**) Carbonhand allowing back-to-work and ADLs assistance. Reproduced with permission from Bioservo Technologies AB [45]. (**b**) ReStore^{TM} for gait training while walking on a treadmill; and (**c**) ReStore^{TM} for gait training outdoors. Reproduced with permission from the Wyss Institute for Biologically Inspired Engineering, Harvard John A. Paulson School of Engineering and Applied Sciences, and Boston University's College of Health and Rehabilitation Sciences Sargent College

stroke. In what follows, we will describe the main features of some of the most relevant soft exoskeletons in the market: CarbonHand and ReStoreTM.

Carbonhand

Carbonhand is a soft exoskeleton designed for upper limb rehabilitation. It consists of a grip-enhancing powered glove designed to facilitate a normal movement pattern of the hand and then aiding back-to-work and ADLs assistance (Fig. 9.7a). It has a total weight of 700 g. When this exoskeleton detects that the user needs more strength to develop a task, it provides additional force and reinforce the patient's hand function without controlling the wearer, being the perfect complement to help a weak grasp. It is an intuitively controllable solution made up of a glove with embedded sensors, transmission cables, and a power and control unit that can be easily adjusted at the maximum added strength for each user by using a smartphone app. The glove and cables are available in multiple sizes. It contains three separate motors, one for each actuated finger, which are housed together with the controller and rechargeable batteries in an adjustable pouch [45].

ReStore^{TM}

ReStore^{TM} is a soft exosuit which is safe and reliable for use during post-stroke gait rehabilitation. In patients with post-stroke hemiparesis, ReStore^{TM} acts to modify the user walking pattern by providing ankle plantarflexion and dorsiflexion assistive forces in parallel with the underlying paretic muscles. This allows the wearer to perform a more symmetrical walking while using less energy, which produces a more efficient walking. The exosuits are anchored to the affected limb through a functional garment. This anchorage provides supplementary force to the ankle joint during walking by transferring mechanical energy via a cable transmission from battery-powered actuators integrated into a belt anchored to the hips or externally placed in a box adjacent to the treadmill. The use of this exoskeleton has demonstrated good results while walking on a treadmill at the physiotherapy clinical room (Fig. 9.7b) and also at patient's home and outdoors (Fig. 9.7c) [46, 47]. This type of exoskeletons has been developed at the Wyss Institute for Biologically Inspired Engineering, the Harvard John A. Paulson School of Engineering and Applied Sciences, and Boston University's College of Health and Rehabilitation Sciences Sargent College.

9.5   Final Considerations and Future Research Directions

Rigid exoskeletons could have the advantage of improving safety avoiding fallings when an unexpected bump or irregular surface is presented in real life. However, this type of exoskeletons may not be as safe as it appears as their spring mechanism could release an unexpected excessive energy when those previous situations occur, not being able to avoid a patient lack of balance and falling. Patients who use these rigid exoskeletons for ambulatory rehabilitation always need technical aids such as two crutches to prevent falls. Furthermore, these devices are usually oversized and heavy, and sometimes they provide an inappropriate physical contact between the surface of the patient limbs and their structure, thus leading to injuries and discomfort.

Because of all these reasons, rigid exoskeletons may be the preferred exoskeletons to train with the patient in a clinical supervised environment, but not at home. So, there is a clear need to develop more comfortable and safer tools that permit rehabilitation training at home, especially during periods such as the current Covid-19 worldwide pandemic, which is causing the isolation of people. In this scenario, soft robotic designs are a promising technology tool to optimize rehabilitation, especially for ambulatory purposes, as they are slighter and more comfortable devices with a more natural appearance. However, some drawbacks need to be addressed before replacing rigid exoskeletons by soft ones. For instance, patients who do not retain some walking ability or do not maintain their own body weight support would not be able to use soft exoskeletons, as they would not be able to provide the stability and walking augmentation required [16, 39]. In the case of gait training rehabilitation exoskeletons for lower limbs, future research should explore soft structures able to assist upper joints at the lower limbs, such as knees and hips, as, for the time being, the current ReStore^{TM} only assists ankle joints [46, 47]. Whenever further investigations find the formula to solve all these limitations, soft exoskeletons could be perfectly customized to fit the patient as a second skin, which is expected to enhance rehabilitation outcomes and patient satisfaction [1].

Another important factor that may impair the correct function of current orthotic and robotic devices is the inexistence of a real connection between the device and the patient CNS. In this sense, future research should focus on the development of systems such as neural implants that directly allow patients to know the kinetic and kinematic information from the wearable device, imitating human proprioception sensation, in order to achieve an immediate feedback that permits to correct posture and motion [1]. Perfect implantable biomaterials acting as artificial sensors of the human nervous system, able to operate in a safe, immediate, and fault-free connection with the orthotic and robotic devices, do not yet exist. Therefore, we believe that it would be beneficial for our scientific and clinical community to review some of the current knowledge about engineered biomaterials. This has been the main rationale for the compilation of this chapter. Likewise, there is a clinical demand to improve the materials that conform external orthotic and robotic devices, to achieve a better connection with implantable biomaterials and to optimize the comfort, external appearance, and functional rehabilitation outcomes that are expected when a patient is wearing these devices. Finally, it is worth noting that this chapter aimed to prove that engineering manufacturing designs and materials for external rehabilitation devices should be considered as important as engineering implantable biomaterials and that their characteristics should be taken into account together with their design, as different parts of the same construction.

Acknowledgements Authors would like to thank Professor Ismael Payo (Department of Electrical Engineering, School of Electrical, Electronic and Aeronautical Engineering, University of Castilla-La Mancha, Spain) for the critical review of the chapter.

References

1. Herr H (2009) Exoskeletons and orthoses: classification, design challenges and future directions. J Neuroeng Rehabil 6:21
2. Garavaglia L, Pagliano E, Baranello G, Pittaccio S (2018) Why orthotic devices could be of help in the management of movement disorders in the young. J Neuroeng Rehabil 15:118
3. Garavaglia L, Pagliano E, Arnoldi MT et al (2017) Two single cases treated by a new pseudoelastic upper-limb orthosis for secondary dystonia of the young. 2017 International Conference on Rehabilitation Robotics (ICORR), 1260–1265
4. Matthews MJ, Payne C, Watson M (2011) The use of a dynamic elastomeric fabric orthosis to manage painful shoulder subluxation: A case study. J Prosthet Orthot 23:155–158
5. Yang K, Isaia B, Brown LJE, Beeby S (2019) E-textiles for healthy ageing. Sensors 19:4463
6. Zhao S, Liu J, Gong Z et al (2020) Wearable physiological monitoring system based on electrocardiography and electromyography for upper limb rehabilitation training. Sensors 20:4861
7. Hesse S, Waldner A, Tomelleri C (2010) Innovative gait robot for the repetitive practice of floor walking and stair climbing up and down in stroke patients. J Neuroeng Rehabil 7:30
8. Krebs HI, Hogan N, Aisen ML, Volpe BT (1998) Robot aided neurorehabilitation. IEEE Trans Rehabil Eng 6:75–87

9. Krebs HI, Conroy SS, Bever CT, Hogan N (2012) Forging mens et manus: The MIT experience in upper extremity robotic therapy. In Dietz V, Nef T, Zev Rymer W (eds) Neurorehabilitation Technology (pp 125–140). Springer, London

10. Colombo G, Joerg M, Schreier R, Dietz V (2000) Treadmill training of paraplegic patients using a robotic orthosis. J Rehab Res Dev 37:693–700

11. Mantone J (2006) Getting a leg up? Rehab patients get an assist from devices such as Health South's AutoAmbulator, but the robot's clinical benefits are still in doubt. Mod Healthc 36:58–60

12. Fleerkotte BM, Koopman B, Buurke JH et al (2014) The effect of impedance-controlled robotic gait training on walking ability and quality in individuals with chronic incomplete spinal cord injury: an explorative study. J Neuroeng Rehabil 11:26

13. Nam KY, Kim HJ, Kwon BS et al (2017) Robot-assisted gait training (Lokomat) improves walking function and activity in people with spinal cord injury: a systematic review. J Neuroeng Rehabil 14:24

14. Esquenazi A, Talaty M, Packel A, Saulino M (2012) The ReWalk powered exoskeleton to restore ambulatory function to individuals with thoracic-level motor-complete spinal cord injury. Am J Phys Med Rehabil 91:911–921

15. Contreras-Vidal JL, Bhagat NA, Brantley J et al (2016) Powered exoskeletons for bipedal locomotion after spinal cord injury. J Neural Eng 13:031001

16. Sánchez-Villamañan MC, Gonzalez-Vargas J, Torricelli D et al (2019). Compliant lower limb exoskeletons: a comprehensive review on mechanical design principles. J Neuroeng Rehabil 16:55

17. Chen Y, Tan X, Yan D et al (2020) A composite fabric-based soft rehabilitation glove with soft joint for dementia in Parkinson's disease. IEEE J Transl Eng Health Med 8:1400110

18. Kumar B, Noor N, Thakur S et al (2019) Shape memory polyurethane-based smart polymer substrates for physiologically responsive, dynamic pressure (re)distribution. ACS Omega 4:15348–15358

19. Nematollahi M, Baghbaderani KS, Amerinatanzi A et al (2019) Application of NiTi in assistive and rehabilitation devices: A review. Bioengineering 6:37

20. Chaudhari R, Vora JJ, Patel V et al (2020) Surface analysis of wire-electrical-discharge-machining-processed shape-memory alloys. Materials 13:530

21. Attard J, Rithalia S (2010) Physiological effects of Lycra® pressure garments on children with cerebral palsy. In: Anand SC, Kennedy JF, Miraftab M, Rajendran S (ed) Woodhead publishing series in textiles, medical and healthcare textiles. Woodhead Publishing, Sawston, pp 300–308

22. Attfield SF, Nicholson J, Morton RE (2009) Evaluation of stability of Lycra soft orthoses using 3D kinematic analysis. Orthopädie Technik, edition IV 1–7

23. Matthews MJ, Watson M, Richardson B (2009) Effects of dynamic elastomeric fabric orthoses on children with cerebral palsy. Prosthet Orthot Int 33:339–347

24. Inoue M, Tange A, Niwa M (2013) Theoretical analysis of biaxial tensile properties of power net. Text Res J 83:1319–1324

25. Matthews M, Blandford S, Marsden J, Freeman J (2016) The use of dynamic elastomeric fabric orthosis suits as an orthotic intervention in the management of children with neuropathic onset scoliosis: A retrospective audit of routine clinical case notes. Scoliosis Spinal Disord 11:14

26. García-Dominguez A, Claver J, Sebastián MA (2020) Integration of additive manufacturing, parametric design, and optimization of parts obtained by fused deposition modeling (FDM). A methodological approach. Polymers 12:1993

27. Telfer S, Pallari J, Munguia J et al (2012) Embracing additive manufacture: implications for foot and ankle orthosis design. BMC Musculoskeletal Disord 13:84

28. Barrios-Muriel J, Romero-Sánchez F, Alonso-Sánchez FJ, Salgado DR (2020) Advances in orthotic and prosthetic manufacturing: A technology review. Materials 13:295

29. Portnova AA, Mukherjee G, Peters KM et al (2018) Design of a 3D-printed, open-source wrist-driven orthosis for individuals with spinal cord injury. PloS One 13:e0193106

30. Yoo HJ, Lee S, Kim J et al (2019) Development of 3D-printed myoelectric hand orthosis for patients with spinal cord injury. J Neuroeng Rehabil 16:162

31. Parker DJ, Nuttall GH, Bray N et al (2019) A randomised controlled trial and cost-consequence analysis of traditional and digital foot orthoses supply chains in a National Health Service setting: application to feet at risk of diabetic plantar ulceration. J Foot Ankle Res 12:2
32. Zuniga JM, Cortes A (2020) The role of additive manufacturing and antimicrobial polymers in the COVID-19 pandemic. Expert Rev Med Devices 17:477–481
33. Li K, Xia C, Qiao Y, Liu X (2019) Dose-response relationships between copper and its biocompatibility/antibacterial activities. J Trace Elem Med Biol 55:127–135
34. Song J, Jin P, Li M et al (2019) Antibacterial properties and biocompatibility in vivo and vitro of composite coating of pure magnesium ultrasonic micro-arc oxidation phytic acid copper loaded. J Mater Sci Mater Med 30:49
35. Cao Q, Li J, Wang E (2019) Recent advances in the synthesis and application of copper nanomaterials based on various DNA scaffolds. Biosens Bioelectron 132:333–342
36. Ahmedova A, Todorov B, Burdzhiev N, Goze C (2018) Copper radiopharmaceuticals for theranostic applications. Eur J Med Chem 157:1406–1425
37. Anitha A, Iswariya K, Karunya S (2016) A survey on next generation in revolution—Leap Motion. Int J Trend Res Develop 3:275–276
38. Wang TX, Renata C, Chen HM, Huang WM (2017) Elastic shape memory hybrids programmable at around body-temperature for comfort fitting. Polymers 9:674
39. Mahon ST, Roberts JO, Sayed ME et al (2018) Capability by stacking: the current design heuristic for soft robots. Biomimetics 3:16
40. Abels C, Mastronardi VM, Guido F et al (2017) Nitride-based materials for flexible MEMS tactile and flow sensors in robotics. Sensors 17:1080
41. Gil-Agudo A, Ama-Espinosa AJ, Lozano-Berrio V et al (2020) Terapia robótica con el exoesqueleto H2 en la rehabilitación de la marcha en pacientes con lesión medular incompleta. Una experiencia clínica. Rehabilitación 54:87–95
42. Robotic exoskeleton Exo-H2 (2021). Technaid. https://www.technaid.com/products/robotic-exoskeleton-exo-exoesqueleto/. Cited 30 Sep 2021
43. Robotic exoskeleton Exo-H3 (2021). Technaid. https://www.technaid.com/products/robotic-exoskeleton-exo-exoesqueleto-h3/. Cited 30 Sep 2021
44. Soft exosuits for lower extremity mobility (2021). Wyss Institute and Harvard University. https://wyss.harvard.edu/technology/soft-exosuits-for-lower-extremity-mobility/. Cited 30 Sep 2021
45. Exoskeleton Report: Carbonhand. https://exoskeletonreport.com/product/carbonhand/. Cited 30 Sep 2021
46. Soft exosuit for post-stroke gait re-training. https://wyss.harvard.edu/media-post/soft-exosuit-for-post-stroke-gait-re-training/. Cited 30 Sep 2021
47. Awad LN, Esquenazi A, Francisco GE, et al (2020) The ReWalk ReStore^{TM} soft robotic exosuit: a multi-site clinical trial of the safety, reliability, and feasibility of exosuit-augmented post-stroke gait rehabilitation. J Neuroeng Rehabil 17:80

Chapter 10
Robotics and Virtual Reality Exer-Games for the Neurorehabilitation of Children and Adults with Traumatic Brain Injury: The IS-BRAIN Model

Pedro A. Serrano, Teresa Criado, Virginia Aranda, Nayra Fernández-Pinedo, Andrea Riendas, Miriam M. Sevilla, Cristina Zafra, Ana Calvo-Vera, and Ignacio Calvo-Arenillas

Abstract In this chapter, we describe the evolving synergistic model IS-BRAIN®, used for a decade in our Robotic Unit at the Brain Damage Unit in the Hospital Beata María Ana (Spain). Conventional approaches with robotics, virtual reality, and exergame-based methodologies are described for their application in children and adults with traumatic brain injury and discussed based on the most current evidence. Our model justifies an adjuvant intervention of balance, stability, lower limbs, gait, upper limbs, and hand in acquired brain injuries.

P. A. Serrano (✉)
Brain Damage Unit, Hospital Beata María Ana, Red Menni Daño Cerebral, Madrid, Spain

Occupational Therapy Deparment, Faculty of Health Sciences, Centro Superior de Estudios Universitarios La Salle, Universidad Autónoma de Madrid, Madrid, Spain

Occupational Thinks Research Group, Institute of Neuroscience and Sciences of the Movement (INCIMOV), Centro Superior de Estudios Universitarios La Salle, Universidad Autónoma de Madrid, Madrid, Spain

Universidad de Castilla La Mancha, Talavera de la Reina, Toledo, Spain

Facultad de Fisioterapia y Enfermería, Universidad de Salamanca, Salamanca, Spain
e-mail: pedrselo@lasallecampus.es

T. Criado · V. Aranda · A. Riendas · M. M. Sevilla
Brain Damage Unit, Hospital Beata María Ana, Red Menni Daño Cerebral, Madrid, Spain

N. Fernández-Pinedo
Children Rehabilitation Unit, Hospital Beata María Ana, Red Menni Daño Cerebral, Madrid, Spain

C. Zafra
Universidad de Castilla La Mancha, Talavera de la Reina, Spain

A. Calvo-Vera · I. Calvo-Arenillas
Facultad de Fisioterapia y Enfermería, Universidad de Salamanca, Salamanca, Spain
e-mail: abcvera@usal.es; calvoreh@usal.es

E. López-Dolado, M. Concepción Serrano (eds.), *Engineering Biomaterials for Neural Applications*, https://doi.org/10.1007/978-3-030-81400-7_10

Keywords Balance · Gait training · Hand robotics · Lower limb robotics ·
Neurorehabilitation · Occupational therapy · Physiotherapy · Traumatic brain
injury · Upper limb robotics · Virtual reality

10.1 Introduction

As a consequence of our increasingly modern urban society with continuous
displacements, the incidence of head injuries (HI) in adults due to work, traffic,
or home accidents is augmenting. In children, there is also an increase in accidents
during play and community trips. Worldwide, HI are considered one of the leading
causes of disability due to neurological disease [1]. Indeed, they are the fourth
cause of death and the second cause of disability among young people [2]. People
who have suffered traumatic brain injury (TBI) require expensive and specialized
interventions, in addition to variable hospital rehabilitation periods [3]. The recovery
time frame from a TBI depends on the severity of the brain trauma and the needs
of care received by the patient. The duration of the rehabilitation process can take
from 6 months to years [1].

TBI can be defined as a disturbance of brain function caused by an external
force, such as a direct impact on the skull, a penetration of a foreign object, a rapid
acceleration or deceleration, or shock waves from an explosion. The severity of the
HI depends on the nature, intensity, duration, and direction of the external force that
causes it, giving rise to three types: mild, moderate, and severe [2]. TBI can cause
physical, cognitive, and even emotional problems in the patient. Therefore, people
who have suffered a TBI will require specific rehabilitation services [1].

Two types of lesions occur right after TBI. First, the primary injury takes place at
the moment of impact, followed by the secondary injury. This latter one gives rise to
functional, structural, cellular, and molecular alterations that lead to disturbances in
the affected area of the brain. After this event, fractures associated to TBI constitute
a serious clinical problem, and their isolated rehabilitation from neurological injury
might result impossible. The consolidation of traumatic fractures is occasionally
concurrent with neurological rehabilitation, so professionals must cope with external
bone fixators, cerclages, and unresolved craniofacial and/or limb fractures.

In this chapter, we present the Interventional Synergistic model Based on exergamed Robotics to Augment Intensity in Neurorehabilitation (IS-BRAIN) as a novel
and customized tool to approach neurorehabilitation for TBI patients. This is an own
model evolved from experimental evidences of clinical practices carried out since
2010 at the Brain Damage Unit (BDU) of the Hospital Beata Maria Ana (HBMA)
toward the current intervention in our Robotic Unit (Fig. 10.1). The IS-BRAIN
model, conceived by expert occupational therapists (OTs) and physiotherapists
(PTs) (Serrano and Criado, authors of this chapter) in neurologic diseases, focuses
on the use of new exer-gamed technologies as an adjunct therapy for rehabilitation in
acquired brain damage (ABD) in order to improve lower limbs (LLs), upper limbs
(ULs), hand dexterity, stability, and balance functions. It is important to note that

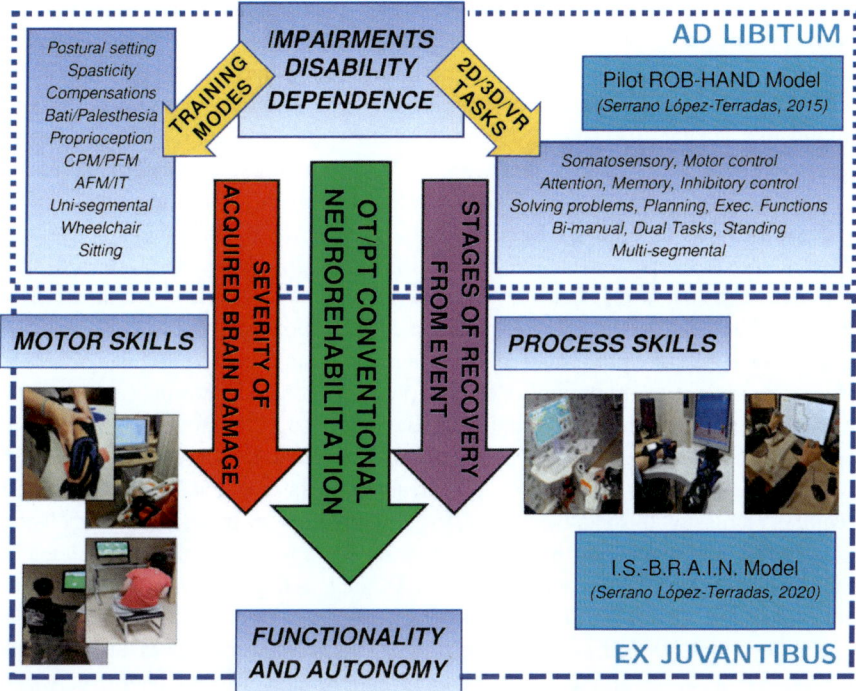

Fig. 10.1 Diagram of the current IS-BRAIN model used in the Robotics Unit of the BDU-HBMA. AFM: Active-facilitated movement, CPM: continuous passive movement, IT: interactive therapy based on exer-games, PFM: passive-facilitated movement, 2D: two dimensions, 3D: three dimensions, VR: virtual reality

this is a model in continuous evolution, in which every year the most innovative virtual reality (VR) and game-based robotic neurorehabilitation techniques and technologies are incorporated and daily implemented by expert professionals in clinical, teaching, and research practices in our Robotics Unit. Adult and child patients, with variable ABD etiology such as stroke, tumors, TBI, or central nervous system (CNS) infections like the recent COVID-19, can benefit from this synergistic neurorehabilitation model. The adaptability, feasibility, and safety of the IS-BRAIN model to multiple etiologies and profiles of patients, like current times of the SARS-COV-2 pandemic, will be extensively discussed in detail below.

10.2 Sensorimotor and Cognitive-Behavioral Disorders After Traumatic Brain Injury

TBI patients often experience multiple and complex sensorimotor and cognitive-behavioral changes as a consequence of the HI, since earliest stages of evolution. Specifically, the miscellany of sequelae produced by TBI are very diverse and can be grouped into cognitive problems (e.g., post-traumatic amnesia, attention problems, neglect, memory and new learning problems, disexecutive functions, agnosia, apraxia, lack of awareness of the deficit), behavioral and neuropsychiatric symptoms (e.g., apathy, lack of initiative or insight, agitation, depression, impulsivity, irritability, aggressiveness), motor problems (e.g., paresis, muscle tone disorders, amyotrophy, spasticity, poor coordination, or dexterity), sensory impairments (e.g., hypoesthesia, dysesthesia, neuropathic pain, anosmy), speech and swallowing disorders (e.g., aphasia, dysphagia), balance and coordination dysfunction (e.g., permanent feeling of dizziness, instability, ataxia), and sleep disorders (e.g., insomnia, sleep apnea), among others. Nonetheless, within all those, the most frequent and disabling sequelae in TBI are motor and behavioral disorders. The most common motor disorders either decrease (i.e., hypotonia, paresis or plegia), increase (i.e., hypertonia), or alter (i.e., spasticity) the normal functioning of muscle strength in the affected limbs [2]. Behaviors such as impulsivity, irritability, aggressiveness, or personality changes, compared to motor ones, can lead to behavioral explosions or akathisia. The presence of these last alterations eventually requires admission to specialized units to avoid involuntary injuries to themselves or their family members until these conditions are solved. These behaviors can lead to the abandonment of the most basic daily tasks, such as hygiene or eating, so hospital admission and the application of multidisciplinary treatments become necessary.

On their turn, sensorimotor manifestations can be very eclectic in TBI. Depending on the location of the lesion and the subsequent affected neural networks, unilateral, bilateral, or crossed sensorimotor affectations can be identified, as well as a combination of central and peripheral lesions, in the same patients. In very early stages, we can find a very severe and flaccid motor impairment or spasticity, depending on the involvement of the reticulo-spinal and cortico-spinal pathways, states of psychomotor restlessness associated with periods of post-traumatic amnesia (PTA), or even delusional body perceptions (e.g., somatoparaphrenias) related to hallucinations or phantom sense of haptic or visual illusory own-body movements [4], including false perceptions of unbelonging of body segments in the body scheme. These patients tend to self-represent using the human figure test (HFT) with multiple arms, with separate segments of the body or to endow themselves with a name or own life (Fig. 10.2). The "hit and kick effect" can produce miscellanea of alterations in voluntary motor control by affecting two contralesional or opposite cortical territories. As a result of apraxia, the spatiotemporal organization of movements can be compromised, affecting whole daily motor behaviors such as swallowing, speech, dressing, grooming, gait, or community movement. Action planning disorders, speech disorders, and upper/lower limb kinetic apraxia are frequent in corticofugal tract injury from the secondary motor area lesions following TBI [5–8]. Moreover, spinal cord injury (SCI) and TBI can co-exist in 11–29% of SCI patients, who are therefore at risk of limb apraxia. People with SCI and TBI, or both, must be trained on powered assistive technologies or robotics, which amplify movements, and on their ability to learn complex motor strategies (i.e., limb praxis), all being critical to future activities of daily living (ADL) performance and function [9].

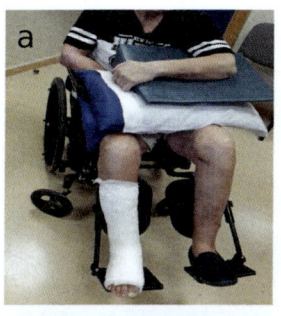
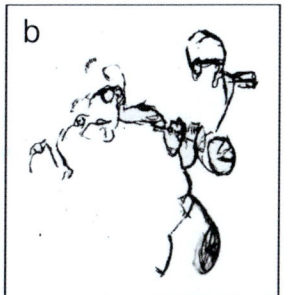

TBI Patient 1, says:

"This is my body draw (...). After the motorcycle accident, my arm and leg are very bad (...). The surgeons chose to put my brother´s arm and leg on my body right half side."

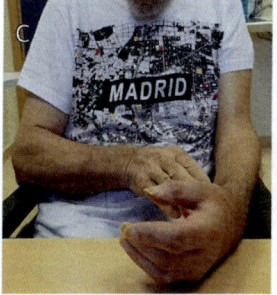
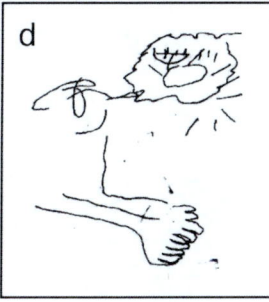

TBI Patient 2, says:

"This is my body draw (...). I have pain in an arm that is not mine (...). I have three arms, I have one left over (...). I lost my left arm 4 months ago, in the trunk of the last car that I trasported."

Fig. 10.2 Exemplary delusional self-referential explanations in patient 1 (**a, b**) and patient 2 (**c, d**) suffering somatoparaphrenia after TBI

The involvement after TBI of the intermediate and cortical connections areas, such as the corpus callosum and the supplementary motor area, can produce the interhemispheric disconnection syndrome [10], alien hand behaviors, or inter-manual diagnostic interference [11–14], compromising the performance of ADL [15, 16].Gripping behaviors, opposite movements to those desired or uncontrolled automatic movements, can appear during the performance of ADL (Fig. 10.3a–c). The same motor behaviors, such as motor magnetization, difficulty in releasing objects, involuntary manipulation, motor oppositionism, or difficulty in executing motor commands voluntarily, can appear during functional and standardized tests with alien hands (Fig. 10.3d–g). In order to improve optimal motor re-learning, the dysfunctional associated reactions after TBI and other neurological diseases must be early and correctly assessed and treated as a main objective of neurorehabilitation [17]. Regardless of the nature of the cognitive, sensory, or motor impairments, their implications for the performance of ADL should be evaluated. Best therapeutic options available in each rehabilitation service should be offered to TBI patients with atypical neurologic hand (Fig. 10.3h–k), as neuroplasticity processes in the brain will never cease to surprise us.

Fig. 10.3 Alien hand syndrome after TBI: functional tasks (**a–c**), standardized assessment (**d–g**), and treatment with the IS-BRAIN model using HandTutor® gloves (**h–j**) and Amadeo® hand robot (**k**)

10.3 Rehabilitation Programs After TBI

TBI rehabilitation includes the need for a specific cognitive-behavioral rehabilitation and the prevention of further complications. Likewise, an early evaluation and an adequate multidisciplinary intervention can contribute to accelerate functional recovery and improve the quality of life of these patients, reducing co-morbidity and the risk of death [18, 19]. This includes emerging assessment and treatment instruments and novel techniques in areas of structural, chemical, and functional neuroimaging and neuropsychology, advances in the realms of cell-based therapies and genetics, promising motor and cognitive restoration techniques, and the use of electronic technologies as coadjutants of OT and PT, including assistive devices, VR, and robotics [20]. For instance, in the earliest stages of recovery, a better prognosis factor is an initial hand distal activation capacity in one or more fingers and a lesser presence of sensitivity affectation [21]. Consequently, treatment and rehabilitation targets must be adapted in the acute, subacute, and chronic state phases of TBI [22]. The theoretical phases through which a person with TBI can evolve are briefly described below, as well as comprehensive, synergic rehabilitation programs adapted to the corresponding phase.

10.3.1 Severe TBI Patients: Disorders of Consciousness State Phase

After acquired severe TBI, disorder of consciousness (DOC) status can be maintained over time beyond the initial phase. If the condition persists after months, they can be considered "hopeless" and "non-rehabilitable" cases. In the last decades, some new sensory techniques and technologies have been developed based on recommendation guidelines from clinical practitioners [23, 24]. Importantly, data exists on the comparison among multiple standardized and validated DOC scales [25]. Within these scales, the Sensory Modality Assessment and Rehabilitation Technique (SMART) is a valid and reliable tool for diagnosis, evolution predictions, and assessment of emergence from vegetative and minimal states [26–30]. The added value of SMART as an extended behavioral assessment and treatment protocol allows its use for the detection of hidden awareness in patients with vegetative state or minimally conscious state [31, 32]. Its practice value and internal construct and predictive validity as a behavioral observation tool and its symbiotic clinical utility with other DOC scales have been already described [33–37]. In this phase, generating opportunities to respond at a highest sensory, communication, and functional level is a priority and SMART becomes a precise tool to assist these objectives. The patient should receive appropriate sensory regulation and stimulation. Sensory deprivation and the effects of immobility due to long periods of bedridden should be controlled. Moreover, adapting a correct positioning system to the wheelchair and educating relatives is essential in these first weeks of accompaniment [32, 38, 39]. Nonetheless, predicting functional recovery from minimal changes in sensory and sensorimotor responsiveness is still a challenge in the neurorehabilitation of these patients [40].

Some tilt-table and verticalization systems that combine muscle functional electrical stimulation (FES) such as ERIGO® , a tilt table with a dynamic foot support, can be safely and effectively used early in TBI, stroke, and SCI patients to minimize the effects of immobility and reduce episodes of (pre)syncope and orthostatic hypotension [41–45]. In patients with extended cortical or sub-cortical lesions, even more in those with DOC, one could ask whether it is really possible to carry out more than these interventions, as robotic gait training, when considered feasible, safe, and recommended in these cases [46]. Some companies and laboratories already provide the latest virtual and augmented reality stimulation and training tools, soft robotics, haptic tools and buttons, and cobotics for assisted minimal participation in ADL care, so that younger survivors with severe TBI can early access the environment by VR headsets, eye gaze, human–computer communicators, and brain-controlled multimodal interfaces [47–49]. Cognitive rehabilitation with electroencephalography (EEG), reach systems, and gait simulators based on VR are currently being used for diagnosis, neurorehabilitation, and research [50–52].

10.3.2 Severe-to-Moderate TBI Patients: Post-Traumatic Amnesia Phase

In this phase, neuropsychological and neurobehavioral rehabilitation becomes the priority. Confused–agitated patients will not be able to benefit from the use of new rehabilitation technologies [38]. Similarly, irritable or aggressive ones should not be included in this type of program, at least until these complications have been solved or controlled by the therapist. As being reverted, and in the judgment of the clinical and therapeutic teams, some technological systems based on exer-games, such as the TYMO® balance platform and the HandTutor® gloves, due to their fast adjustment, could be used with these patients as a catalyst for akathisia and psychomotor restlessness to improve hand coordination problems, early weight distribution, and trunk control [53–55].

10.3.3 Mild-to-Moderate TBI Patients: Subacute to Chronic Phases

10.3.3.1 Current Conventional Approaches for Neurorehabilitation

For the upper limb rehabilitation by PTs and OTs [56, 57], it is common to use an eclectic approach combining theoretical frameworks such as biomechanics, neurodevelopment, and re-learning and motor control. Within these, techniques such as the Bobath concept, motor imaginary, action observation, mirror therapy [58], healthy side constrained-induced restriction movements therapy (CIMT), forced use (FU), increased error-augmented techniques, cognitive exercise therapy, Rood's principles for neuro-sensitive re-training, proprioceptive neuromuscular facilitation, Brunnstrom movement therapy, and task-oriented motor re-learning (TOMR) can be highlighted. Based on the best evidences, CIMT, FU, and TOMR stand out from the rest for UL rehabilitation after stroke [59, 60]. The abovementioned techniques have also been applied synergistically as specific hand therapies, depending on the individual needs of each patient [61]. New technologies based on robotics and VR and serious exer-games turn out to be a valid co-adjunct to these approaches for the evaluation and treatment of the UL [62–67]. This synergy, together with a more intensive therapy and the use of the most advanced electrophysiological and electromechanical monitoring systems, would allow a greater optimization of the rehabilitation outcomes [59, 65, 68]. An updated version of the OT protocol used in the IS-BRAIN model is shown in Table 10.1.

Table 10.1 Standardization of synergistic 2020 OT UL protocol adapted for the BDU-HBMA (unpublished final version revised by Serrano, Aranda, González, Rodríguez, Baños, and Riendas, December 2019), based on the original synergistic OT protocol by Serrano (unpublished, March 2018). This protocol has been used in the current research studies HandBoost-TMS® 2019 (Romero-Muñoz) and REHandMAP® 2020 (Serrano). Protocol modified in September 2020 for this chapter by Serrano and Criado

	INTERVENTION TECHNIQUE	INTERVENTION SUB-TECHNIQUE	INTERVENTION TARGETS	
COADJUANTS FOR REHABILITATION: rTMS, TDCS, BioFEEDBACK by EEG / qEEG, sEMG, …	□ SENSORIMOTOR PREPARATION	Postural Setting	□ Lying on □ In a wheelchair □ Supported sitting □ Unsupported sitting □ High-Sitting □ Standing □ With displacements	ROBOT-ASSISTED THERAPY, VIRTUAL REALITY AND SERIOUS GAMES FOR REHABILITATION
		UL Setting	□ Hands-on upper-limb mobilizations (ROM, biomechanic parameters, …) □ Aedema treatment □ Sensitive regulation □ Tone modulating	
	□ SENSITIVE REEDUCATION	Motor Imaginary and Action-Observation Therapy	□ Motor Imaginary □ Sensory Imaginary □ Augmented Combined Sensory Feedback Techniques □ Augmented Mirror Visual Feedback □ Passive Sensitive training □ Active Sensitive training □ Mirror Therapy	
	□ MOTOR EXECUTION		□ CIMT (original and modified) □ Forced Use Techniques □ Repetitive task-oriented training □ Motor Shaping □ Controlled Task Practice □ Technical-specific activities	
	□ TASK-ORIENTED MOTOR RELEARNING (ADL) □ NEUROMOTOR CONTROL	Affected UL Use And UL Integration in Contextual Functional Tasks	□ Functional movements training (necessary for ADL performance) □ Real-objects use training □ Real environment training □ Use of objects in specific environment training □ Complete-practice of activity training □ Fractional-practice of activity training □ Structured-practice of activity training □ Random-practice of activity training □ Distributed-practice of activity training □ Unilateral practice training □ Crossing midline practice training □ Bimanual practice training	

For standardized assessment and monitoring of UL recovery, based on the latest systematic reviews for neurological disease [69–72], the Fugl-Meyer for UL scale can be used as a global measure of recovery, although not sensitive to detect small changes in hand functionality. In addition, other short scales such as the Box and Block test and the Nine Hole Peg test can be used providing a global measurement of gross and fine motor function, although not sensitive to detect small changes in sensorimotor function in the subacute phase or during shorter hand rehabilitation processes. For UL clinical evaluation in ABD, professionals can also make use of the Wolf Motor Function test, the Action Research Arm test, the Motor Assessment scale, Jebsen–Taylor Hand Function test, and JAMAR® dynamometry [73–78]. In these patients, the quality of movements, postural setting, and coordination must also be evaluated [79, 80], in addition to the presence and evolution of associated motor reactions [17, 81] and spasticity [82].

10.3.3.2 New Technologies Based on Robotics-Assisted Therapy, VR, and Serious Exer-Games

Despite the increasing evidence on the use of robotics in stroke and other neurological diseases, very few studies have explored patients with TBI. In the next section, we discuss unique cases and case series reported to date that evaluate the efficacy of robotics-assisted therapy (RAT), VR systems, and exer-games-based technologies for improving the functionality of traumatic or neurological hands in adults [21, 38, 62] and children [68, 83, 84] with subacute or chronic TBI.

In the current generations of children, the usability of adjuvant new technologies for head trauma is evident. VR environment, classroom, and scenarios could help to improve attention deficits after ABD [85]. With the IREX® VR system, for instance, children with the greatest motor deficits showed the greatest gains in motor skills, thus creating an opportunity for intensive practice of motor skills [84]. In this TBI population, the motivation, adherence to treatment and cognitive skills, can be specially increased using new VR, RAT, and game-based technologies such as LEGO® [86]. From the clinical practice with ABD children, including obstetrical brachial plexus injury, it has been observed that movement velocity, fluidity, and motion precision in kinematic parameters improve after Armeo® Spring UL rehabilitation [87–92]. Combining this robotic therapy with CIMT can be useful to improve UL recovery in hemiplegic and diplegic children [93, 94]. For instance, VR exer-games TYROMOTION technologies like PABLO® and AMADEO® have been proposed, together with functional tasks, to improve UL motor and functional performance [95] and to enhance UL rehabilitation in neuromotor disorders [96], respectively. In a different approach, HandTutor® gloves, with their augmented biomechanical feedback, have been used to treat language disability in speech therapy in preschool children with dysarthria [97].

For children with neurological gait disorders, the evidence suggests that robot-assisted gait training has functional, temporospatial, and kinematic gait performance benefits [98]. In the case of children with acute ABD, a wearable robotic device for ankle passive and active movement training could help to promote neuroplasticity and prove improvements in motor performance related to isometric torque generation, stretching, and active movement [99]. In line with these findings, preliminary results with a robotic exoskeleton gait trainer on a teenager with ABD suggest that high dose, repetitive, and consistent gait training by using robotic exoskeletons has the potential to induce function recovery [100]. Robotically driven orthoses are promising for treating gait impairment in children with hemiplegia after ABD. They impose a proximal-to-distal differential effect on LLs, increasing proximal segments and hip joint stimulation, quality of gate, and gate velocity [88]. In this context, TYMO® balance platform and balance treadmills with interactive computers are feasible and safe for children and teenagers with cerebral palsy and

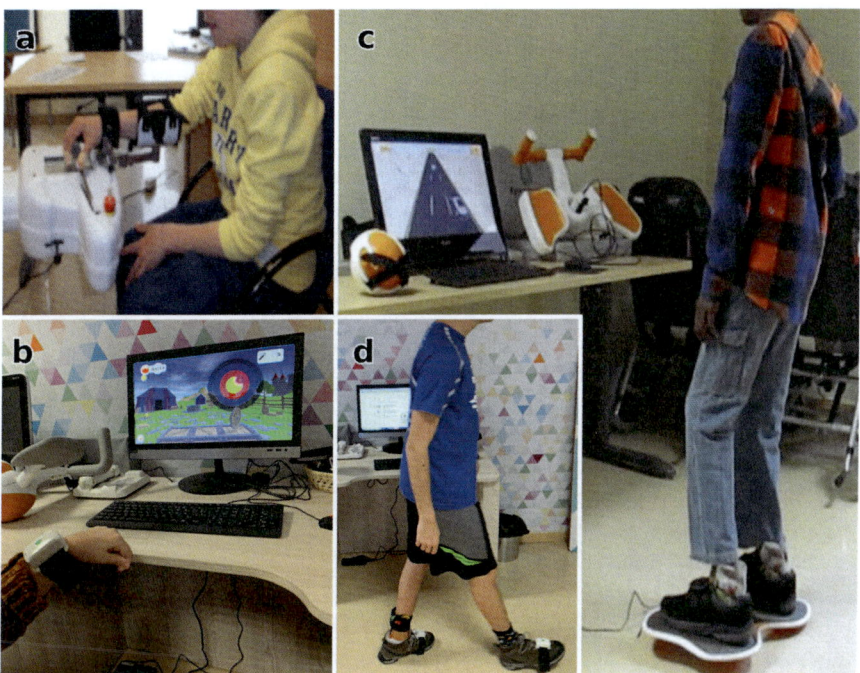

Fig. 10.4 Amadeo® hand robot (R5 version) (**a**) and PABLO® X2 sensors (**b**) for UL rehabilitation in TBI children. TYMO® plate (**c**) for static and dynamic balance treatment and PABLO® X2 gait assessment (**d**) used in TBI teenagers

other neurological diseases, allowing for significant improvements in gross motor function and balance [101] (Fig. 10.4).

In adults with TBI, the specific application of RAT models, like IS-BRAIN, has proven benefits for the recovery of UL, LL, gait, and balance functions as they provide repeated, high-intensity, and well-defined exercises [54, 62]. Specifically, the main objective of RAT is to obtain a better motor recovery motivating the user to carry out different tasks. In this sense, robotic devices allow to record performance by monitoring important bio-psycho-physiological parameters and to offer feedback to patients, thus becoming a more attractive therapy, simulating videogames [102]. For robotic systems, it is necessary to work with different degrees of freedom of movement, programs, and types of assistance on the most distal components of the hemiparetic UL, hands, and fingers and to introduce bio-feedback systems so that, in the short term, a functional training through its use can be translated into an increase in motor control and functional grip capacity. Likewise, this technology has positive effects on motivation, improving body-somatosensory awareness, own-body perception, and the complexity of motor learning that can be achieved through repetition [54, 62].

To date, there are two main robotic systems designed for the rehabilitation of UL and LL: exoskeletons and end-effectors. The exoskeletons control each joint of the limbs to be treated, while the end-effectors guide the most distal part of the affected limbs [103]. End-effectors generate forces at the interface, holding the forearm/hand at a specific point. Their design allows the path of the end-effector to coincide with the natural path of the hand in space for a required task. These systems are suitable for patients with sufficient residual motor skills to control their movements, since they do not allow inter-segmental control of the UL [104, 105]. In what follows, we describe major features and accomplishments in neurorehabilitation of the most relevant robotic end-effectors, including Amadeo® hand robot, HandTutor® gloves, and PABLO® device, and robotic exoskeletons such as the ARMEO® SPRING. Consequently, approaches concerning UL rehabilitation will be first exposed, with later focus on those related to LL.

The Amadeo® hand robot is a feasible, safe, and adjustable type of robotic end-effector device, whose assessment and treatment goal is the hand (Fig. 10.5). This hand robot helps to train various aspects of the grasping movement and to work on targeted finger extension training. The entire hand or individualized finger movements can be controlled through isometric force, range of motion, and surface electromyography (sEMG) trigger signals [106]. Overall, it is considered a good augmented feedback device for improving hand motor function and cognitive skills rehabilitation [54, 62, 107–109]. The Amadeo® robot is the hand device with the highest level of clinical and research evidences, including functional brain mapping studies that suggest neuroplasticity changes by rebalancing interhemispheric connectivity [62, 110–112]. A hand-centered intervention has been found necessary when the proximal improvements do not migrate to distal segments or vice versa [113, 114]. This robot can be used in all phases of finger–hand rehabilitation for multiple pathologies and can be adapted to the needs of each patient in combination with EEG, sEMG, and traditional UL scales [115–117]. When properly used, the Amadeo® hand robot can detect small changes in biomechanical kinetics and kinematics parameters such as range of motion, hand grip strength, and finger flexor and extensor forces [21]. Despite its intrinsic augmented feedback, motor re-learning should be reinforced by increased extrinsic feedback such as verbal instructions, modeling, and rhythmical auditory stimulation [118]. Intensive therapies with robotic systems such as Amadeo® [119] are beneficial in the recovery of manual motor function in association with PT [112] and/or OT [107, 110]. Other own specific synergistic hand rehabilitation models, like the Robotics for Hand (RobHand2015), base of the current IS-BRAIN concept, have found evidence of better recovery of manual function in TBI patients compared with stroke, being the ability to activate the fingers/hand in the first days after stroke a positive prognostic factor for ulterior recovery [120].

On its turn, the HandTutor® is a portable ergonomic glove and a laptop with rehabilitation software designed to allow finger, wrist, and hand rehabilitation in TBI, stroke, and rheumatologic patients. It is wearable, flexible, and adaptable to different residual motor abilities for adult and children subjects [121–123]. It provides augmented feedback and can be useful to detect small changes in senso-

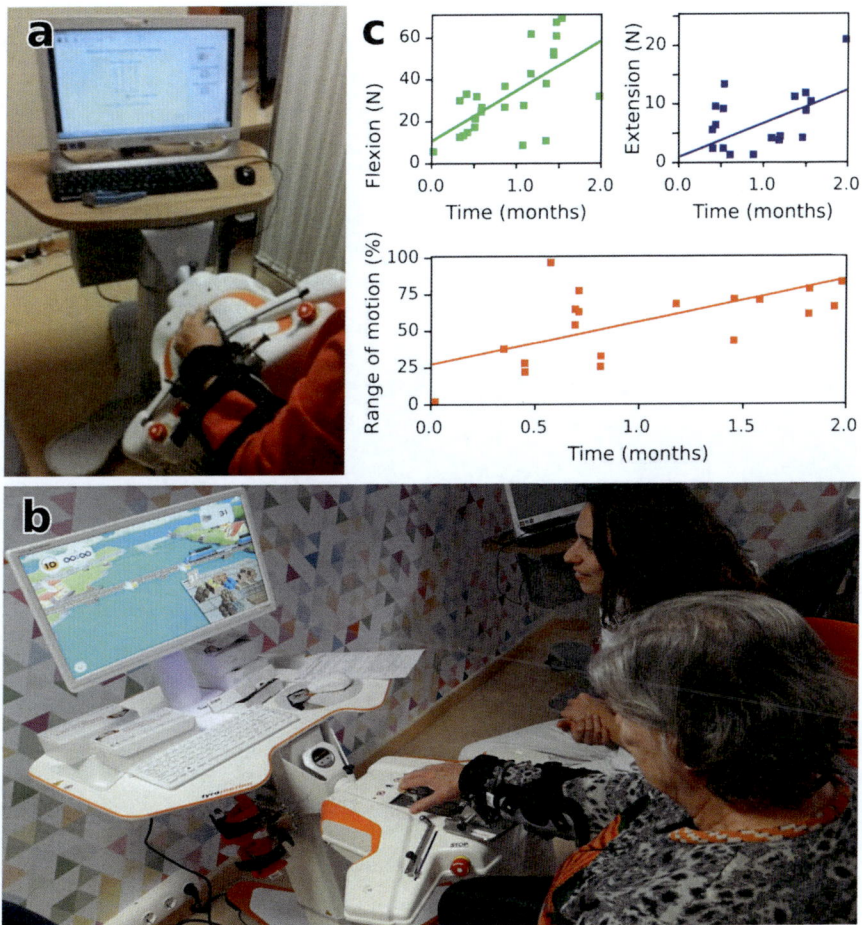

Fig. 10.5 Amadeo® hand robot (R5 and R7 versions) can be used in adult TBI patients in passive (**a**) and interactive (**b**) mode therapies providing a monitoring graphic report (**c**)

rimotor recovery associated to individualized finger/wrist active range of motion and grip frequencies [108, 124]. In adults with Colles fracture, improvements in wrist frequency and mobility were found [125]. Moreover, a case study reported significant improvements in passive and active range of motion of the injured hand, as well as in the speed of the movements, after the use of this device in patients with surgical treatment after finger flexor tendons injury [126]. A home-based tele-rehabilitation program with on-line distant control appeared to be efficacious according to the neurological Disability of the Arm, Shoulder and Hand, Rivermead Motor, and Barthel Index scales in stroke patients using HandTutor® gloves, Virtual Rehab® with Kinect® and Re-Joyce® devices [127]. The integration of VR

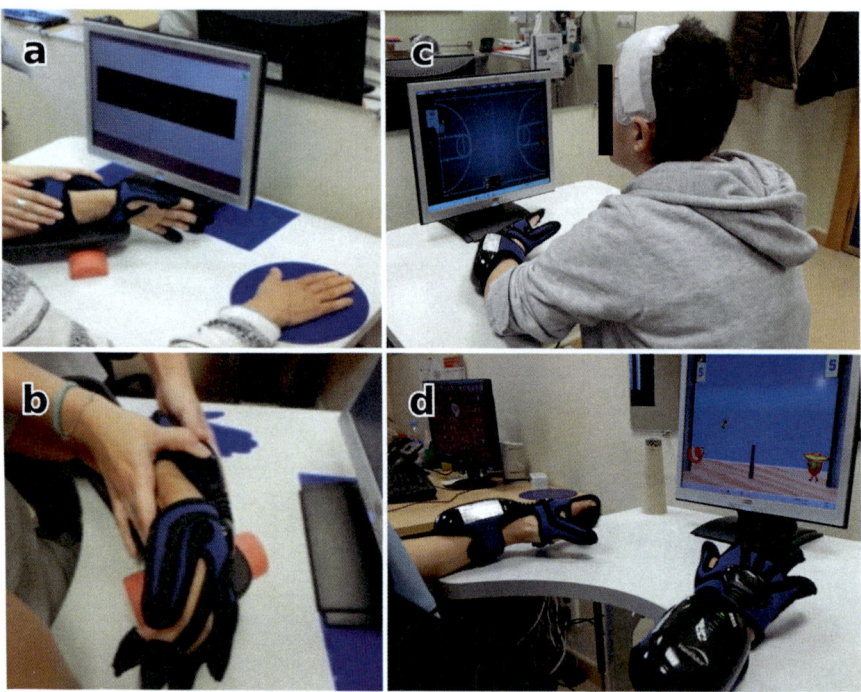

Fig. 10.6 HandTutor® gloves applied to TBI patients allow a quick adjustment and increase the extrinsic feedback (**a**) to control motor compensations (**b**) from both its one-handed (**c**) and bimanual (**d**) approaches

exer-games in patients post-stroke is used to analyze the success degree of factors involved in hand recovery [128] (Fig. 10.6).

Alternatively, the PABLO® device for the entire UL rehabilitation offers the possibility of interactive games based on the VR concept with application in OT programs (Fig. 10.7). Moreover, it represents a modern option for the evaluation of hand deficits and dysfunctions, with computerized data base, monitoring of kinetics and kinematics parameters and better muscular and neuro-cognitive feedback. The handle sensor of PABLO® shows its therapeutic benefit in various UL pathologies (e.g., hand surgery for traumatic UL lesions with or without neurological problems after TBI) [129–131], subacute and chronic strokes [119, 132], Parkinson's disease, other neurological diseases, healthy adult subjects [133, 134], and elderly [135]. The impact of brain-derived neurotrophic factor (BDNF) administration and intensive therapies with a set of robotic and sensor-based devices [106, 107, 136] have been investigated in stroke patients [137]. With very sensitive sensors, the PABLO® device has a complete evaluation of all the joints of the UL, including finger pinch force, hand grip strength, and wrist, elbow, and shoulder range of motion. In the clinical practice, these sensors could get out of calibration, requiring frequent recalibration, so the therapists must monitor the frequent presence of

Fig. 10.7 The PABLO® final-effector device can be used with TBI patients even in times such as those of the COVID19 pandemic. Handle sensor (**a**) and Multiball (**b**) allow working the hand and wrist, while motion sensors help to redirect the arm/foot (**c**) motion. Bimanual and high-cognitive tasks (**d**) can also be implemented with PABLO® . Exemplary use of the ARMEO® SPRING exoskeleton for entire UL neurorehabilitation (**e**)

motor compensations. Recently, Rogelj and Zajc, from the University Rehabilitation Institute of Slovenia, have elaborated a novel protocol with the PABLO® device for TBI and other neurological conditions as multiple sclerosis. By using this device as an effective additional treatment, the authors found increased range of motion, gripping ability, and occupational performance (e.g., cutting bread), as evaluated with the Canadian Occupational Performance Measure [138]. Finally, PABLO® Sensors in X2 version can also be used for gait assessment and LL training [139].

Other end-effector devices can also be implemented for neurological neurorehabilitation in TBI patients. For instance, DIEGO® has proven its feasibility and efficiency for UL rehabilitation in neurological, orthopedic, and stroke patients [134, 136]. Intensity-dependent clinical effects have been found using this individualized technology-supported task-oriented UL training program in patients with multiple sclerosis and ABD [140, 141]. Finally, the application of MYRO® has benefitted fitness, balance, and UL function in multiple sclerosis and stroke individuals [106, 142–144].

Within exoskeletons, ARMEO® SPRING is used at intermediate stages of recovery (Fig. 10.7e). Weight compensation characteristics and implications for clinical practice and research have been described and analyzed [145]. In the acute and chronic stroke phases, this device is effective in biomechanical and functional parameters, but it does not substitute the presence of the therapists and it is not significantly more effective than common UL OT programs [146, 147]. Nonetheless, zero-gravity arm training can help to improve poor energy-efficient movements and reduce the formation of abnormal muscle synergies [148]. A 2-week training program in combination with the Kinect® system found positive effects and significantly recovered post-stroke functional levels in self-care, UL motor ability, visual constructive abilities, cognitive skills, and decreased anxiety [149]. By using this exoskeleton, patterns of functional recovery have been compared between stroke and SCI patients [150]. In cervical SCI, conclusions from the use of this device suggest that individuals with some preserved hand function may be better candidates for UL rehabilitation training [151]. Dose and staffing have been compared in neurological injuries [152] and traumatic fractures [153]. Other encouraging findings by using this device include positive gains in motor control in a bilateral elbow transplanted patient with the restoration of motor and tactile sensation [154].

RobHand2015 and IS-BRAIN2020 models allow that all of these devices can be implemented in the neurorehabilitation of typical traumatic and neurological hand syndromes, but also in atypical hand syndromes like the alien hand, UL somatoparaphrenias or with Gerstmann's syndrome, in which the use of new technologies based on robotics, VR, and exer-games can become a feasible and effective therapeutic option [120]. Also, for improving traditional assessment methods, a virtual tabletop workspace could optimize UL function evaluation [155]. Additionally, novel therapies using myoelectric orthoses in TBI patients improve motor control, coordination, and motor learning [156]. Synchronously to sensorimotor impairments, cognitive deficits such as visuospatial attention, working memory, and executive function can be rehabilitated through the use of these novel technologies [109, 157].

To what concerns to LL rehabilitation, the correct coordination of its movements, postural control, and balance are essential to regain ambulation in TBI patients. Gait deficits can increase the risk of falls, reduce mobility, and worsen quality of life [158], then making these targets a priority in the treatment plan. Specifically, the key objectives to assess and treat in gait neurorehabilitation are: speed, balance, traveled distance, and functionality. Typically, patients with moderate ABD show difficulties in activities such as walk and balance, which largely affect ADL (e.g., having a shower, community mobility). These impairments could be produced by many factors such as muscular strength reduction, asymmetric weight distribution, spasticity, and loss of motor control. From the perspective of LL rehabilitation, the recovery of TBI patients is variable, as a result of balance deficits, availability of motor residue, and compensation mechanisms such as lack of weight transfer during walking, foot drop, knee hyperextension, and others, which may appear. With this clinical picture, clinicians and therapists cannot always provide sufficient repetitions and massive practice in individuals with moderate or severe deficits that require enormous physical assistance during motor practice [159]. Clinical experience demonstrates that 1–3 therapists are needed to facilitate the walking of subjects with moderate CNS lesions [46].

To alleviate these limitations, the use of LL robotic devices appears as an attractive solution to perform more precise movements and repetitions with less effort and to control different deficits, thus increasing the efficiency of traditional interventions. Tools that can be used to rehabilitate LL and balance deficits are robotics, VR systems, and game-based balance platforms. While the first one gives feedback to the patient when making exercises related to the recovery of neuromotor elements with balance strategies and postural adjustments before movement, the second and third tools focus on keeping specific training postures.

Gait speed, commonly measured in meters per second, is an important parameter for clinicians and researchers [160]. It is established as a predictive parameter of functional capacity after stroke. In the absence of a similar scale for TBI patients, clinicians keep this parameter as a standardized reference due to the similarities, within the heterogeneity in sensorimotor deficits that may appear. Perry and colleagues affirm that (a) with a speed lower than 0.4 m/s, the walk should be carried out indoors, (b) with a speed between 0.4 and 0.8 m/s, a limited deambulation can be done in the community environment, and (c) a speed greater than 0.8 m/s is needed to carry out the gait process in community. Along these same lines, Carr and Shepherd [161] consider necessarily a speed greater than 0.5 m/s to carry out the march in the community safely.

Control stability is very different during standing and gait, as the latter requires a much more complex control than the first one due to the idiosyncrasies of the task itself [162–164]. During neurorehabilitation, knowledge on neuromotor control and neuroplasticity is applied through the principle of acquisition by training specific skills and tasks such as walking. In order to evaluate and train the aspects of speed, cadence, and control and to provide measurable assistance to make easier for a patient to walk, the systems currently available in clinical practice are: suspension of body weight, suspension on a treadmill or on the ground itself, robotic exoskeletons, and end-effector robots. Specifically, recent research with robotic exoskeletons in TBI patients has found changes in cortical activity after gait robotic training, evidencing an increase in cortical activation in the prefrontal cortex, bilateral premotor cortex, and primary motor cortex while walking with robotic exoskeletons [165].

For years, gait training with partial weight bearing has been common in the clinical practice, based on the principle of producing a physiological gait pattern that is as normalized as possible and taking into account the ideal kinetics and kinematics aspects of human gait. Gait training can be done on a treadmill or directly on the ground, with suspension systems that are anchored to the ceiling and move through rails. These suspension systems can be combined with a manual-drive-assist or a robot-drive-assist. Working with the walking belt allows modifying the pathological cadence, which refers to the number of steps per minute. This is the main objective of the BIODEX® gait trainer. In a subject with ABD, training on a treadmill with a speed selected by the therapist after evaluation has been shown to be just as effective as the use of the speed that the patient chooses, with which he/she feels the most comfortable [166, 167].

One of the most used robots for gait assistance following stroke and TBI is the LOKOMAT® by Hocoma [168–171]. It is an exoskeleton for LLs and trunk. Although, in the first versions, this exoskeleton kept the pelvis fixed in the coronal and sagittal planes, which guaranteed balance through external elements, this has been corrected in the latest versions. LOKOMAT® consists of a computer-controlled linear system monitored treadmill with sensors built for each hip and knee joints and a weight-bearing system. The gait pattern and motive force can be individually adjusted according to the patient needs to optimize performance and working in the sagittal, frontal, and transverse planes. This system can also assess physiological stiffness of the patient hip and knee and the isometric force exerted, respectively, for hip and knee extension [171].

Following TBI, the spatiotemporal characteristics of the gait are affected. Particularly, the ability to coordinate flexion and extension of both LLs and to modulate the contraction of the different muscle groups is impaired. The patient is commonly unable to access normal movement patterns that facilitate a smooth transition between the different phases of the gait and that allow the limb to slide and advance during the oscillation phase [172]. This is the target where electromechanical devices such as end-effector robots have an important role, although they are less kinematically precise than exoskeletons, which leave the knee and ankle totally free in a more natural approach to gait training. Differently to exoskeletons, end-effector robots only come into contact with the patient body through the support of the sole of the foot, working the two feet separately. Both feet are connected through a contact plate that simulates the stance phase and the gaiting swing phase or other movements such a gliding. Scientific evidence has demonstrated that an increase in maximum gait speed occurs in subjects with TBI when gait is trained with partial weight bearing and assisted with an exoskeleton-type robot [173]. However, these devices have had only a modest clinical impact to date [174]. Robot assistance seems to statistically improve stride symmetry, while manual assistance significantly improves speed [166, 172]. When the three interventions (i.e., partial weight bearing on treadmill, exoskeleton, and robot end-effector) are compared in TBI patients, an increase in maximum speed is reported for treadmill training and exoskeletons. Besides, stride symmetry is greater when working with a robot, either an end-effector or an exoskeleton. Improvements in symmetry appear in both phases of gait in patients trained with exoskeletons. Finally, all the interventions are equally valid to improve gait functionality (i.e., meters covered per unit of time).

Another important aspect of gait rehabilitation is the generalization of motor learning and the capacity to perform functional walking outside or in different environments. This is where VR technologies intervene, helping to simulate potentially dangerous scenarios and unsafe situations and allowing the design of familiar virtual contexts for the patient. In these virtual or augmented scenarios, patients can be controlled by therapists, reducing the fear to fall in safe conditions and enhancing the motivation of the patient, which usually translates into a greater number of repetitions of the trained task and longer treatment sessions to optimize motor re-learning [172]. An important parameter is the degree of virtual immersion of the therapy, which can be: fully immersive (where the VR system eliminates the perception of the real world, like a VR headset), semi-immersive, and non-immersive (i.e., both worlds are perceived and what varies is the size of the monitor in front). Some studies associate a larger degree of immersion with a greater benefit of the VR therapy [166]. VR in HI and TBI patients helps to improve cognitive functions, attention, and executive functions that directly intervene in motor planning and could facilitate and improve care processes and patient autonomy. In TBI patients, the use of exoskeletons such as LOKOMAT® in synergy with VR systems activates a reinforced re-learning process through the perception of a stimulus, the motor plan cognition to be implemented, and the selected motor plan execution, which helps these subjects to improve the performance of proposed motor tasks [171, 175, 176].

From all the different interventions performed to increase walking speed and distance covered in TBI subjects, only interventions with sufficient scientific evidence recommended by the American Physical Therapy Association, according to its practice guide, are considered for use in clinical practice [177]. These are the use of static and dynamic balance platforms together with augmented visual feedback or VR.

Another key feature in gait rehabilitation is maintaining balance and stability. Postural stability and standing balance are defined as the ability to leave the center of gravity of the subject within the boundaries of the support base [161, 162]. Balance is a complex function necessary in static and dynamic positions, for which postural adjustment strategies are needed to maintain stability. In static status, the ankle and hip strategy is used to maintain and, incidentally, to recover from instability; while in dynamic status, the trunk and feet control is responsible for maintaining the stability. Systematic reviews with meta-analyzes report benefits in the prevention of falls through the application of balance platforms and treadmills that provide disturbances in ABD patients [178].

Assessment and rehabilitation of balance present limitations when working in a conventional way. Regarding evaluation, conventional scales such as Time Up and Go and Stability Assessment by Pull-Test have a good fall prediction validity to date but provide limited and, sometimes even, subjective information [179, 180]. A complete balance protocol based on these screening scales takes a long time to be carried out and do not measure reactive control during gait [175]. Alternatively, robotic devices allow an objective and quantitative assessment of instability, which can be monitored, while the sensory transduction of the patient is stimulated in a multisensory way [171].

In the case of posturography, a standing platform that can be moved from 1 to 6 degrees of freedom, depending on the model, offers a controlled and safe environment to test the balance while standing. These devices provide quantitative measures of the ability to maintain the body center of mass within the base of support in static standing. Its use has been traditionally linked to assessment and research, but, in recent times, these systems have evolved by reducing the platform that serves as base and adding a screen with software that allows the patient to work following the principles of intensive, repetitive, and task-oriented practice.

Fig. 10.8 The BIODEX® Balance Trainer allows the assessment of the postural stability, the limits of stability (**a**), the risk of falls (**b**), and the sensory integration of balance in four different conditions (**c**). After TBI, dynamic balance and gait can be assessed and treated by either using a traditional PT/OT approach (**d**) or a technological evaluation using the Balance Tutor® treadmill (**e**)

In this context, the BIODEX® Balance Trainer (Fig. 10.8) combines a complete assessment protocol with a good task-oriented training program that can be adjusted based on evaluation findings. It assesses postural stability, risk of fall, limits of stability, and sensory integration processes of balance. Results by our research team have shown that the use of virtual platforms with exercises to enhance stability could be effective to improve the balance of ABD patients in an isolated and immediate way. Exercises to work on the reaction speed and load transfer performed with the TYMO® and BIODEX® trainers seem to have a positive influence in the improvement of latero-lateral and antero-posterior stabilities [53]. Also, both devices have shown effectiveness in the evaluation and rehabilitation of patients with stroke, TBI, and other ABD resulting from brain infections (e.g., malaria, staphylococcus) and tumors [53, 54]. They are also good coadjutants to detect hidden stability problems in the evaluation of balance in other neurodegenerative pathologies, such as polyneuropathies, multiple sclerosis, and Parkinson's disease [179–181].

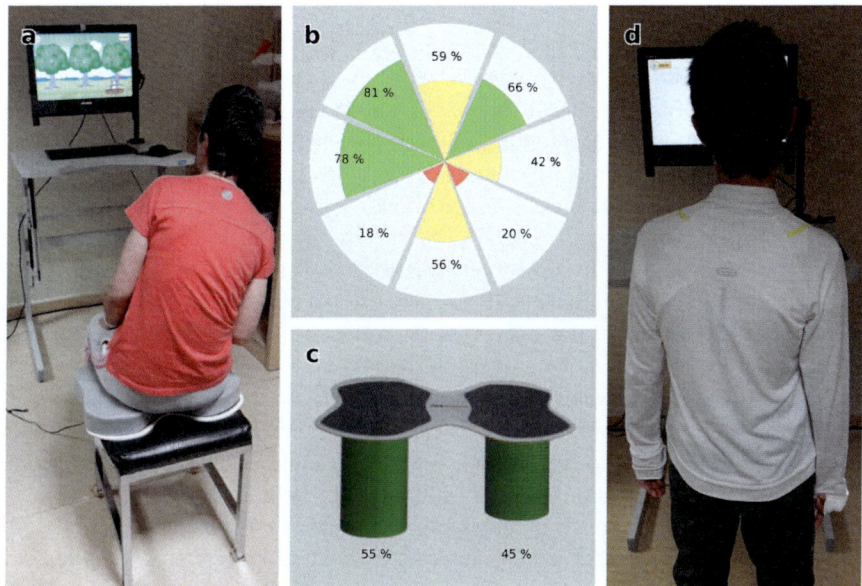

Fig. 10.9 After a pelvic fracture in adults with subacute TBI, the TYMO® platform can be used for improving sitting (**a**), active weight transfer in all balance direction controls (**b**), asymmetries in weight distribution (**c**), and motor compensations or abnormalities in proprioception or batiesthesia (**d**)

The TYMO® plate (Tyromotion GmbH, Austria), which stands out for its small size (Fig. 10.9), is another example of these electromechanical systems. Some of the most remarkable advantages of this system are its portability, easy adjustment, and task-oriented approach provided by the TYRO-s® software, which emphasizes the most playful component of rehabilitation. It favors an increase in the number of repetitions and helps to gain adherence to the treatment in children and adults with TBI and other neurologic diseases [53–55, 182].

A third type of device with interest in this context is the Balance Tutor® Rehab treadmill by Meditouch (Ltd.), roughly the equivalent of placing a gait tape on top of the posturography, which makes it possible to work and assess reactive balance and gait reactions during walk tasks. This device also allows the generation of omnidirectional, antero-posterior and latero-lateral disturbances (Fig. 10.9e). It also incorporates two screens, one for the therapist and another for the patient, to visualize the disturbances that are going to be generated and then work the reactive or proactive balance in dynamics in a differentiated way. This system, feasible and effective, incorporates a suspension harness as a safety element, interesting for patients with TBI, or cardiovascular accidents, with significant stability deficits that generate aversion to work on a treadmill. In addition, this support generated by the restraint harness helps TBI patients to perform different tasks and facilitates the assessment of activities that the patient could not carry out otherwise [183, 184].

There are some other types of devices to work on dynamic balance during various activities of daily life. For instance, the SENLY® system [185] divides the platform/treadmill into two parts to work independently each LL, with mono or bilateral disturbances in two planes of movements, and incorporates sensors to measure the force generated in activities such as running or walking. The disturbances generated by these robotic tools are precisely measured and quantified and can be used as many times as desired. Their sensors provide detail information on the equilibrium reactions generated by specific disturbances. This information is of great relevance as it can serve to unravel the cause of the reduced ability to balance and can potentially help in the assessment, analysis, and design of a personalized training [175]. Other specific robotics include the HIROB® robot horse by REDOX Game Labs (Austria, GmbH), which enables unique movement therapies for the improvement of trunk control and stability in patients with neurological deficits. This effective and fun device simulates the path of motion as it would result from riding a horse and thus enables automated hippotherapy to be conducted. The platform uses an engaging, unlimited virtual terrain with adjustable levels of training difficulty to motivate and encourage patients to perform therapeutic movements. All patient movements are tracked by using the Microsoft Kinect software, and VR movements are controlled through specific motion sensors that identify the relative position of the device. In stroke and TBI patients, it provided significant improvements in trunk stability, selective pelvis movements, ability to maintain the pelvis and upper body in a straight posture, and back muscles activity, as demonstrated in a prospective case series with 20 subjects [186, 187].

Finally, we would like to devote some attention to the wide variety of electrophysiological-based technologies that are emerging in the last decades and their remarkable potential for the neurorehabilitation of TBI patients. Brain–computer interface technologies are being progressively explored in research and incorporated in current clinical practices for predicting functional outcomes in neurological diseases. High-quality non-invasive EEG signals during robotic rehabilitation has been used in research projects such as BCI4REHAB® [188] and ARTE® with AMADEO® [116] to analyze the activation of the motor cortex, the level of engagement, and the brain (a)symmetry index [116].

Intention-to-move signals are more pronounced during more engaging activities, such as playing games, both in healthy and stroke subjects [189]. Electrophysiological assessment using parameters like the patient heart rate, inter-beat interval, galvanic skin response, task-related coherence, and short-latency afferent inhibition suggests remodeling of sensorimotor plasticity and interhemispheric inhibition between sensorimotor cortices after the application of intensive hand robot-assisted therapies [110]. At the BDU-HBMA, EEG activation patterns have been studied currently in both neurological patients and healthy subjects for the establishment of assessment protocols (Fig. 10.10). Results showed that there were no difference between assisted and non-assisted movements in healthy subjects, while desynchronizated associated to movements (ERD) pattern in neurological patients during non-assisted programs was altered and associated to potent ability to functional

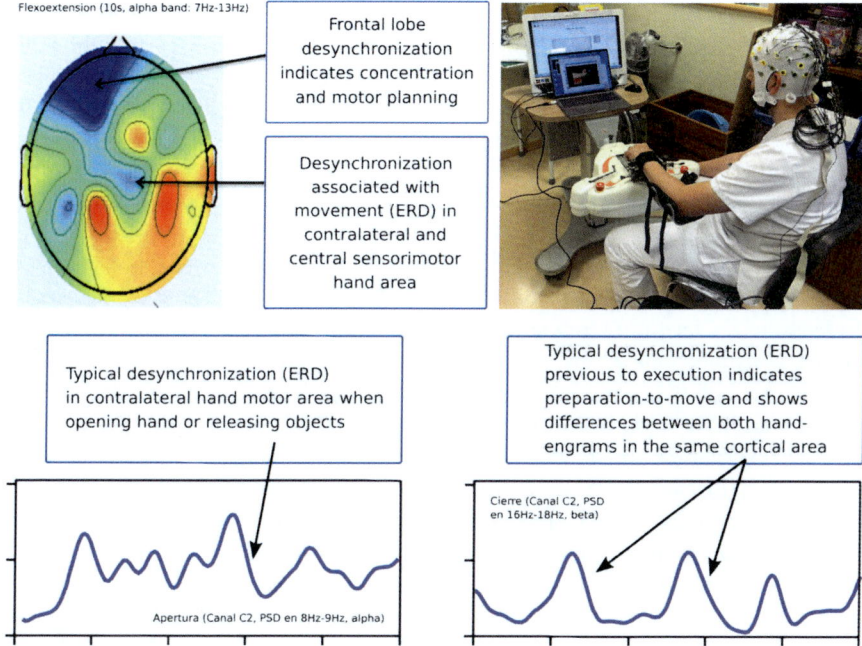

Fig. 10.10 REHandMAP® project by Serrano at the BDU-HBMA (2020–2022) combining brain–computer interface, EEG, and Amadeo® hand robot in ABD patients. Pilot results of subject HC02 at 2019 are shown

recovery [190]. The effect of hand robot-assisted therapy was also enhanced by sEMG, thus establishing as best predictors for UL recovery [115, 191].

Finally, the effect of the combination of transcranial direct current stimulation and a robot-assisted therapy to promote UL motor recovery is still unknown, but preliminary findings point out toward beneficial effects at the subacute phase using ARMEO® [192]. The combined use of the ARMEO® passive device to support the arm weight against gravity with functional electric stimulation to improve the reaching task also increases the feedback error-based learning mixed with adaptive sliding mode control [141, 193]. For ABD patients, applying direct transcranial magnetic stimulation, either to decrease excitability in the healthy hemisphere or to increase cortical excitability in the injured one, improves motor function from 1–3 weeks to 3 months after the last session received [194–196].

10.4   Future Perspectives

Nowadays, the development and effectiveness of novel technologies and robotic devices are increasing. Some of these tools have already achieved great specificity and sensitivity in neurorehabilitation, while having equaled their effectiveness to conventional therapies, or even exceeded it. Many of them have obtained higher positive clinical outcomes, thanks to their capacity to focus on real and functional sensorimotor engrams and neurocognitive objectives. In this sense, the effectiveness of rehabilitation therapies could be significantly enhanced by using interactive and engaging exercise protocols such as the IS-BRAIN model described in this chapter.

The future of neurorehabilitation could be found at the bridge between bioengineering materials and the use of the most modern and updated robotic technologies. By doing so, this synergy could allow a more objective assessment of the efficacy of rehabilitation processes, from both bottom-up and top-down approaches, and discover the real effects produced in the peripheral and CNS. These changes could be measured in terms of reorganization of the functional brain map and the creation of new synaptic connections, beyond the activation of pre-existing inactive neural connections and the concurrent natural processes of neuroplasticity. We could then hypothesize how these neural changes might improve not only the lost cognitive and/or sensorimotor functions, but also the autonomy of patients with TBI and other neurological diseases on ADL performance.

Our clinical experience using some of these new technologies with ABD patients along 10 years demonstrates the feasibility and effectiveness of the IS-BRAIN concept by improving the range of motion, strength, and muscle activity, reducing maladaptative hypertony and spasticity and increasing UL and LL functions, balance, and gait abilities. The IS-BRAIN rehabilitation model shows that augmenting traditional rehabilitation with robotics, VR, and exer-games therapies is cost-effective and efficient in children and adults with ABD, including TBI.

Nonetheless, important questions remain unanswered and will constantly arise. Could these new technologies also help to rehabilitate our patients with COVID19 neurological sequelae as they have already proved with ABD of different etiologies? Where is the limit for the use of these new evolving exer-gamed robotic models? Is it time to affirm, after 2 decades of clinical practice, that the clinical application of these new technologies in neurorehabilitation should not be longer novel *ad libitum* but evidence-based *ex juvantibus*?

Acknowledgments The authors first thank Hospital Beata Maria Ana (HBMA) for the professional trust during these decades work. Special posthumous thanks are given to Dr. Aurelio Capilla for his generous support of research at HBMA. The authors would also like to thank Dr. Marcos Ríos and Dr. Paulina Oliva for their support on the acquisition of these new technologies at the Brain Damage Unit (BDU) of the HBMA. They also acknowledge OT and PT colleagues in BDU. PAS thanks the OT College of Castilla-La Mancha for a 2016 grant, which undoubtedly contributed to the growth of some of the clinical evidences presented in this chapter. PAS also acknowledges Dr. Ignacio Serrano from the G-Nec Group at the *Consejo Superior de Investigaciones Científicas* (CSIC) for fruitful discussions and support in the context of the current REHandMAP® project. PAS also thanks colleagues and Master students in neurologic therapies based on robotics and VR from the University of Castilla-La Mancha since 2014. Finally, PAS acknowledges colleagues at the Occupational Thinks Research Group and the Ethics Committee of the *Centro Superior de Estudios Universitarios La Salle*.

References

1. Andelic N, Forslund MV, Perrin PB et al (2020) Long-term follow-up of use of therapy services for patients with moderate-to-severe traumatic brain injury. J Rehabil Med 52:jrm00034
2. Pérez GAR, Perdomo HA, García MA (2015) Factores pronósticos de muerte en pacientes con traumatismo craneoencefálico. Rev Cub Med Int Emerg 14:61–69
3. Demir Y, Köroğlu Ö, Tekin E et al (2018) Factors affecting functional outcome in patients with traumatic brain injury sequelae: Our single-center experiences on brain injury rehabilitation. Turk J Phys Med Rehabil 65:67–73
4. Heydrich L, Blanke O (2013) Distinct illusory own-body perceptions caused by damage to posterior insula and extrastriate cortex. Brain 136:790–803
5. Jang SH, Seo JP (2017) Limb-kinetic apraxia in a patient with mild traumatic brain injury. Medicine 96:e9008
6. Buchmann I, Dangel M, Finkel L et al (2020) Limb apraxia profiles in different clinical samples. Clin Neuropsychol 34:217–242
7. Choi EB, Kim JY, Jang SH (2020) Motor recovery of hemiparetic leg by improvement of limb-kinetic apraxia in a chronic patient with traumatic brain injury: A case report. Medicine (Baltimore) 99:e20144
8. Shriberg LD, Strand EA, Jakielski KJ et al (2019) Estimates of the prevalence of speech and motor speech disorders in persons with complex neurodevelopmental disorders. Clin Linguist Phon 33:772–789
9. McKenna C, Thakur U, Marcus B et al (2013) Assessing limb apraxia in traumatic brain injury and spinal cord injury. Front Biosci (Schol Ed) 5:732–742
10. Falchook AD, Porges EC, Nadeau SE et al (2015) Cognitive-motor dysfunction after severe traumatic brain injury: A cerebral interhemispheric disconnection syndrome. J Clin Exp Neuropsychol 37:1062–1073
11. Brugger F, Galovic M, Weder BJ, Kägi G (2015) Supplementary motor complex and disturbed motor control—A retrospective clinical and lesion analysis of patients after anterior cerebral artery stroke. Front Neurol 6:1–12
12. Ridley B, Beltramone M, Wirsich J et al (2016) Alien hand, restless brain: Salience network and interhemispheric connectivity disruption parallel emergence and extinction of diagonistic dyspraxia. Front Hum Neurosci 10:1–9
13. Alfaro A, Bernabeu Á, Badesa FJ et al (2017) When playing is a problem: An atypical case of alien hand syndrome in a professional pianist. Front Hum Neurosci 11:1–6
14. Gao X, Li B, Chu W et al Alien hand syndrome following corpus callosum infarction: A case report and review of the literature. Exp Ther Med 12:2129–135
15. McBride J, Sumner P, Jackson SR et al (2013) Exaggerated object affordance and absent automatic inhibition in alien hand syndrome. Cortex 49:2040–2054
16. Olszewska DA, McCarthy A, Murray B et al (2017) A wolf in sheep's clothing: An "alien leg" in corticobasal syndrome. Tremor Other Hyperkinet Mov 7:455
17. Kahn MB, Mentiplay BF, Clark RA et al (2016) Methods of assessing associated reactions of the upper limb in stroke and traumatic brain injury: A systematic review. Brain Inj 30:252–266
18. Hanafi MH (2017) Acute rehabilitation in traumatic brain injury. Malays J Med Sci 24:101–103
19. Mollayeva T, Xiong C, Hanafy S et al (2017) Comorbidity and outcomes in traumatic brain injury: Protocol for a systematic review on functional status and risk of death. BMJ Open 7:e018626
20. Flanagan S, Cantor J, Ashman T (2008) Traumatic brain injury: Future assessment tools and treatment prospects. Neuropsychiatr Dis Treat 4:877–892
21. López-Terradas PAS, Rosendo DM, Lago MR (2013) Hand functional recovery in sub-acute brain injury stage patients using AMADEO® robotic-assisted therapy—A pilot clinical study with apraxic and neglect patients. In: Londral AR, Encarnação P, Pons JL,

editors. Neurotechnix 2013—International Congress on Neurotechnology, Electronics and Informatics (Special Session on Virtual and Augmented Reality Systems for Upper Limbs Rehabilitation); Vilamoura, Algarve, Portugal. Scitepress Digital Library; Short paper 6:1–4

22. Marklund N, Bellander BM, Godbolt AK et al (2019) Treatments and rehabilitation in the acute and chronic state of traumatic brain injury. J Intern Med 285:608–623

23. Seel RT, Sherer M, Whyte J et al (2010) Assessment scales for disorders of consciousness: Evidence-based recommendations for clinical practice and research. Arch Phys Med Rehabil 91:1795–813

24. Crawford T (2013) Low awareness conditions: their assessment and treatment. Pract Neuropsychol Rehabil Acquir Brain Inj. 1st ed, Routledge, New York

25. Chatelle C, Schnakers C, Bruno MA, et al (2010) The sensory modality assessment and rehabilitation technique (SMART): A behavioral assessment scale for disorders of consciousness. Rev Neurol (Paris) 166:675–682

26. Gill-Thwaites H (1997) The sensory modality assessment rehabilitation technique—A tool for assessment and treatment of patients with severe brain injury in a vegetative state. Brain Inj 11:723–734

27. Gill-Thwaites H, Munday R (1999) The sensory modality assessment and rehabilitation technique (SMART): A comprehensive and integrated assessment and treatment protocol for the vegetative state and minimally responsive patient. Neuropsychol Rehab 9:305–320

28. Wilson SL, Gill-Thwaites H (2000) Early indication of emergence from vegetative state derived from assessments with the SMART—A preliminary report. Brain Inj 14:319–331

29. Gill-Thwaites H, Munday R (2004) The sensory modality assessment and rehabilitation technique (SMART): A valid and reliable assessment for vegetative state and minimally conscious state patients. Brain Inj 18:1255–1269

30. Kempny A, Teixeira LC, Gill-Thwaites H et al (2012) What component of the sensory modality assessment and rehabilitation technique (SMART) is the best predictor of diagnosis? Brain Inj 26:343

31. Godbolt AK, Stenson S, Winberg M, Tengvar C (2012) Disorders of consciousness: Preliminary data supports added value of extended behavioural assessment. Brain Inj 26:188–193

32. López-Terradas PAS, Requejo LC, Gill-Thwaites H (2015) Técnica SMART para la evaluación y el tratamiento de personas con disfunción física y sensorial grave. In: Terapia Ocupacional en disfunciones físicas: teoría y práctica. Editorial Médica Panamericana, 2nd ed, España, 295–309

33. Garlick G (2016) Assessment for disorders of consciousness—more than just a diagnostic tool? Br J Occup Ther, COT Conference 79

34. Tennant A, Gill-Thwaites H (2017) A study of the internal construct and predictive validity of the SMART assessment for emergence from vegetative state. Brain Inj 31:185–192

35. Teixeira LC, Gill-Thwaites H, Reynolds F, Duport S (2018) Can behavioural observations made during the SMART assessment detect the potential for later emergence from vegetative state? Neuropsychol Rehabil 28:1340–1349

36. Gill-Thwaites H, Elliott KE, Munday R (2018) SMART-Recognising the value of existing practice and introducing recent developments: leaving no stone unturned in the assessment and treatment of the PDOC patient. Neuropsychol Rehabil 28:1242–1253

37. Morrissey AM, Gill-Thwaites H, Wilson B et al (2018) The role of the SMART and WHIM in behavioural assessment of disorders of consciousness: clinical utility and scope for a symbiotic relationship. Neuropsychol Rehabil 28:1254–1265

38. Noreña D, Calderón CA, Antonio MG et al (2012) Caso clínico desde un enfoque multidisciplinar. In: Daño Cerebral Adquirido, 1st ed, Síntesis, Madrid, España, 341–74

39. López-Terradas PS (2010) Terapia ocupacional en los desórdenes de conciencia. Revista de Terapia Ocupacional, Colegio Profesional de Terapeutas Ocupacionales de Castilla y León 0:22–33

40. Billeri L, Naro A, Leo A et al (2019) Looking toward predicting functional recovery in disorders of consciousness: can sensorimotor integration help us? Brain Inj 33:364–369

41. Kumar S, Yadav R, Aafreen (2020) Comparison between Erigo tilt-table exercise and conventional physiotherapy exercises in acute stroke patients: a randomized trial. Arch Physiother 10:1–9

42. Calabrò RS, Naro A, Russo M et al (2015) Do post-stroke patients benefit from robotic verticalization? A pilot-study focusing on a novel neurophysiological approach. Restor Neurol Neurosci 33:671–681

43. Ancona E, Quarenghi A, Simonini M et al (2019) Effect of verticalization with Erigo® in the acute rehabilitation of severe acquired brain injury. Neurol Sci 40:2073–2080

44. Daunoraviciene K, Adomaviciene A, Svirskis D et al (2018) Necessity of early-stage verticalization in patients with brain and spinal cord injuries: Preliminary study. Technol Health Care 26:613–623

45. Moineau B, Brown A, Brisbois L et al (2019) Lessons learned from the pilot study of an orthostatic hypotension intervention in the subacute phase following spinal cord injury. J Spinal Cord Med 42:176–185

46. Williams K, Christenbury J, Niemeier JP et al (2019) Is robotic gait training feasible in adults with disorders of consciousness? J Head Trauma Rehabil 35:E266–E270

47. Goodman L, Schaler R, Schaler P (2017) Life and living: co-designing real and virtual spaces for survivors of severe acquired brain injury (sABI). Twenty third International Conference on Virtual System and Multimedia (VSMM), Dublin, 1–4

48. Czyzewski A, Kostek B (2016) A study in experimental methods of human-computer communication for patients after severe brain injuries. In: Ortuño F, Rojas I (eds) Bioinformatics and Biomedical Engineering. IWBBIO 2016. Lecture Notes in Computer Science, vol 9656. Springer, Cham

49. Kara DD, Ring M, Hennig FF, Michelson G (2020) Effects of mild traumatic brain injury on stereopsis detected by a virtual reality system: attempt to develop a screening test. J Med Biol Eng 40:639–647

50. Maggio MG, Naro A, La Rosa G et al (2020) Virtual reality based cognitive rehabilitation in minimally conscious state: A case report with EEG findings and systematic literature review. Brain Sci 10:414

51. Moraes T, Paiva WS (2019) Immersive virtual reality cognitive training for patients with moderate to severe traumatic brain injury. Brain Stimul 12:e79–e80

52. Aiello E, Peruzzi A, Fadda P et al (2012) New virtual reality protocol for gait training in patients with multiple sclerosis. Gait Posture 35:S46–S47

53. Mendigutía A, Ferrer TC, Balsera LA et al (2014) Virtual reality: Immediate effect in the improvement of the balance. In: Londral AR, Encarnação P (eds). NEUROTECHNIX 2014: 2nd International Congress on Neurotechnology, Electronics and Informatics. Rome, Italy, Scitepress 121–125

54. López-Terradas PS, Ferrer TC, Balsera LA et al (2014) *Terapia asistida por robot y neurorrehabilltación de la mano y el equilibrio: Evidencia y experiencia clínica con* AMADEO© *y* TYMO© . Revista de Terapia Ocupacional, Colegio Profesional de Terapeutas Ocupacionales de Castilla y León (5):4–7

55. NCT02975804 (2016) RCT on Interactive Computer Play on Trunk Control in CP. https://clinicaltrials.gov/show/NCT02975804. Cited 20 Nov 2021

56. Bennett S, McKenna K, McCluskey A et al (2007) Evidence for occupational therapy interventions: Effectiveness research indexed in the OTseeker database. Br J Occup Ther 70:426–430

57. Case-Smith J (2008) Building the evidence for occupational therapy interventions. OTJR-Occup Part Heal 28:98–99

58. Zeng W, Guo Y, Wu G et al (2018) Mirror therapy for motor function of the upper extremity in patients with stroke: A meta-analysis. J Rehabil Med 50:8–15

59. Wattchow KA, McDonnell MN, Hillier SL (2018) Rehabilitation interventions for upper limb function in the first four weeks following stroke: a systematic review and meta-analysis of the evidence. Arch Phys Med Rehabil 99:367–382

60. Subramanian S, Fountain M, Hood A (2018) Interventions to augment upper extremity motor improvement in individuals with a traumatic brain injury: A systematic review. Neurorehabil Neural Repair 32:1110–1111

61. Roe C, Tverdal CB, Howe EI, Andelic N (2018) Effective rehabilitation services in the post-acute phase of moderate and severe traumatic brain injury. Ann Phys Rehabil Med 61:e233

62. López-Terradas PS, Navarrete PO, Barbás JM (2015) *Terapia asistida por robot con Amadeo para la rehabilitación de la mano.* In: Medica Panamericana, *Terapia ocupacional en las disfunciones físicas: teoría y práctica*, 2nd ed, Madrid. España 449–457

63. Basteris A, Amirabdollahian F, Nijenhuis S et al (2014) Training modalities in robot-mediated upper limb rehabilitation in stroke: A framework for classification based on a systematic review. J Neuroeng Rehabil 11:1–15

64. Debert CT, Herter TM, Scott SH, Dukelow S (2012) Robotic assessment of sensorimotor deficits after traumatic brain injury. J Neurol Phys Ther 36:58–67

65. Mehrholz J, Pohl M, Platz T et al (2015) Electromechanical and robot-assisted arm training for improving activities of daily living, arm function, and arm muscle strength after stroke. Cochrane Database Syst Rev 2015:CD006876

66. Archambault PS, Norouzi-Gheidari N, Kairy D et al (2019) Upper extremity intervention for stroke combining virtual reality, robotics and electrical stimulation. International Conference on Virtual Rehabilitation (ICVR), Tel Aviv, Israel 1–7

67. Norouzi-Gheidari N, Archambault P, Fung J (2012) Effects of robot-assisted therapy on stroke rehabilitation in upper limbs: Systematic review and meta-analysis of the literature. J Rehabil Res Dev 49:479–496

68. Tay EL, Lee SWH, Yong GH, Wong CP (2018) A systematic review and meta-analysis of the efficacy of custom game based virtual rehabilitation in improving physical functioning of patients with acquired brain injury. Technol Disabil 30:1–23

69. Burridge J, Murphy MA, Buurke J et al (2019) A systematic review of international clinical guidelines for rehabilitation of people with neurological conditions: What recommendations are made for upperlimb assessment? Front Neurol 10:567

70. Kelly G, Moys R, Burrough M et al (2020) Rehabilitation in practice: improving delivery of upper limb rehabilitation for children and young people with acquired brain injuries through the development and implementation of a clinical pathway. Disabil Rehabil 20:1–8

71. McCreary JK, Rogers JA, Forwell SJ (2018) Upper limb intention tremor in multiple sclerosis: An evidence-based review of assessment and treatment. Int J MS Care 20:211–223

72. Eraifej J, Clark W, France B et al (2017) Effectiveness of upper limb functional electrical stimulation after stroke for the improvement of activities of daily living and motor function: A systematic review and meta-analysis. Syst Rev 6:40

73. Abedi S, Akbarfahimi N (2020) The effect of modified constraint-induced movement therapy on upper extremity function of a patient with severe acquired brain injury. Archiv Rehabil 21:106–119

74. Turtle B, Porter-Armstrong A, Stinson M (2020) A systematic review of the application and psychometric properties of the graded Wolf Motor Function Test. Brit J Occup Ther 83:285–296

75. Cunha BP, de Freitas SMSF, de Freitas PB (2017) Assessment of the ipsilesional hand function in stroke survivors: the effect of lesion side. J Stroke Cerebrovasc Dis 26:1615–1621

76. Behrendt F, Schuster-Amft C (2018) Using an interactive virtual environment to integrate a digital Action Research Arm Test, motor imagery and action observation to assess and improve upper limb motor function in patients with neuromuscular impairments: A usability and feasibility study protocol. BMJ Open 8:e019646

77. Pike S, Lannin NA, Wales K, Cusick A (2018) A systematic review of the psychometric properties of the Action Research Arm Test in neurorehabilitation. Aust Occup Ther J 65:449–471

78. Rahman HA, Yeong CF, Khor KX, Su ELM (2017) Important parameters for hand function assessment of stroke patients. TELKOMNIKA 15:1501–1511

79. Ustinova KI, Chernikova LA, Dull A, Perkins J (2015) Physical therapy for correcting postural and coordination deficits in patients with mild-to-moderate traumatic brain injury. Physiother Theory Pract 31:1–7

80. Nordin N, Xie SQ, Wünsche B (2014) Assessment of movement quality in robot-assisted upper limb rehabilitation after stroke: a review. J Neuroeng Rehabil 11:137

81. Kahn MB, Clark R, Bower K et al (2015) Measurements scales for associated reactions of the upper limb in stroke and traumatic brain injury (TBI): a systematic review. Physiotherapy 101:e703–e704

82. De-La-torre R, Oña ED, Balaguer C, Jardón A (2020) Robot-aided systems for improving the assessment of upper limb spasticity: A systematic review. Sensors 20:5251

83. Vallejo EG, Rosendo DM (2017) Terapia ocupacional y terapia robótica asistida con amadeo en la atención sostenida y el nivel de consciencia y alerta en un niño con traumatismo craneoencefálico infantil. TOG (A Coruña) 14:80–96

84. Galvin J, McDonald R, Catroppa C, Anderson V (2011) Does intervention using virtual reality improve upper limb function in children with neurological impairment: a systematic review of the evidence. Brain Inj 25:435–442

85. Rizzo AA, Buckwalter JG, Bowerly T et al (2000) The virtual classroom: A virtual reality environment for the assessment and rehabilitation of attention deficits. Cyberpsychol Behav 3:483–499

86. Barco A, Albo-Canals J, Garriga-Berga C et al (2014) A drop-out rate in a long-term cognitive rehabilitation program through robotics aimed at children with TBI. The twenty third IEEE International Symposium on Robot and Human Interactive Communication, Edinburgh 186–192

87. Biffi E, Maghini C, Cairo B et al (2018) Movement velocity and fluidity improve after Armeo® Spring rehabilitation in children affected by acquired and congenital brain diseases: an observational study. Biomed Res Int 2018:1537170

88. Beretta E, Molteni E, Biffi E et al (2018) Robotically-driven orthoses exert proximal-to-distal differential recovery on the lower limbs in children with hemiplegia, early after acquired brain injury. Eur J Paediatr Neurol 2:652–661

89. Cesareo A, Beretta E, Biffi E et al (2016) A comparative study among constraint, robot-aided and standard therapies in upper limb rehabilitation of children with acquired brain injury. IXIV Mediterranean Conference on Medical and Biological Engineering and Computing. IFMBE Proceedings 57. Springer

90. Colegate J, Ward R, Valentine J (2016) The robotic arm in activity based rehabilitation for children. Dev Med Child Neurol 58:69

91. NCT03780322 (2018) Effectiveness of Armeo Spring Pediatric in Obstetric Brachial Plexus Injury. https://clinicaltrials.gov/show/NCT03780322. Cited 20 Nov 2021

92. Cimolin V, Germiniasi C, Galli M et al (2019) Robot-assisted upper limb training for hemiplegic children with cerebral palsy. J Dev Phys Disabil 31:89–101

93. Roberts H, Shierk A, Clegg NJ et al (2020) Constraint induced movement therapy camp for children with hemiplegic cerebral palsy augmented by use of an exoskeleton to play games in virtual reality. Phys Occup Ther Pediatr 7:1–16

94. Turconi AC, Biffi E, Maghini C et al (2016) Can new technologies improve upper limb performance in grown-up diplegic children? Eur J Phys Rehab Med 52:672–681

95. Dorich J, Lowe A, Harpster K (2014) Using virtual reality technologies combined with functional activities to improve upper extremity motor and functional performance. AACPDM 68th Annual Meeting San Diego, CA

96. Falzarano V, Marini F, Morasso P, Zenzeri J (2019) Devices and protocols for upper limb robot-assisted rehabilitation of children with neuromotor disorders. Appl Sci 9:2689

97. Medvedeva E, Olkhina E (2017) Using the appliance hand tutor in speech therapy to children with dysarthria. Perspectives Sci Educ 6:92–96

98. Lefmann S, Russo R, Hillier S (2014) What evidence exists on the effectiveness of the use of robotic-assisted gait training in children with neurological gait disorders? Dev Med Child Neurol 56:93

99. Chen K, Xiong B, Ren Y et al (2018) Ankle passive and active movement training in children with acute brain injury using a wearable robot. J Rehabil Med 50:30–36

100. Karunakaran KK, Ehrenberg N, Cheng J, Nolan KJ (2019) Effects of robotic exoskeleton gait training on an adolescent with brain injury. Annu Int Conf IEEE Eng Med Biol Soc 4445–4448

101. Pin TW, Butler PB (2019) The effect of interactive computer play on balance and functional abilities in children with moderate cerebral palsy: a pilot randomized study. Clin Rehabil 33:704–710

102. Esquenazi A. Robotics in neurorehabilitation, Professor Stefan Hesse Memorial Lecture. Neurol und Rehabil 23:S14–S15

103. Mazzoleni S, Duret C, Grosmaire AG, Battini E (2017) Combining upper limb robotic rehabilitation with other therapeutic approaches after stroke: current status, rationale, and challenges. Biomed Res Int 2017:8905637

104. Bertani R, Melegari C, De Cola MC et al (2017) Effects of robot-assisted upper limb rehabilitation in stroke patients: a systematic review with meta-analysis. Neurol Sci 38:1561–1569

105. Chien W, Chong Y, Tse M, Chien C, Cheng H (2020) Robot-assisted therapy for upper-limb rehabilitation in subacute stroke patients: a systematic review and meta-analysis. Brain Behav 10:e01742

106. Jakob I, Kollreider A, Germanotta M et al (2018) Robotic and sensor technology for upper limb rehabilitation. PM R 10:S189–S197

107. Aprile I, Germanotta M, Cruciani A et al (2020) Upper limb robotic rehabilitation after stroke: a multicenter, randomized clinical trial. J Neurol Phys Ther 44:3–14.

108. López-Terradas PS (2018) *Rehabilitación de la mano neurológica con Amadeo Hand Robot y Hand Tutor Glove: Resultados sensoriomotores y funcionales de 5 años de investigación.* APETO Journal (number 63), APTO, editors. II Congreso Ibérico de Terapia Ocupacional. Madrid, España, 95

109. Fasoli SE, Adans-Dester CP (2019) A paradigm shift: Rehabilitation robotics, cognitive skills training, and function after stroke. Front Neurol 10:1088

110. Calabrò RS, Accorinti M, Porcari B et al (2019) Does hand robotic rehabilitation improve motor function by rebalancing interhemispheric connectivity after chronic stroke? Encouraging data from a randomised-clinical-trial. Clin Neurophysiol 130:767–780

111. Sale P, Lombardi V, Franceschini M (2012) Hand robotics rehabilitation: Feasibility and preliminary results of a robotic treatment in patients with hemiparesis. Stroke Res Treat 2012:820931

112. Sale P, Mazzoleni S, Lombardi V et al (2014) Recovery of hand function with robot-assisted therapy in acute stroke patients: A randomized-controlled trial. Int J Rehabil Res 37:236–242

113. Hwang CH, Seong JW, Son D (2012) Individual finger synchronized robot-assisted hand rehabilitation in subacute to chronic stroke: a prospective randomized clinical trial of efficacy. Clin Rehabil 26:696–704

114. Takahashi CD, Der-Yeghiaian L, Le V et al (2008) Robot-based hand motor therapy after stroke. Brain 131:425–437

115. Baldan F, Turolla A, Pregnolato G et al (2019) Rehabilitation robotics of hand function, after stroke: diagnostic criteria for reference to therapy. WCPT Congress. Geneva PLR5-1048

116. Calcagno A, Coelli S, Tacchino G et al (2020) ARTE project: EEG analysis during robotic rehabilitation. In: Henriques J, Neves N, de Carvalho P (eds) XV Mediterranean Conference on Medical and Biological Engineering and Computing—MEDICON 2019. IFMBE Proceedings, 76. Springer, Cham

117. Jung JH, Lee HJ, Cho DY et al (2019) Effects of combined upper limb robotic therapy in patients with tetraplegic spinal cord injury. Ann Rehabil Med 43:445–457

118. Speth F, Wahl M (2014) Specifying rhythmic auditory stimulation for robot-assisted hand function training in stroke therapy. The twentieth International Conference on Auditory Display (ICAD-2014), New York, USA

119. Ward NS, Brander F, Kelly K (2019) Intensive upper limb neurorehabilitation in chronic stroke: outcomes from the Queen Square programme. J Neurol Neurosurg Psychiatry 90:498–506
120. López-Terradas PS (2021) Quo vadis, Amadeo? PhD Dissertation. In preparation
121. Carmeli E, Vatine J, Peleg S et al (2009) Upper limb rehabilitation using augmented feedback: Impairment focused augmented feedback with HandTutor. 2009 Virtual Rehabilitation International Conference, Haifa. IEEE 220–220
122. Carmeli E, Peleg S, Bartur G et al (2011) HandTutor^{TM} enhanced hand rehabilitation after stroke—a pilot study. Physiother Res Int 16:191–200
123. Hernández MR, Panadero CF, Martín OL, López BP (2017) Hand rehabilitation after chronic brain damage: effectiveness, usability and acceptance of technological devices: a pilot study. Physical Disabilities—Therapeutic Implications, IntechOpen
124. López-Terradas PS, Gonzalez A (2016) Advanced sinergic neurorehabilitation based on robotic and gamed systems with augmented feedback: functional improvements in activities of daily living. In: APTO, APETO (number 61), editors. I *Congreso Ibérico de Terapia Ocupacional*. Beja, Portugal
125. Krajczy M, Łuniewski J, Bogacz K et al (2014) Impact of elastic therapeutic tape on final effects of physiotherapy in patients with Colles' fracture. Fizjoterapia Pol 14:42–49
126. Krajczy M, Krajczy E, Szczegielniak A et al (2015) Physiotherapy after a hand surgery performed with a diagnostic and functional therapy device. Case study. Fizjoterapia Pol 15:46–54
127. Lyadov KV, Snopkov PS, Shapovalenko TV et al (2015) Distantly controlled rehabilitation in chronic stroke. Int J Stroke 10:1
128. Perez C, Kaizer F, Archambault P, Fung J (2017) A novel approach to integrate VR exer-games for stroke rehabilitation: Evaluating the implementation of a "games room". International Conference on Virtual Rehabilitation (ICVR), Montreal, QC, 2017:1–7
129. Brăilescu C, Scarlet R, Nica A et al (2013) A study regarding the results of a rehabilitation program in patients with traumatic lesions of the hand after surgery. Palestrica Third Millennium—Civilization and Sport 14:263–270
130. Nica AS, Brailescu CM, Scarlet RG (2013) Virtual reality as a method for evaluation and therapy after traumatic hand surgery. Stud Health Technol Inform 191:48–52
131. Germanotta M, Cruciani A, Di Sipio E et al (2018) Effects of a robotic rehabilitation treatment in a patient with traumatic lesion of the right brachial plexus measured by means of motion analysis: A case stud. Gait Posture 66:S17–S18
132. Brackenridge J, Bradnam LV, Lennon S et al (2016) A review of rehabilitation devices to promote upper limb function following stroke. Neurosci Biomed Eng 4:25–42
133. Seitz JR, Kammerzell A, Samartzi M et al (2014). Monitoring of visuomotor coordination in healthy subjects and patients with stroke and Parkinson's disease: an application study using the PABLO^R-Device. Int J Neurorehabil 1:1000113
134. Aprile I, Pecchioli C, Loreti S et al (2019) Improving the efficiency of robot-mediated rehabilitation by using a new organizational model: An observational feasibility study in an Italian rehabilitation center. Appl Sci 9:5357
135. Chaudhary P, Hamdani N, Sharma P (2019) The effects of visuomotor training using Pablo System on hand grip strength and wrist movements in adults and elderly. Iran Rehabil J 17:215–224
136. Aprile I, Cruciani A, Germanotta M et al (2019) Upper limb robotics in rehabilitation: An approach to select the devices, based on rehabilitation aims, and their evaluation in a feasibility study. Appl Sci 9:3920
137. NCT03701035 (2018) Moderate intensity aerobic training in sub-acute and chronic stroke patients—the influence on brain derived neurotrophic factor (BDNF) and upper-limb rehabilitation. A protocol for a randomized control trial and health economic evaluation. https://clinicaltrials.gov/show/NCT03701035. Cited 20 Nov 2021
138. Rogelj P, Zajc D (2019) Effectiveness of robotics or sensory-supported training in improving upper extremity functions among people with multiple sclerosis—Case study. Mult Scler J 25:1063

139. Jocham A, Laidig D, Kastnbauer E, Seel T (2019) Evaluation of gait parameters in patients with neurological diseases using inertial measurement units. Neurol Rehabil S17-06:20–21

140. Lamers I, Raats J, Spaas J et al (2019) Intensity-dependent clinical effects of an individualized technology-supported task-oriented upper limb training program in Multiple Sclerosis: A pilot randomized controlled trial. Mult Scler Relat Disord 34:119–127

141. Passon A, Seel T, Massmann J et al (2018) Iterative learning vector field for FES-supported cyclic upper limb movements in combination with robotic weight compensation. IEEE/RSJ International Conference on Intelligent Robots and Systems (IROS), Madrid, 5169–5174

142. Feys P, Straudi S (2019) Beyond therapists: Technology-aided physical MS rehabilitation delivery. Mult Scler J 25:1387–1393

143. Mace M, Guy S, Hussain A et al (2017) Validity of a sensor-based table-top platform to measure upper limb function. IEEE Int Conf Rehabil Robot 652–657

144. Ščurić I, Blažinčić V, Klepo I et al (2018) Hand function recovery with sensor-based task specific feedback training in patients after acute stroke or traumatic brain injury: preliminary results. Neurorehabilitation Conference Proceedings

145. Perry BE, Evans EK, Stokic DS (2017) Weight compensation characteristics of Armeo® Spring exoskeleton: implications for clinical practice and research. J Neuroeng Rehabil 14:14

146. Bocanová R, Gueye T, Švestková O et al (2018) Efficiency of robot-assisted therapy through the device Armeo Spring in patients after stroke in acute phase of early rehabilitation. Rehabil Fyz Lek 25:119–125

147. Colomer C, Baldoví A, Torromé S et al (2013) Efficacy of Armeo® Spring during the chronic phase of stroke. Study in mild to moderate cases of hemiparesis. Neurol (English Ed) 28:261–267

148. Aziatskaya G, Kovyazina M, Khizhnikova A et al (2017) Virtual reality efficacy during zero gravity arm training in post stroke. Brain Inj 31:792

149. Adomavičienė A, Daunoravičienė K, Kubilius R et al (2019) Influence of new technologies on post-stroke rehabilitation: A comparison of Armeo Spring to the Kinect system. Medicina (Kaunas) 55:98

150. NCT04383873 (2020) Effectiveness analysis of Armeo Spring device as a rehabilitation treatment in spinal cord injured patients. https://clinicaltrials.gov/show/NCT04383873. Cited 20 Nov 2021

151. Zariffa J, Kapadia N, Kramer JLK et al (2012) Feasibility and efficacy of upper limb robotic rehabilitation in a subacute cervical spinal cord injury population. Spinal Cord 50:220–226

152. Wuennemann MJ, Mackenzie SW, Lane HP et al (2020) Dose and staffing comparison study of upper limb device-assisted therapy. Neurorehabilitation 46:287–297

153. Nerz C, Schwickert L, Becker C et al (2017) Effectiveness of robot-assisted training added to conventional rehabilitation in patients with humeral fracture early after surgical treatment: protocol of a randomised, controlled, multicentre trial. Trials 18:589

154. Filippi L (2019) Early implementation of sensorimotor retraining for cortical reintegration in postoperative rehabilitation in an bilateral above elbow allotransplantation. J Hand Ther 32:560

155. Wilson PH, Duckworth J, Mumford N et al (2007) A virtual tabletop workspace for the assessment of upper limb function in Traumatic Brain Injury (TBI). Virtual Rehabilitation. Venice, Italy, 14–19

156. Pundik S, McCabe J, Kesner S et al (2020) Use of a myoelectric upper limb orthosis for rehabilitation of the upper limb in traumatic brain injury: A case report. J Rehabil Assist Technol Eng 7:2055668320921067

157. Logan LM, Semrau JA, Debert CT et al (2018) Using robotics to quantify impairments in sensorimotor ability, visuospatial attention, working memory, and executive function after traumatic brain injury. J Head Trauma Rehabil 33:E61–E73

158. Niedermeier M, Ledochowski L, Mayr A et al (2017) Immediate affective responses of gait training in neurological rehabilitation: A randomized crossover trial. J Rehabil Med 49:341–346

159. Nolan KJ, Karunakaran KK, Ehrenberg N, Kesten AG (2018) Robotic exoskeleton gait training for inpatient rehabilitation in a young adult with traumatic brain injury. Annu Int Conf IEEE Eng Med Biol Soc 2809–2812

160. Perry J, Garrett M, Gronley JK, Mulroy SJ (1995) Classification of walking handicap in the stroke population. Stroke 26:982–9

161. Shepherd RB, Carr JH (2006) Neurological rehabilitation. Disabil Rehabil 28:811–812

162. Horak FB, Henry SM, Shumway-Cook A (1997) Postural perturbations: new insights for treatment of balance disorders. Phys Ther 77:517–533

163. Fino PC, Mancini M, Curtze C et al (2018) Gait stability has phase-dependent dual-task costs in Parkinson's disease. Front Neurol 9:373

164. Horak FB (2006) Postural orientation and equilibrium: What do we need to know about neural control of balance to prevent falls? Age Ageing 35 Suppl 2:ii7–ii11

165. Karunakaran KK, Nisenson DM, Nolan KJ (2020) Alterations in cortical activity due to robotic gait training in traumatic brain injury. Second Annual International Conference of the IEEE Engineering in Medicine and Biology Society (EMBC), Montreal, QC, Canada 3224–3227

166. Esquenazi A, Lee S, Wikoff A et al (2016) A randomized comparison of locomotor therapy interventions: partial body weight supported treadmill, Lokomat® and G-Eo® Training in traumatic brain injury. PM R 8:S154

167. Kim TW, Kim YW (2014) Treadmill sideways gait training with visual blocking for patients with brain lesions. J Phys Ther Sci 26:1415–1418

168. van Kammen K, Boonstra AM, van der Woude LHV et al (2017) Differences in muscle activity and temporal step parameters between Lokomat guided walking and treadmill walking in post-stroke hemiparetic patients and healthy walkers. J Neuroeng Rehabil 14:32

169. van Kammen K, Boonstra AM, van der Woude LHV et al (2020) Lokomat guided gait in hemiparetic stroke patients: the effects of training parameters on muscle activity and temporal symmetry. Disabil Rehabil 42:2977–2985

170. Medvedev IN (2019) Place and possibilities of the robotic system Lokomat in the rehabilitation of patients after ischemic stroke. Biomed Pharmacol J 12:131–140

171. Maggio MG, Torrisi M, Buda A et al (2020) Effects of robotic neurorehabilitation through Lokomat plus virtual reality on cognitive function in patients with traumatic brain injury: A retrospective case-control study. Int J Neurosci 130:117–123

172. Esquenazi A, Lee S, Packel AT, Braitman L (2013) A randomized comparative study of manually assisted versus robotic-assisted body weight supported treadmill training in persons with a traumatic brain injury. PM R 5:280–90

173. O'Brien A, Adans-Dester C, Scarton A et al (2016) Robotic-assisted gait training as part of the rehabilitation program in persons with traumatic and anoxic brain injury. Arch Phys Med Rehabil 97:e117

174. Durandau G, Farina D, Asín-Prieto G et al (2019) Voluntary control of wearable robotic exoskeletons by patients with paresis via neuromechanical modeling. J Neuroeng Rehabil 16:1–18

175. Shirota C, Van Asseldonk E, Matjačić Z et al (2017) Robot-supported assessment of balance in standing and walking. J Neuroeng Rehabil 14:80

176. Calabrò RS, Naro A, Russo M et al (2017) The role of virtual reality in improving motor performance as revealed by EEG: a randomized clinical trial. J Neuroeng Rehabil 14:53

177. Hornby TG, Reisman DS, Ward IG et al (2020) Clinical practice guideline to improve locomotor function following chronic stroke, incomplete spinal cord injury, and brain injury. J Neurol Phys Ther 44:49–100

178. Mansfield A, Wong JS, Bryce J et al (2015) Does perturbation-based balance training prevent falls? Systematic review and meta-analysis of preliminary randomized controlled trials. Phys Ther 95:700–9

179. Romero J, Arroyo A, Andreo J et al (2019) Virtual reality EEG guided motor neurofeedback training has mid-term effects in Parkinson's disease patients limits of stability. International Parkinson and Movement Disorders Society, International Congress of Parkinson Disease and Movement Disorders (MDS Congress 2019). Nice, France 34(Suppl S2):896

180. Andreo J, Arroyo A, Periañez I et al (2019) Relationship of cognitive processing speed with postural instability in non-demented Parkinson's disease patients. International Parkinson and Movement Disorders Society, International Congress of Parkinson Disease and Movement Disorders (MDS Congress 2019). Nice, France 34(Suppl S2):483

181. Karimi N, Ebrahimi I, Kahrizi S, Torkaman G (2008) Reliability of postural balance evaluation using the Biodex Balance System in subjects with and without low back pain. J Postgrad Med Inst. 22:95–101

182. Soto J (2019) *Realidad virtual y robótica para reparar el cerebro tras una lesión*. ABC Salud, Periódico ABC Dec 14, 3, 16–7, 32

183. Volovets SA, Sergeenko EY, Darinskaya LY et al (2018) The modern approaches to the restoration of postural balance in the patients suffering from the consequences of an acute cerebrovascular accident (CVA). Vopr Kurortol Fizioter Lech Fiz Kult 95:4–9

184. Sergeenko EY, Volovets SA, Darinskaya LY et al (2017) The use of the balance tutor rehabilitation treadmill for balance and gait recovery in poststroke patients. Bull Russ State Med Univ 6:58–64

185. Luciani LB, Genovese V, Monaco V et al (2012) Design and Evaluation of a new mechatronic platform for assessment and prevention of fall risks. J Neuroeng Rehabil 9:51

186. IM-Hirob: robotic hippotherapy for improvement of impaired trunk function. Intelligent Motion (IM) Switzerland—Austria—Germany (GmbH). https://www.intelligentmotion.at/?page_id=56&ref=steemhunt&lang=en. Cited 2 Nov 2020

187. Hirob VR rehabilitation robot uses virtual environments to enhance neurological rehabilitation. Health Rehab News. https://www.fitness-gaming.com/news/health-and-rehab/hirob-vr-rehabilitation-robot-uses-virtual-environments-to-enhance-neurological-rehabilitation.html#.WByPnB8o-SM. Cited 20 Nov 2021

188. Scherer R, Grieshofer P, Enzinger C et al (2012) Predicting functional stroke-rehabilitation outcome by means of brain-computer interface technology: The BCI4REHAB Project. World Congress for NeuroRehabilitation, 26:772

189. Butt M, Naghdy G, Naghdy F et al (2019) Investigating the detection of intention signal during different exercise protocols in robot-assisted hand movement of stroke patients and healthy subjects using EEG-BCI system. Adv Sci Technol Eng Syst 4:300–307

190. Caimmi M, Visani E, Digiacomo F et al (2016) Predicting functional recovery in chronic stroke rehabilitation using event-related desynchronization-synchronization during robot-assisted movement. Biomed Res Int 7051340

191. Dziemian K, Kiper A, Baba A et al (2017) The effect of robot therapy assisted by surface EMG on hand recovery in post-stroke patients. A pilot study. Rehabil Med 21:4–10

192. Triccas LT, Burridge JH, Hughes A et al (2015) A double-blinded randomised controlled trial exploring the effect of anodal transcranial direct current stimulation and uni-lateral robot therapy for the impaired upper limb in sub-acute and chronic stroke. NeuroRehabili 37:181–91

193. Barbouch H, Resquín F, Gonzalez-Vargas J et al (2019) Feedback error learning with sliding mode control for functional electrical stimulation: Elbow joint simulation. Int J Innov Technol Explor Eng 8:2971–2982

194. Sandrini M, Cohen L (2013) Non invasive brain stimulation in neurorehabilitation. Handb Clin Neurol 116:499–524

195. Tanaka S, Sandrini M, Cohen LG (2011) Modulation of motor learning and memory formation by non-invasive cortical stimulation of the primary motor cortex. Neuropsychol Rehabil 21:650–75

196. Zhang L, Xing G, Shuai S et al (2017) Low-frequency repetitive transcranial magnetic stimulation for stroke-induced upper limb motor deficit: a meta-analysis. Neural Plast 2758097

Chapter 11
Current Implantable Devices in Human Neurological Surgery

Raquel Madroñero-Mariscal, Ana de los Reyes Guzmán,
Fernando García-García, Antonio García Peris, José Luis Polo, Ángel
Rodríguez de Lope, and Elisa López-Dolado

Abstract The different nosological entities that lead to central nervous system (CNS) damage often produce neurological deficits that drastically impair the level of functional independence and quality of life of the individuals who suffer from them. Traditionally, neurosurgeons have played an important role in the acute phase of CNS lesions by carrying out different interventions aimed at their stabilization and the prevention of further progression. However, the constant appearance of new

R. Madroñero-Mariscal
Fundación del Lesionado Medular, Cmo., Madrid, Spain

Hospital Universitario Infanta Leonor, SERMAS, Madrid, Spain

Laboratory of Interfaces for Neural Repair, Hospital Nacional de Parapléjicos, SESCAM, Toledo, Spain
e-mail: medico.reh@medular.org

A. R. Guzmán
Biomechanics and Technical Aids Department, Hospital Nacional de Parapléjicos, SESCAM, Toledo, Spain
e-mail: adlos@sescam.jccm.es

F. García-García · A. G. Peris
Department of Radiology, Hospital Nacional de Parapléjicos, SESCAM, Toledo, Spain

J. L. Polo
Department of Electrical Engineering, Escuela de Ingeniería Industrial y Aeroespacial, Universidad de Castilla-La Mancha, Toledo, Spain
e-mail: joseluis.polo@uclm.es

Á. R. de Lope
Department of Neurosurgery, Complejo Hospitalario de Toledo, SESCAM, Toledo, Spain

E. López-Dolado (✉)
Rehabilitation Department, Hospital Nacional de Parapléjicos, SESCAM, Toledo, Spain

Department of Medicine and Medical Specialities, School of Medicine, Universidad de Alcalá (UAH), Madrid, Spain

Laboratory of Interfaces for Neural Repair, Hospital Nacional de Parapléjicos, SESCAM, Toledo, Spain
e-mail: elopez@sescam.jccm.es

E. López-Dolado, M. Concepción Serrano (eds.), *Engineering Biomaterials for Neural Applications*, https://doi.org/10.1007/978-3-030-81400-7_11

implantable devices offers new ways of treatment while providing novel strategies to induce plastic changes in the damaged circuits. On top of that, the new generations of neurosurgeons have on hand novel and more advanced cellular and sub-cellular tools that permit the manipulation of cells, molecules, and genes and the application of specific immunotherapy techniques in order to restore CNS morphology and function. The role of neurosurgery is, therefore, changing, and there is now an imminent need to redesign and develop less invasive, precise, and safe surgical interventions to apply these new therapies to a greater number of patients and at different stages of their diseases. This chapter reviews the main neurological devices for implantation in the CNS and the tools and methodologies used for their indication, diagnosis, and monitoring. Some concepts on deep brain stimulation and brain–machine interfaces are also discussed.

Keywords Brain–machine interface · Central nervous system implants · Deep brain stimulation · Electrical impedance tomography · Electrical stimulation therapies · Electrode–tissue interface · Manganese-enhanced magnetic resonance imaging · Neurosurgical restorative techniques · Tractography

11.1   The Role of Neurosurgery in Central Nervous System Restoration

Injuries at the central nervous system, being either a traumatic brain injury (TBI) or a spinal cord injury (SCI), usually leave permanent sequelae with a drastic impact on patients. Typically, they prevent them from carrying out activities of daily living (ADLs) autonomously and deteriorate their quality of life. Moreover, these pathologies deeply affect society as a whole, as they entail high human and economic costs [1]. Traditionally, neurosurgeons have played an important role at the acute stage of the CNS damage. For instance, patients with a traumatic SCI commonly needed a prompt surgery to get their spinal column stabilized as soon as possible in order to avoid posterior complications [2]. After such surgical procedures, in most cases, patients used to go through a rehabilitation program with other staff physicians, scientists, and health professionals such as physiotherapists and occupational therapists. At that point, appointments with neurosurgeons were no longer scheduled, as there were no such repairing surgical procedures available. Nowadays, the emergence of new technological devices able to provide more functional responses to patients with permanent sequelae makes neurosurgeons to actively participate in the functional recovery process, so playing an important role not only at the acute phase of the neurological disease, but also at more chronic stages.

11.1.1 Traditional Neurological Functional Recovery after CNS Injury: A "Macroscopical–Morphological Approach"

The functional recovery that occurs mostly during the first year after CNS lesions is supported by some accepted systems-circuitry principles, whose validity extends throughout the rehabilitation period from sub-acute to chronic stages. This potential of functional improvement is based on three accepted by consensus facts that sustain the functional optimization of the spontaneous recovery [3]:

1. There is a primary functional neurological recovery due to the initial decrease of neuroinflammation features such as oedema inside the CNS, but also because of the plasticity phenomena that sustain collateral sprouting of the intact and some of the axotomized neurons. This leads to remarkable changes in synaptic sites and increases neurotrophic factors levels, all of the above providing a more suitable environment for neural connections reorganization and thus for neurofunctional improvement. However, this promising scenario is subsequently replaced, 6–12 months after the injury, by the formation of a glial scar that hampers functional recovery and may even worsen it. Rehabilitation programs applied during the acute and sub-acute period in these patients aim to stimulate plasticity and focus on those circuits that generate more functional behaviours (for more details, see Chaps. 9 and 10). At this stage, various methods for evoking functionally useful motor responses through electrical stimulation (e.g., cortex electrical stimulation, vagus nerve stimulation, and paired peripheral nerve stimulation) and transcranial magnetic stimulation (e.g., brain state-dependent stimulation [4]) are available and have shown promising results. While electrical stimulation has been already used in humans, transcranial magnetic stimulation has been exclusively used in animal models and is not yet suitable for human application.
2. There is a reinforcement of the synaptic network to enhance functionalities due to Hebbian plasticity, which can be improved by neural interference and deep stimulation strategies.
3. The use of two different electrical stimulation systems: (I) functional electrical stimulation (FES) of distal musculature, and (II) epidural electrical stimulation (EES) of distal spinal elements attempts to develop a bypass or brain–computer interface (BCI) to stimulate the intact neuromuscular elements distal to the CNS lesion.

Based on these facts, the "macroscopic morphological approach" aims to stimulate the plasticity phenomena that occur during the recovery period by providing a rehabilitation program with oriented tasks to the patient (for further details, please refer to Chaps. 9 and 10), especially during the first year after injury. The electrical stimulation therapies mentioned above also intend to preserve the motor function below the level of the lesion.

11.1.2 New Approaches for Neurological Functional Recovery after CNS Injury: A "Microscopic Approach"

There are some cellular and molecular events that can influence neural regeneration processes, which could be considered as potential therapeutic targets. Concretely, studies in animal models have identified some molecular factors—such as the factor 10 and growth-associated protein 43 (GAP43)—that contribute to provide a suitable environment for axonal regeneration [5]. Also, some genes such as the proto-oncogene bcl-2 avoid neural cell death [6]. Moreover, the initial inflammation and the alteration of the immune system after a CNS injury play a critical role in axonal regeneration [7–9]. In this context, the transplantation of stem and progenitor cells is currently being investigated to enhance the chances of functional recovery due to their demonstrated capacity to improve neurological deficits in animal models [10–12]. Based on these facts, the possibility of manipulating all these cellular and molecular factors and the application of immunotherapies aided by neurosurgical interventions may provide promising and revolutionary avenues for the treatment for these patients. They might make the healing of CNS lesions at the end of the rehabilitation period a reality in the near future, thus relieving patients from dreaded functional sequelae. Nonetheless, it will be of huge importance to determine the best combination of traditional rehabilitation programs with the new possibilities of microscopic neurosurgical interventions to establish the appropriate timing for each particular procedure based on the most favourable synergistic action among them [3].

Given the vast degree of recent developments that fields such as BCI, FES, EES, gene therapy, stem cell therapy, and immunotherapy are experiencing, the role of neurosurgery has moved from just taking an action at the initial stabilization of these patients to play a more active role in their posterior recovery outcomes, closely intertwined with the multidisciplinary neurorehabilitation team. In this chapter, we will focus on describing the main neurological implantable devices that are being used nowadays as a treatment for different neurological disorders, as well as some of their follow-up tools. We will also introduce some concepts on deep brain stimulation (DBS) and BCI systems.

11.2 CNS Implantable Devices Commonly Used in Neurosurgical Practice

Since ancient times, cranial and nervous system surgery has used different materials for the repair of injuries. For instance, cranioplasty, defined as the surgical repair of a defect in the skull, is one of the oldest neurosurgical procedures, with archaeological remains in the area of Peru showing procedures to repair defects in the cranial vault with gold plates of 1 mm in thickness 2000 years before Christ. Along the centuries, the daily work of neurosurgeons has evolved based on previously described neurosurgical techniques and the success and failure of new materials and modified procedures. The evolution of neurosurgical techniques to approach increasingly deeper areas in the CNS has driven the design and development of novel materials and tools throughout history [13]. Later on, the discovery of the treatment of hydrocephalus by Dandy and coworkers with a study on experimental animals led to the design of cerebrospinal fluid shunt catheters placed intraventricularly as a treatment that revolutionized this previously fatal disease [14]. It was then realized that the insertion of these catheters not only provided release of intracranial pressure, but also allowed the instillation of intraventricular medication. The administration of contrasts for the radiological study of intracranial lesions was also made possible through this route, at a time in which diagnostic systems such as computed tomography and magnetic resonance imaging were not available. Intraspinal catheters were also designed for diagnosis and drug administration, giving rise later to intrathecal infusion systems.

In 1820, Charles Babbage pioneered early mechanical computing devices. Around the 1950s, work by Penfield and collaborators in the study of cortical functions led to the design of systems for recording cortical electrical activity. Then, the first electrocorticographic systems appeared, which later led to implantable cortical electrode arrays. These devices found the way to interact with the cerebral cortex by means of electrical stimulation, recording its activity and inducing specific responses. Experts in Otolaryngology, most notably William House, began working in the early 1960s with cochlear electrical stimulation. These early attempts later evolved, in collaboration with William Hitselberger, to the first auditory brainstem implant, which was placed in a patient in 1979 for the first time [15]. Nowadays, small and more energy-efficient microelectronic devices are enabling the transition from portable to wearable to implantable ones. We are facing a quick succession of advances in the field, from cortical stimulation and recording electrode blankets for drug-resistant epilepsy, to retinal visual prostheses, paddle electrodes, vestibular prostheses for balance recovery and intraparenchymal electrodes for DBS. In the spinal cord, neuroprosthesis for chronic pain management is worth mentioning (Fig. 11.1). Table 11.1 summarizes standard procedures and CNS implants in the current neurosurgical practice according to their location.

There is a clear parallelism between the functioning of nervous tissues and that of microelectronic semiconductors. However, in the CNS, ions in solution are the elements that move across neuronal membranes charging it, while microelectronic semiconductors transport electrons for the same purpose. Hybridizing the two systems to create neurobionic interfaces seems a logical goal to pursue, regardless of its difficulty and the many biological and engineering challenges to face. In this context, current implantable micro-electromechanical systems combine miniaturized mechanical and electromechanical elements. Figure 11.2 shows the different components of these systems. About sensors, there is a wide range of microsensors available in the market and capable of monitoring various physiological variables such as temperature, blood pressure, magnetic fields, electrode–tissue impedance, inertial forces acting around the sensor and certain chemicals. On their turn, microactuators include instruments for tissue ablation (e.g., by heating, light

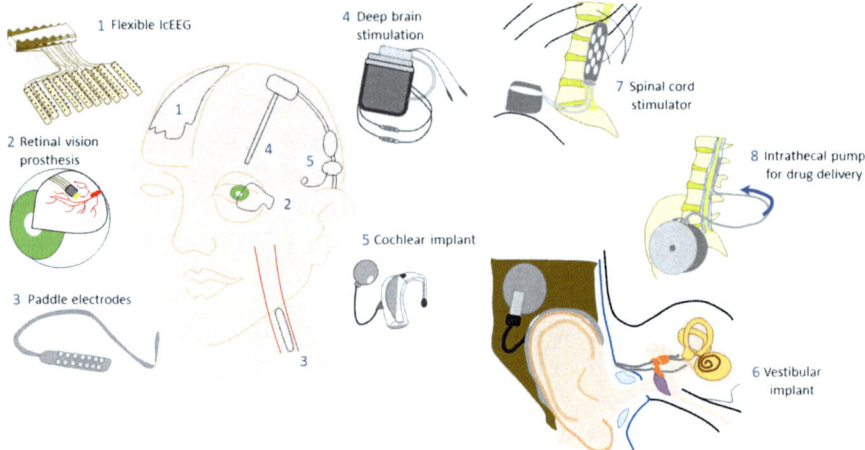

Fig. 11.1 Current neural implants and their location in the CNS [16]: (1) electrode blankets for EEG recording; (2) retinal implant as a visual prosthesis; (3) paddle electrodes; (4) DBS electrode; (5) cochlear implant as a hearing prosthesis; (6) vestibular implant as a balance prosthesis; (7) implantable spinal cord stimulator for pain control; and (8) intrathecal pump for drug delivery. Modified from [16] with permission

Table 11.1 Current CNS implants in neurosurgical clinical practice classified by location

Implant location	Procedure	Type of material/device used for implantation
Calotte	Craneoplasty	Titanium mesh Biocompatible materials
Cortex	Intraoperative recording of epileptiform activity	Electrode cortical blankets
Dura mater	Duroplasty	Striated muscles Bovine pericardium Biocompatible synthetic materials
Epidural space	Intracranial pressure recording Implantation of stimulation electrodes at spinal level for pain control	Silicone Biocompatible materials Epidural electrodes
Intraparenchymal location	Brachytherapy for local radiation release Local chemotherapy release Intraparenchymal pressure recording sensors Deep brain stimulation	Iodine-125/131 seeds Carmustine implants Intracranial pressure sensors Brain electrodes
Intraventricular	Cerebrospinal fluid shunt systems for intraventricular drug administration	Silicone catheters Ommaya reservoir
Subdural space	Pressure recording catheters Spinal catheters for drug administration	Subdural electrodes Intrathecal pumps

Fig. 11.2 Main components of micro-electromechanical systems

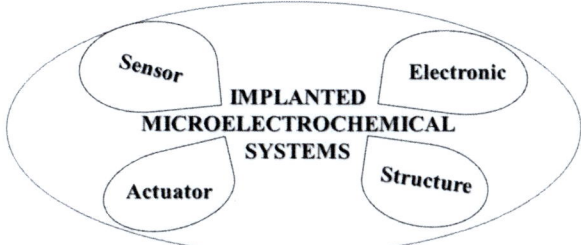

radiation, or ultrasound) and bioactive molecules such as chemotherapy agents, shunt systems for fluid flow control, optical switches to modulate or redirect light beams, and micro-resonators, among many others [17].

Supported by an extensive progress in the design of neuro-microelectronic interfaces and a better understanding of the functioning of the human brain, current microelectronic systems are enabling new forms of CNS control, interaction, and intervention, both invasive and noninvasive. At present, there are many different types of prosthetic devices available, ranging from electric wheelchairs to innervated robotic limbs, synthetic exoskeletons, and artificial sphincters. Importantly, recent technological developments are even permitting that effector devices and prostheses do not interact with the actual physical world but can do so in a virtual environment, in which the effector appears as an electronic avatar. The feasibility of this novel concept has been demonstrated by using BCI experiments [18–20]. In what follows, we will briefly describe general concepts on DBS and BCI techniques applied in the context of neurological diseases.

11.2.1 Deep Brain Stimulation (DBS)

Deep brain stimulation (DBS) is a functional and reversible neurosurgical technique capable of modulating the response of certain neuronal populations, either by stimulating or inhibiting them. These actions then modify the responses of the neural circuits leading to networks reorganization. Based on its potential and results obtained to date, DBS is considered one of the most promising clinical therapies for neurological disorders nowadays. It was first approved by the Food and Drug Administration of the United States of America (FDA) in 1995, for the treatment of essential tremor, then for the treatment of Parkinson's disease in 2002, and for the management of dystonia in 2003. Although its main indication is movement disorders, encouraging results are being reported from clinical trials in other diseases such as Tourette's syndrome, obsessive-compulsive disorder, certain types of epilepsy, major depression, and Alzheimer's disease. Recently, it has also been applied for the recovery of dynamic stability during locomotion in post-stroke patients and for the management of certain addictive disorders [21].

The surgical implantation technique required for DBS usually involves the placement of two intraparenchymal brain electrodes introduced into the previously chosen relay nucleus using a stereotaxic frame, a guided neuronavigation system, or even a stereotaxic arm robot. Each of the electrodes connects to the neurostimulator via an external cable, which, in a second step, is subcutaneously implanted together with the wires in the chest or the abdomen. The electrodes implantation procedure can be carried out, while the patient is anesthetized—always in the paediatric population—or awake. If the patient is anaesthetized during the surgical procedure, it will be necessary to perform intraoperative neurophysiological recordings to check the correct location of the electrode. The initial programming of the stimulation parameters and their periodic checking and readjustment is a very important goal for the optimal effect of DBS, and it is performed by telemetry. Post-surgical complications are not frequent and, in most of the cases, mild. Between 14 and 50 % of cases with undesirable effects related to the implantation also experience some hardware-related minor problems [22].

In this context, magnetic resonance imaging (MRI) is the modality of choice for morphological diagnosis at the surgical site, as well as for monitoring the correct position of DBS electrodes and assessing post-implantation complications and subsequent controls. In some particular cases, computerized axial tomography (CT) and the fusion of MRI and CT images could also be used [23].

11.2.2 Brain–Machine Interface (BMI)

BCI consists of the creation of a direct communication path between the brain, in this case a wired brain, and an external device. BCI provides a continuous association between brain activity and peripheral stimulation, with the necessary and sufficient potential to induce neuroplastic changes in the CNS. This technique is used during rehabilitation programs with the aim to restore human cognitive or sensory-motor functions in patients who suffered paralysis, such as in SCI patients. In this context, electroencephalographic (EEG) signals from the patients serve to decode the intention of voluntary movements.

At the Hospital Nacional de Parapléjicos of Toledo (Spain), BCI has been applied in two different research approaches. The first one relates to gait restoration. In this case, BCI was used to decode the gait intention from the patient and trigger movements of the lower-limb exoskeleton (Fig. 11.3) [24]. The second one refers to the combination of BCI and FES for hand rehabilitation in incomplete cervical SCI patients. In this case, the patient's movement intention was used to trigger a grasping movement induced by FES, while the patient simultaneously received visual feedback from a virtual hand simulating this movement [25]. These studies analysed the feasibility of using BCI in SCI patients. Based on the encouraging results obtained, further experiments with a larger patients population and a greater number of experimental sessions to evaluate the therapeutic potential of BCI in combination with traditional therapies are envisioned [19, 20].

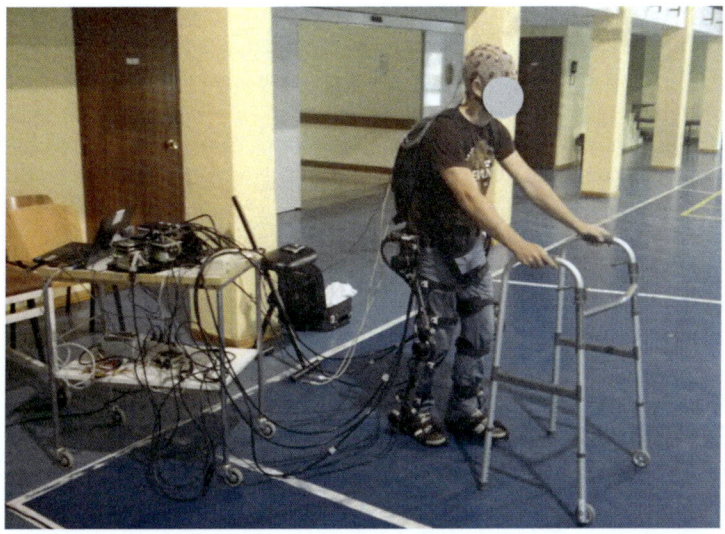

Fig. 11.3 BCI in combination with a lower-limb exoskeleton for gait rehabilitation at the Hospital Nacional de Parapléjicos of Toledo (Spain)

Surveys of potential end users have identified key characteristics of BCIs, including high precision, fast response times, and multifunctionality. All these BCI performance characteristics are fundamentally dependent on the neural decoding algorithm used, which is trained to associate neural activation patterns with the intended actions of the user. For instance, Schwemmer and colleagues reported intracortical recordings from a 27-year-old SCI patient with AIS A C5 implanted with a 96-channel microelectrode array in the hand and arm area of his left primary motor cortex. In this chapter, authors demonstrated that the patient was able to maintain the performance of a BCI implanted in the cerebral cortex and synchronized with FES to allow object reaching, grasping, and subsequent release with one of his paralysed arms for more than a year after implantation. No signs of explicit daily retraining were found, and functionality could even increase with minimal retraining using a technique known as transfer learning. They also showed that the participant could use the deep neural network decoder in real time to move his forearm using FES, all of which being a solid step towards the translation of this implantable BCI technology to the clinical practice [18].

11.3 Diagnostic Tools for Monitoring Patients with CNS Implants

The incorporation of neuroimplants in the neurosurgical practice is clearly driving the progress of medical imaging techniques. Within those, MRI is, without a doubt, the gold standard technique in the field. The increasing prevalence of neurodegenerative diseases (e.g., Parkinson's and Alzheimer's diseases and macular degeneration), treatment-refractory temporal epilepsy, and neuromodulation with the need of treatment for chronic neuropathic pain has increased the number of patients with implantable bioelectronic devices to restore function and improve their quality of life. In this context, some parameters become essential. First, the study of the interactions with the neuroimaging equipment for disease diagnosis and monitoring. Second, the evaluation of the histological environment in which the implant is immersed. Third, the investigation of the neuroimplant itself. Another important matter is safety when carrying an implantable CNS device, as it involves not only medical assessments, but also periodic monitoring by engineers and physicists responsible for its design. The conditions needed for optimal implant functioning should also be carefully considered and reviewed, including system tightness, batteries, Wi-Fi connection microantennas, wiring, electrodes, tissue interfaces, and, simultaneously, the physiological conditions of the targeted tissue area (e.g., a minimal and sensitive area of brain and/or spinal cord neural tissue).

11.3.1   MRI Interactions with the CNS and the Implantable Devices

The wide variety of neuroimplants to be used (e.g., cochlear stimulators, deep brain stimulators, flexible intracranial electroencephalography antennas, retinal vision prostheses, and spinal cord stimulators), as well as the constant increase in the capacity of MRI equipments, now reaching magnetic fields of 8 T for clinical use, is a constant challenge for multidisciplinary teams in the field. As far as the implant materials are concerned, the periodical performance of monitoring and follow-up MRI exposes them to a high-electromagnetic-field environment, radio frequencies, and rapid changes in secondary magnetic fields such as gradients. This could lead to high material stress due to mechanical forces, induced voltages, heat and eventual device magnetization (Fig. 11.4), circumstances that could therefore hamper the use of the different radiological techniques, especially MRI. Among major advantages of MRI, we could note the avoidance of ionizing energy, the high contrast obtained among tissues—especially soft ones—the versatility of its use for multiplanar and 3D anatomical imaging, its angiographic capacity and the possibility of performing functional studies (i.e., BOLD effect), in vivo studies of diffusion, diffusion tensor, and tractography; all of those by using the same unique equipment while scanning [16].

It is worth noting the tremendous increase in potential complications associated with the performance of exploratory radiological procedures if the subject under examination has an implant. The magnitude of these complications depends on the type of implant, the particular MRI equipment to be used, and the acquisition sequence applied (Fig. 11.5). For these reasons, physicians, physicists, and engi-

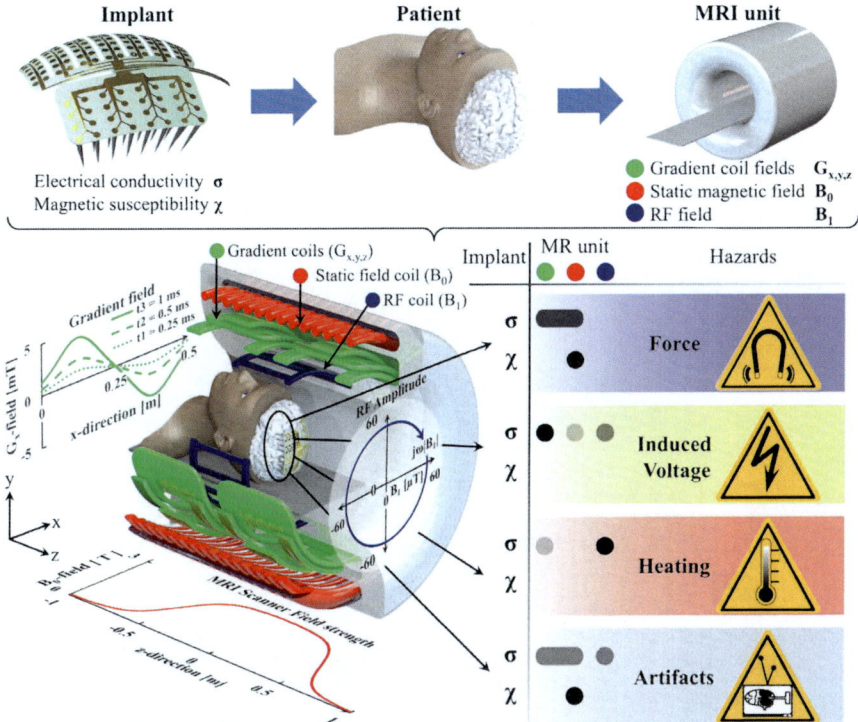

Fig. 11.4 The complex magnetic resonance environment can interact with an implant in many different ways that could cause hazardous conditions. Forces may occur from gradient coil and static fields interacting with the implant. Voltages may be induced by the time-variant gradient and radio-frequency fields or by the rapid transportation of the patient towards the scanner. Heating effects originate from the electric fields of the time-variant magnetic fields, where the main contribution is produced by the powerful radio-frequency pulses. Artefacts are mainly attributed to the B0-field by virtue of its deviation caused by susceptibility mismatch and to the radio-frequency field depending on the sequence used. The correlation between the hazard and the magnetic resonance component is displayed by grey-scale dots in a table format, where the intensity gives an indication of how severe this interaction can become. Reprinted by permission from [16], https://doi.org/10.1088/1741-2552/aab4e4

neers need a shared study framework to jointly understand these interactions. This will ensure patient safety, as well as appropriate implant and bioelectronic device preservation, thus maintaining optimal diagnostic and therapeutic capabilities of radiological procedures.

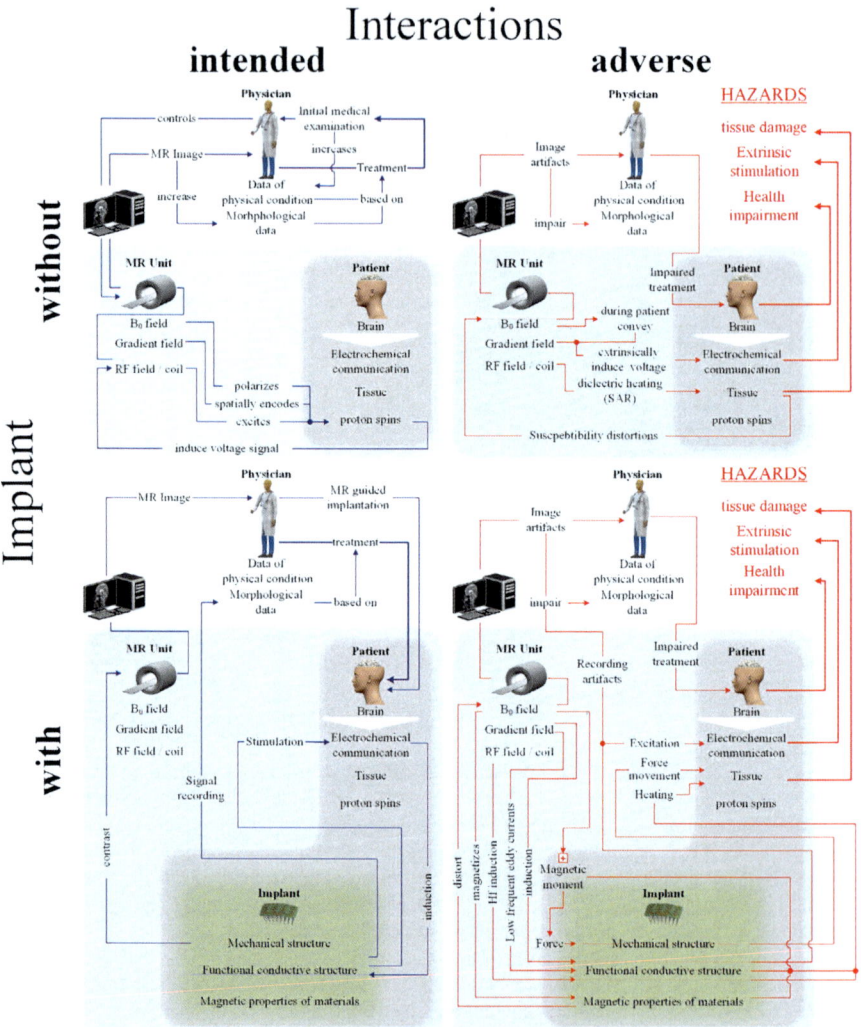

Fig. 11.5 A detailed map of interaction pathways among the patient, the brain implant, the MR environment, and the responsible physician. Those pathways that only occur in the presence of an implant sum up to those occurring in patients without implants, displayed in the top part of the graphic. Reprinted by permission from [16], https://doi.org/10.1088/1741-2552/aab4e4

11.3.1.1 Feasibility of MRI for the Assessment of Morphological Changes in SCI Animal Models

Despite MRI is widely used in human injuries to identify qualitative and quantitative pathological changes after TBI and SCI, its use is not so widespread in preclinical research with animal models. In the particular case of the rat spinal cord, a higher MRI resolution might be required to detect damage without errors due to its smaller size [26–28]. Although a good correlation between MRI findings and histological results has been reported in some experimental studies in rats using different MRI resolutions [26–28], the best resolution remains unclear yet. In order to elaborate MRI protocols capable of monitoring morphological changes in the follow-up of rat SCI models, López-Dolado and colleagues developed a protocol, extracted from the daily clinical MRI practice with cervical SCI patients, to be applied in rats with cervical hemisection.

In this chapter, five adult male Wistar rats received a C6 right spinal cord hemisection at the age of 20 ± 3 weeks (360 ± 36 g in weight). Four months after injury, animals were sacrificed, their spinal cords extracted, and the spinal fragments between C4 and T1 prepared for ex vivo MRI analysis [29]. Simultaneously, MRI results from 5 young human adults with incomplete chronic traumatic SCI due to C6 myelopathy were chosen. The same clinical MRI protocol was used to analyse both human and rat images in order to verify whether it could be possible to detect morphological changes in both types of samples in a comparable way. In the case of rat samples, MRI images were acquired by using a 7 T Bruker Biospin MRI machine (Ettlingen, Germany) BioSpec70/30. Human images were acquired by using a 3 T Siemens (Erlangen, Germany) Magnetom Trio MRI machine. The different measurements were made on images of a turbo spin echo sequence enhanced in T2 (RARE turbo) and axial acquisition. A Weasis 2.6.0 viewer was used for the visualization and evaluation of the images. Importantly, samples of SCI patients whose MRI was analysed were selected to mimic in age, location, and severity those of rat SCI samples (i.e., adult young age, traumatic SCI, lesion epicentre at the right C6 level, complete or almost complete preservation of the left hemicord at the lesion site, above and below). This specific effort of selection was expected to improve the comparability of the parameters under investigation between both groups, making the clinical scenarios more suitable for comparison (Fig. 11.6).

Prior to analysis, a size normalization (3.571:1) was applied to MRI slices to make them comparable [30]. No statistically significant differences were found for any morphological parameters studied between human and rat MRI myelopathies (i.e., maximal anteroposterior diameter, $p = 0.15$; maximal transverse diameter, $p = 0.84$; lesion epicentre length, $p = 0.56$; total SCI volume, $p = 0.06$). Perilesional gliosis showed a significantly higher hyperintensity in humans ($p = 0.008$), but it was unclear that this finding related to a more severe gliosis as MRI resolution was higher in rat samples [31]. It is worth noting that rat models represent precise and controlled laboratory injuries (Fig. 11.6d–f), while humans suffer uncontrolled SCI lesions and second traumas due to the stabilization surgery (Fig. 11.6a–c). This fact could explain, at least partially, the higher perilesional gliosis hyperintensity found

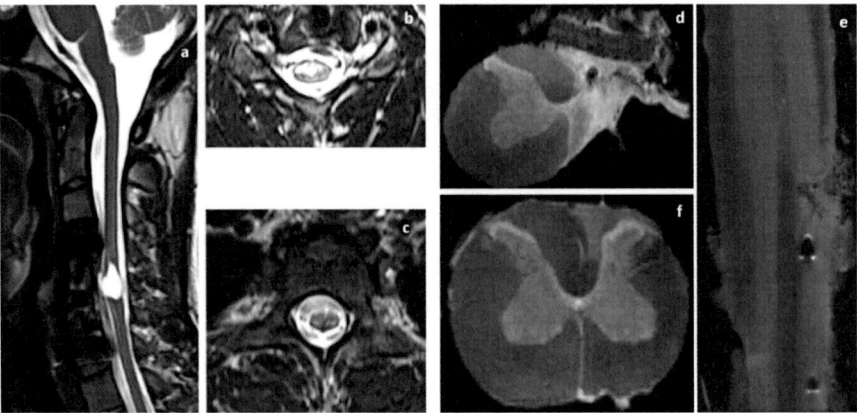

Fig. 11.6 MRI example of a young human adult with chronic incomplete traumatic SCI due to C6 myelopathy. Lesion epicentre in sagittal (**a**) and axial (**b**) sections. Perilesional damage in axial section (**c**). MRI example of a rat C6 right spinal cord hemisection. Lesion epicentre in axial (**d**) and coronal (**e**) sections. Perilesional damage in axial section (**f**). (**d**)–(**f**) Reproduced from [29], Copyright Elsevier (2019)

in human MRI images. Based on these findings, MRI is a useful technique to analyse SCI lesions in animal models as a complementary method, even a replacement, to other more expensive and laborious methods such as classical anatomopathological histology.

11.3.1.2 Other Emerging MRI Functional Imaging Methods: Manganese-Enhanced MRI and Tractography

In addition to conventional T1- and T2-weighted imaging, the continuous development of functional MRI methods is enabling a better understanding and investigation of the occurrence, development, and prognosis of CNS diseases. To the already known diffusion-weighted imaging (DWI), diffusion tensor imaging (DTI), and blood-oxygen-level-dependent (BOLD) imaging, manganese-enhanced magnetic resonance imaging (MEMRI) has been recently developed. MEMRI relies on the strong paramagnetism of Mn^{2+}, a calcium ion analog that can enter excitable cells through voltage-gated calcium channels. Mn^{2+} can be transported along the axons via microtubule-based fast axonal transport, allowing to describe neuroanatomical structures, to monitor neural activity and to evaluate axonal transport rates [32]. At present, MEMRI is mainly used in three areas: studies of anatomy and cellular structure, tracing of neural connections, and brain function monitoring. Although it provides unquestionable higher resolution and image quality than DWI, DTI, and BOLD, it has toxicity issues. The excess of Mn^{2+} often accumulates in organs such as the liver, pancreas, bones, kidneys, and brain, so causing liver damage, neurotoxicity, impaired cardiovascular function, and even death, what highly limits its clinical use [33].

It is well established that MRI biomarkers provide a noninvasive method to visualize abnormalities in TBI and SCI patients. Particularly, DTI can accurately detect changes in white matter microstructure [34]. Alternatively, MRI angiography is the best technique to detect changes in cerebral blood flow, while magnetic resonance spectroscopy (MRS) is the one of selections to observe changes in neuronal metabolism and susceptibility-weighted imaging (SWI) shows micro-bleeding in patients with diffuse axonal injury. MEMRI could also be potentially useful for TBI because of its high-contrast and detailed information on structural and functional changes in the brain for continuous and dynamic explorations, as Mn^{2+} remains in the body for several days [32]. In the context of SCI, MEMRI has been used to assess the type, extent, and conditions of the lesion epicentre. As Mn^{2+} signal intensity significantly correlates with motor function, MEMRI has been proposed to determine SCI severity and to monitor the efficacy of some regenerative therapies, such as cell transplantation [35].

The purpose of tractography is to make neural tracts visible and able to follow. Images are acquired using diffusion tensor MRI, assessing the degree of anisotropy of the CNS region under study through mathematical algorithms presented in 2D and 3D images. This technique enables mapping of structural connections in the human and animal CNS and has great potential for estimating complete maps of neuronal connections in the brain known as "connectomes." However, its resolution is suboptimal in white matter areas. Recently, complex methods such as restricted spherical deconvolution and global tractography have been developed to improve tracts tracing. Besides, the reconstruction of tractography images greatly benefits from having clearly identifiable anatomical landmarks and local features such as the proximity of the analysed areas to the grey matter [36]. Because tractography conditions also differ between intra- and interhemispheric connections at different connection distances, the algorithms must be chosen on purpose. Indeed, the more precise the algorithm choice, the better the connectome reconstruction [37].

11.3.2 Electrical Impedance: A Tool to Monitor Electrodes, Tissues, and Interfaces

From an electrical point of view, when an electrode is implanted in a biological tissue, three regions are created: the electrode itself, the electrode–tissue interface (ETI), and the bulk tissue [38]. All of these regions take part in both biopotential measurements and the propagation of electrical signals for the stimulation. Although the biointegration (into the surrounding tissue) of implanted electrodes has already become an attainable goal [39], the ETI plays a significant role in the signal transfer by acting as a transducer, that is, the electronic current of the external circuit is converted into an ionic current flow in the tissue (or vice versa). This is of special interest in chronically implanted electrodes where the electrical behaviour of the ETI changes progressively over time due to a wide variety of mechanisms, such as the foreign body reaction (glial scar in a CNS lesion), electrochemical reactions, and the physiological state of the tissues involved, among other considerations [39–41]. This could imply difficulties in measuring biopotentials (recording electrode) or in maintaining an appropriate propagation of the applied stimuli (stimulation electrode), thus compromising the therapeutic benefits. Therefore, it is useful to monitor the electrical behaviour of the ETI to apply an optimal, efficacious, and safe protocol over chronic timescales [42].

11.3.2.1 Electrode–Tissue Interface

The electrical behaviour of the ETI can be monitored using impedance measurements [40, 41, 43]. Impedance is defined as the opposition that a material/tissue presents to an alternating current flow as a function of the frequency. Therefore, electrode faults, as well as changes in the physiological states of the tissues, can be detected by analysing deviations in the impedance [40]. The concept of impedance is defined in the context of sinusoidal signals and linear time-invariant circuits. The impedance, $Z(j\omega)$, is a complex number (j is the imaginary unit) whose real and imaginary parts are the resistance $R(\omega)$ and the reactance $X(\omega)$, respectively; ω is the angular frequency ($\omega = 2\pi f$, where f is the frequency) [44, 45].

Figure 11.7a shows an electrical equivalent circuit (EEC) including a very simple model of ETI, that is, an electrode–electrolyte interface consisting of an ideal double-layer capacitance C_{DL} in parallel with a charge-transfer resistance R_{CT} (linearization of the Butler–Volmer equation for small overpotentials) [42, 44]. A high (low) value of R_{CT} is related to a polarizable (non-polarizable) electrode. R_{EL} is introduced to model the resistance of the electrode itself. This EEC is a frequency-dependent circuit: at sufficiently high (low) frequencies, the capacitor impedance is very small (large) compared with R_{CT} and then C_{DL} (R_{CT}) will carry almost all the current. Figure 11.7b shows the Nyquist plot, $-X(\omega)$ vs. $R(\omega)$, of the EEC of Fig. 11.7a. The semicircle intersects the real axis at $R_{EL}(\omega = \infty)$ and $R_{EL} + R_{CT}(\omega = 0)$ [46]. The frequency ω_{top} at which $-X(\omega)$ reaches a maximum has also been labelled. It should be mentioned that the experimental impedance data can also be graphically displayed using a Bode plot (i.e., magnitude $|Z(j\omega)|$ and phase $arg[Z(j\omega)]$ vs. frequency) [46].

A more realistic study of ETIs requires the introduction of other circuit elements, such as constant phase elements and diffusion impedances [40, 41, 44], to describe the experimental impedance measurements. In the time domain, these elements involve non-integer order derivatives (i.e., fractional calculus) [45, 46]. Specifically, the well-known depressed semicircle is obtained in the Nyquist plot if the capacitance C_{DL} (refer to Fig. 11.7a) is replaced by a constant phase element. The branch R_{CT} is also more general involving all the Faradaic electrode processes, and it is often described (in the frequency domain) by replacing R_{CT} by the Faradaic impedance [42, 44].

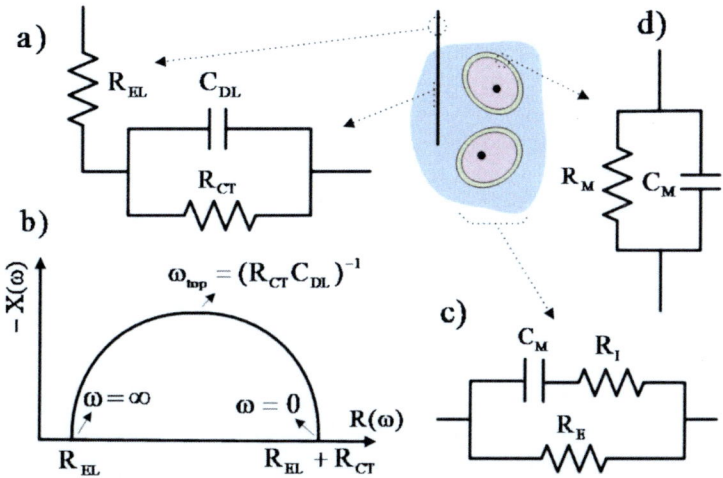

Fig. 11.7 Electrical equivalent circuits (EECs). The inset (electrode implanted in tissue) illustrates the different regions being modelled by the EEC. (**a**) EEC of an electrode—electrolyte interface in which R_{EL} has been added to model the electrode resistance. (**b**) Nyquist plot for the EEC shown in panel (**a**). (**c**) Fricke–Morse circuit model. (**d**) EEC modelling the cell membrane

11.3.2.2   Tissue and Cells

In general, the electrical behaviour of the tissue itself involves capacitance (such as that of cell membranes) and resistance (intra- and extra-cellular compartments) [45]. Figure 11.7c shows the Fricke–Morse model (for passive electrical behaviour) [47], where C_M, R_I, and R_E describe the cell membrane capacitance, intra- and extra-cellular media, respectively. At sufficiently low (high) frequencies, the current will flow through the extra-cellular (both intra- and extra-cellular) media. The Nyquist plot (not sketched here) is similar to that of Fig. 11.7b. Now, the semicircle intersects the real axis at $R_I // R_E$ ($\omega = \infty$) - $//$ denotes parallel - and R_E ($\omega = 0$). Nyquist plot sketches the well-known depressed semicircle when C_M (refer to Fig. 11.7c) is replaced by a constant phase element [45]. In that case, the electrical impedance of that new EEC is similar to that proposed by Cole in his famous formula [48]. Impedance spectroscopy has been used to monitor tissue reaction to inserted neural implants. A resistance was included in the EEC to model the tissue encapsulation adjacent to the implant and formed after implantation [49, 50]. Figure 11.7d illustrates an EEC that models the passive electrical behaviour of a cell membrane. The resistor R_M is related to the permeabilities of the membrane to the different ionic species [51]. For simplicity, the voltage source (in series with R_M) that models the resting transmembrane potential has not been included. A more general EEC can be obtained by replacing C_M (refer to Fig. 11.7d) by a constant phase element. This EEC has been used to interpret patch-clamp recordings from neuronal membranes exhibiting fractional dynamics [51]. In excitable cells, the membrane potential (i.e., voltage across C_M) can be actively modulated, so that a threshold level for action potential triggering could eventually be reached. In that case, the linear time-invariant assumption, and thus the EEC of Fig. 11.7d, is not valid. An EEC of an excitable cell membrane was proposed by Hodgkin and Huxley [52].

Although not highlighted before, the ETI system is complex. Typically, it involves non-linearities and a time-variant behaviour. Specifically, the electrical behaviour of the neural tissue is non-linear, non-homogeneous, anisotropic, and time-variant [53]. One must be careful when linear time-invariant EECs (small-signal models) are used. In general, voltages and currents are the variables used in circuit theory. To study the current or potential distributions in the electrode, the ETI, or the bulk tissue, the field theory should be used. In many cases, neural tissue is modelled by a resistor (circuit theory), while in field theory, it is considered as a volume conductor and the potential distribution is obtained from a particular type of stimulation pattern [54]. On the theoretical basis abovementioned, a new imaging technology has emerged, the electrical impedance tomography (EIT). In EIT, small currents are injected through surface electrodes attached to the body. Simultaneously, the boundary voltages are collected to reconstruct an impedance-based image. Since the impedance properties of biological tissues differ for physiological and pathological states, EIT has demonstrated potential in several clinical scenarios, the most consolidated of which is the assessment of pulmonary ventilation. Its use in the CNS, in particular in the brain, arouses great interest to continuously monitor cerebral oedema and intracranial haemorrhage, providing timely and invaluable information for diagnosis and clinical management of TBI patients [55]. EIT could be of special value to evaluate those patients with CNS implantable devices safely and accurately, as electrical impedance spectroscopy could differentiate the electrode, the tissue, and the interface responses. Although very promising, EIT offers a relatively low spatial resolution compared with CT and MRI. Nonetheless, it provides several unique advantages in terms of low cost, non-invasion, dynamic image monitoring, and functional imaging [56], and it is expected to improve in a foreseeable future.

11.4   Future Perspectives

As it has been exposed in each of the chapters of this book, the combination of new materials designed for reparative purposes with a much deeper understanding of CNS damage mechanisms and more powerful and versatile technological tools brings the possibility of finding more efficient therapies for TBI and SCI. While neural interfaces, FES, and EES are starting to prove usefulness in patients, most cellular therapies, immunotherapies, molecular interventions, and optogenetics, although promising, are not suitable for translation yet. Future clinical trials should focus on the synchronization between cellular and behavioural interventions in order to achieve evident and pursued effects to restore lost neurological functions [3].

Being one of the most promising and innovative treatments in neurology, DBS has great potential for clinical neuroscience research. As MRI helps to understand the neurocircuitry changes after effective DBS, novel developments of 3 T MRI-compatible DBS devices are needed. Indeed, most of available ones, while safe under 1.5 T MRI, are still considered to be unsafe under 3 T MRI, thus preventing those patients to benefit from its more accurate diagnosis. The latest DBS devices for LFP recordings would be expected to include large-capacity batteries or a wireless charging technology, continuous high-precision data acquisition capability, wireless real-time external transmission that can be applied in freely moving conditions, and multiple differential signal acquisition channels functioning simultaneously with DBS.

Looking ahead to the future, there is a natural evolution towards the application of machine learning algorithms for motor control evaluation and objective output readings. As in almost all biomedical fields, artificial intelligence has great potential in DBS. The application of these strategies could help to better understand how DBS influences brain networks, uncovering underlying DBS mechanisms not yet completely understood. Interestingly, recent advances in MRI-compatible DBS devices are allowing for the acquisition of neuroimages during stimulation. Undoubtedly, personalization of the DBS device implantation process and the use of stimulation programs created with artificial intelligence algorithms will increase the safety and efficacy of this type of treatment and thus improve the quality of life of DBS-carrying patients. It is also expected that the combination of advanced imaging techniques and artificial intelligence algorithms will facilitate the individualization of surgical targets and then the final outcome of these procedures [21].

Finally, there is no doubt that a more extensive use of MEMRI and tractography would provide new clues to better understand lesion mechanisms and propose new therapeutic avenues. Further developments in material science are expected to give rise to safer manganese-containing contrast agents that could provide broader prospects and endless possibilities for MEMRI clinical applications [32, 35]. Similarly, more accurate choices of diffusion MRI tractography algorithms could improve connectome reconstructions [37].

References

1. National Spinal Cord Injury Statistical Center (2014) Spinal cord injury facts and figures at a glance. J Spinal Cord Med 37:355–356
2. Fariña MM, Barrera SS, Marqués AM et al (2017) Actualización en lesión medular aguda postraumática. Parte 2. Med Intensiva 41:306–315
3. Krucoff MO, Miller JP, Saxena T et al (2019). Toward functional restoration of the central nervous system: A review of translational neuroscience principles. Neurosurgery 84:30–40
4. Gharabaghi A, Kraus D, Leão MT et al (2014) Coupling brain-machine interfaces with cortical stimulation for brain-state dependent stimulation: enhancing motor cortex excitability for neurorehabilitation. Front Hum Neurosci 8:122

5. Li S, Nie EH, Yin Y et al (2015) GDF10 is a signal for axonal sprouting and functional recovery after stroke. Nat Neurosci 18:1737–1745

6. Chen DF, Schneider GE, Martinou JC, Tonegawa S (1997) Bcl-2 promotes regeneration of severed axons in mammalian CNS. Nature 385:434–439

7. Hu X, Leak RK, Shi Y et al (2015) Microglial and macrophage polarization—new prospects for brain repair. Nat Rev Neurol 11:56–64

8. Sofroniew MV (2015) Astrocyte barriers to neurotoxic inflammation. Nat Rev Neurosci 16:249–263

9. Brommer B, Engel O, Kopp MA et al (2016) Spinal cord injury-induced immune deficiency syndrome enhances infection susceptibility dependent on lesion level. Brain 139:692–707

10. Dinsmore JH, Martin J, Siegan J et al (2002) CNS grafts for treatment of neurologic disorders. In: Methods of Tissue Engineering. 1st Ed, San Diego: Academic Press

11. Borlongan CV, Tajima Y, Trojanowski JQ, Lee VM, Sanberg PR (1998) Transplantation of cryopreserved human embryonal carcinoma-derived neurons (NT2N cells) promotes functional recovery in ischemic rats. Exp Neurol 149:310–321

12. Ahuja CS, Mothe A, Khazaei M (2020) The leading edge: Emerging neuroprotective and neuroregenerative cell-based therapies for spinal cord injury. Stem Cells Transl Med 9:1509–1530

13. Jea A, Al-Otibi M, Rutka JT et al (2007) The history of neurosurgery at the Hospital for Sick Children in Toronto. Neurosurgery 61:612–625

14. Dandy WE, Blackfan KD (1914) Internal hydrocephalus: An experimental, clinical and pathological study. Am J Dis Child 8:406–482

15. Otto SR, Moore J, Linthicum F et al (2012) Histopathological analysis of a 15-year user of an auditory brainstem implant. Laryngoscope 122:645–648

16. Erhardt JB, Fuhrer E, Gruschke OG et al (2018) Should patients with brain implants undergo MRI? J Neural Eng 15:041002

17. Hughes MA (2016) Insinuating electronics in the brain. Surgeon 14:213–218

18. Schwemmer MA, Skomrock ND, Sederberg PB et al (2018) Meeting brain-computer interface user performance expectations using a deep neural network decoding framework. Nat Med 24:1669–1676

19. Hochberg LR, Bacher D, Jarosiewicz B et al (2012) Reach and grasp by people with tetraplegia using a neurally controlled robotic arm. Nature 485:372–375

20. Bouton CE, Shaikhouni A, Annetta NV et al (2016) Restoring cortical control of functional movement in a human with quadriplegia. Nature 533:247–250

21. Sui Y, Tian Y, Ko WKD et al (2021). Deep Brain Stimulation Initiative: Toward innovative technology, new disease indications, and approaches to current and future clinical challenges in neuromodulation therapy. Front Neurol 11:597451

22. Cantó SC, Dueñas BP, Arboix JR (2017) Estimulación cerebral profunda como tratamiento de los trastornos del movimiento en la edad pediátrica. In: Martín JO, Pisón JL. Neurocirugía Pediátrica. Editorial Ergon, Madrid, 1ª (ed), vol 2017

23. Barnaure I, Pollak P, Momjian S et al (2015) Evaluation of electrode position in deep brain stimulation by image fusion (MRI and CT). Neuroradiology 57:903–908

24. López-Larraz E, Trincado-Alonso F, Rajasekaran V et al (2016) Control of an ambulatory exoskeleton with a brain–machine interface for spinal cord injury gait rehabilitation. Front Neurosci 10:359

25. Trincado-Alonso F, López-Larraz E, Resquín F et al (2018) A pilot study of brain-triggered electrical stimulation with visual feedback in patients with incomplete spinal cord injury. J Med Biol Eng 38:790–803

26. Sandner B, Pillai DR, Heidemann RM et al (2009) In vivo high-resolution imaging of the injured rat spinal cord using a 3.0T clinical MR scanner. J Magn Reson Imaging 29:725–730

27. Cunha L, Horvath I, Ferreira S, et al (2014) Preclinical imaging: an essential ally in modern biosciences. Mol Diagn Ther 18:153–73

28. Weber T, Vroemen M, Behr V et al (2006) In vivo high-resolution MR imaging of neuropathologic changes in the injured rat spinal cord. Am J Neuroradiol 27:598–604

29. Domínguez-Bajo A, González-Mayorga A, Guerrero CR et al (2019) Myelinated axons and functional blood vessels populate mechanically compliant rGO foams in chronic cervical hemisected rats. Biomaterials 192:461–474

30. Kjell J, Olson L (2016) Rat models of spinal cord injury: from pathology to potential therapies. Dis Model Mech 9:1125–1137

31. Felder J, Celik AA, Choi CH et al (2017) 9.4 T small animal MRI using clinical components for direct translational studies. J Transl Med 15:264

32. Yang J, Li Q (2020) Manganese-enhanced magnetic resonance imaging: application in central nervous system diseases. Front Neurol 11:143

33. Chen P, Bornhorst J, Aschner M (2018) Manganese metabolism in humans. Front Biosci 23:1655–79

34. Hoogenboom WS, Rubin TG, Ye K et al (2019). Diffusion tensor imaging of the evolving response to mild traumatic brain injury in rats. J Exp Neurosci 13:1179069519858627

35. Martirosyan NL, Turner GH, Kaufman J et al (2016) Manganese-enhanced MRI offers correlation with severity of spinal cord injury in experimental models. Open Neuroimag J 10:139–147

36. Donahue CJ, Sotiropoulos SN, Jbabdi S et al (2016) Using diffusion tractography to predict cortical connection strength and distance: a quantitative comparison with tracers in the monkey. J Neurosci 36:6758–6770

37. Sinke MRT, Otte WM, Christiaens D et al (2018) Diffusion MRI-based cortical connectome reconstruction: dependency on tractography procedures and neuroanatomical characteristics. Brain Struct Funct 223:2269–2285

38. Hernández-Balaguera E, López-Dolado E, Polo JL (2018) In vivo rat spinal cord and striated muscle monitoring using the current interruption method and bioimpedance measurements. J Electrochem Soc 165:G3099–G3103

39. Salatino JW, Ludwig KA, Kozai TDY, Purcell EK (2017) Glial responses to implanted electrodes in the brain. Nat Biomed Eng 1:862–877

40. Sawan M, Mounaim F, Lesbros G (2008) Wireless monitoring of electrode-tissues interfaces for long term characterization. Analog Integr Circ Sig Process 55:103–114

41. Lempka SF, Miocinovic S, Johnson MD et al (2009) In vivo impedance spectroscopy of deep brain stimulation electrodes. J Neural Eng 6:046001

42. Merrill DR, Bikson M, Jefferys JG (2005) Electrical stimulation of excitable tissue: design of efficacious and safe protocols. J Neurosci Methods 141:171–198

43. Krukiewicz K (2020) Electrochemical impedance spectroscopy as a versatile tool for the characterization of neural tissue: A mini review. Electrochem Commun 116:106742

44. Barsoukov E, Macdonald JR (2018) Impedance Spectroscopy. Theory, Experiment, and Applications, 3rd edn. Wiley, New York

45. Hernández-Balaguera E, López-Dolado E, Polo JL (2016) Obtaining electrical equivalent circuits of biological tissues using the current interruption method, circuit theory and fractional calculus. RSC Adv 6:22312–22319

46. Hernández-Balaguera E, Vara H, Polo JL (2016) An electrochemical impedance study of anomalous diffusion in PEDOT-coated carbon microfiber electrodes for neural applications. J Electroanal Chem 775:251–257

47. Fricke H, Morse S (1925). The electric resistance and capacity of blood for frequencies between 800 and 4(1/2) million cycles. J Gen Physiol 9:153–167

48. Cole KS (1940) Permeability and impermeability of cell membranes for ions. Cold Spring Harbor Symp Quant Biol 8:110–122

49. Williams JC, Hippensteel JA, Dilgen J et al (2007) Complex impedance spectroscopy for monitoring tissue responses to inserted neural implants. J Neural Eng 4:410–423

50. Mercanzini A, Colin P, Bensadoun JC, Bertsch A, Renaud P (2009) In vivo electrical impedance spectroscopy of tissue reaction to microelectrode arrays. IEEE Trans Biomed Eng 56:1909–1918

51. Hernández-Balaguera E, Vara H, Polo JL (2018) Identification of capacitance distribution in neuronal membranes from a fractional-order electrical circuit and whole-cell patch-clamped cells. J Electrochem Soc 165:G3104–G3111
52. Hodgkin AL, Huxley AF (1952) A quantitative description of membrane current and its application to conduction and excitation in nerve. J Physiol 117:500–544
53. van Dongen M, Serdijn W (2016) Design of efficient and safe neural stimulators: a multidisciplinary approach. Springer, Switzerland
54. Hernández-Labrado GR, Polo JL, López-Dolado E, Collazos-Castro JE (2011) Spinal cord direct current stimulation: finite element analysis of the electric field and current density. Med Biol Eng Comput 49:417–429
55. Yang L, Li H, Ding J et (2018) Optimal combination of electrodes and conductive gels for brain electrical impedance tomography. Biomed Eng Online 17:186
56. Bayford RH (2006) Bioimpedance tomography (electrical impedance tomography). Biomedical Sciences, vol. 8. Middlesex University, London

Glossary

3D bioprinting Recent top technology for the production of biologically complex three-dimensional microstructures providing control over shape and microarchitecture across conventional manufacturing techniques. It allows to precisely position biological elements, including living cells and molecular components, in a specified 3D hierarchical organization and to even create artificial multi-cellular tissues/organs through computer-aided design/computer-aided manufacturing.

Bacterial translocation Process by which certain bacteria move from their usual location in the body, such as the bowel, to other sterile locations, such as the blood.

Biocompatibility Ability of a biomaterial to perform its desired function with respect to a medical therapy without eliciting any undesirable local or systemic effects in the recipient or beneficiary of that therapy, but generating the most appropriate beneficial cellular/tissue responses in that specific situation and optimizing the clinically relevant performance of that therapy (David F. Williams, 2008).

Biodegradability Ability of a given material to break into smaller elements by the action of biological agents. It implies the disassembly and dissociation of the biomaterial in fragments of micron and submicron size that can be either degraded by phagocytic cells around, then entering metabolic routes, or transported to the blood stream for posterior elimination through detoxification routes at the liver or, preferentially, discarded in the urine by the kidneys, if water soluble.

Biomaterial Natural, synthetic, or hybrid materials suitable to directly interact with living cells, tissues, or organisms for diagnostic and/or therapeutic purposes. Each biomaterial presents a specific set of properties (e.g., biocompatibility, biodegradability, mechanical compliance) that can be used or adapted to match the characteristics of a particular biological microenvironment.

Bipolar electrochemistry Phenomena by which unconnected conducting materials immersed in an ionic media show a dipole between their edges when a remote external field is applied with external driving electrodes. If the potential is sufficiently large, the induced dipole may yield electrochemical reactions at any of the poles, negative or positive.

Brain–computer/machine interface Direct communication path between the brain, a wired brain, and some external device. It provides a continuous association between brain activity and peripheral stimulation, with the necessary and sufficient potential to induce neuroplastic changes in the central nervous system.

Capacitance Ability of a system (e.g., an electrode) to store an electric charge. Electrostimulation protocols for functional therapies in the nervous system mainly rely on capacitive and relatively high-impedance electrodes including platinum and its alloys, steel, and titanium nitride.

Carbon-based nanomaterials Materials fabricated by combining different allotropic forms of carbon at the nanoscale. For instance, carbon nanotubes, nanofibers, and nanowires have allowed the development of scaffolds characterized by electrical conductivity while enabling axonal interfacing and growth.

Cell transplantation Use of stem/progenitor cells from the patient itself (i.e., autologous) or from donors that are genetically similar (i.e., allogenic) or identical (i.e., syngeneic) to be located in a receptor patient. These cells can derive from either fetal tissues (e.g., umbilical cord and placenta) or adult tissues (e.g., bone marrow, adipose tissue, olfactory mucosa, and dental pulp) and have been tested over the last 10 years for the treatment of a wide variety of diseases.

Constant phase element Constant phase provided by an electric circuit element, used to model, for instance, the non-ideal capacitance exhibited by double layers or cell membranes, in the frequency domain.

Contusion/cavity Type of spinal cord injury in which a blunt force deforms the cord without invading it. The continuity of the spinal cord surface is preserved, and no adhesions are found between this surface and the dura mater. This type of lesion usually affects more severely the innermost aspects of the cord in which evident areas of hemorrhage and necrosis are formed. After weeks to months, these areas end up forming cysts (fluid-filled cavities).

Deep brain stimulation Functional and reversible neurosurgical technique capable of modulating the response of certain neuronal populations, either by stimulating or inhibiting them. It modifies the responses of neural circuits, thus leading to network reorganization. Nowadays, it is considered one of the most promising clinical therapies for the treatment of neurological disorders.

Direct current/alternating current Direct current is a type of current in which the direction of the electron flow is consistent in a single direction. In contrast, when the current is alternating, the direction of electron flow changes back and forth at regular intervals or cycles.

Discomplete spinal cord injury Type of spinal cord injury that is clinically complete, but anatomically incomplete. Discompleteness was first uncovered by the finding of electrophysiological transmission of signals across the lesion in patients who were clinically complete. It was afterward confirmed by histological observations of anatomical continuity of the white matter across the lesion in this type of patients.

Electrical equivalent circuit Circuit modeling a process that has an electrical equivalence. It is an interconnection of circuit elements (e.g., resistors, capacitors) approximating the electrical behavior of a physical system.

Electrical impedance tomography New imaging technology in which small currents are injected through surface electrodes attached to the body, while the boundary voltages are simultaneously collected to reconstruct an impedance-based image. This technique could be of special value to evaluate those patients with central nervous system implantable devices safely and accurately. Nonetheless, it still offers a relatively low spatial resolution compared with computerized axial tomography and magnetic resonance imaging. Within its advantages, we could cite low-cost, non-invasion, and dynamic image monitoring.

Electrically active materials Materials able to generate electric fields in the extracellular matrix that could act as signals to promote and control growth, remodeling, and protein adsorption. Some examples of widely investigated electrically active materials include poly(tetrafluoroethylene), polyvinylidene fluoride, and polypyrrole.

Electrodeposition Well-known method for in situ coatings by applying an electric current to a conductive material immersed in a solution containing the salt of the metal to be deposited.

Electrode Electrical conductor used to make contact with a non-metallic part of a circuit (e.g., an electrolyte, a biological tissue). The application of an electric field involves capacitive charging at the electrodes and a faradaic electronic–ionic transfer between the electrode and the ionic electrolyte medium of variable charge depending on electrode materials and potentials used.

Electrode–tissue interface Region located between the electrode and the tissue in which the electrode is implanted. It plays a significant role in signal transfer by acting as a transducer. The electronic current of the external circuit is converted into an ionic current flow in the tissue (or vice versa).

E-Textiles Soft orthoses or platforms used to build wearable devices not considered orthotic, yet playing an important role as a tool for monitoring rehabilitation.

Exergame therapies Rehabilitative therapies based on serial games.

Extracellular matrix Non-cellular 3D network mainly composed of macromolecules, such as fibrous proteins (e.g., collagen and integrin) and glycosaminoglycans that are responsible for supporting cells and providing biochemical and biomechanical gradients able to organize and influence cell response.

Fibrous scaffold Biomimetic fiber-based construct widely used to recreate biochemical and biophysical gradients analogous to each specific cellular microenvironment present in the nervous system. There is a wide range of natural (e.g., collagen, chitosan) and synthetic (e.g., PCL, PLGA) biomaterials that can be successfully fabricated in this fashion, although desirable mechanical strength and biodegradability are usually achieved combining both of them.

Glial cells Group of cells generally smaller than neurons responsible for supporting them. Their number exceeds that of neurons 5–10 times. The main glial cell types are astrocytes, oligodendrocytes, ependymal cells, Schwann cells, and satellite cells.

Graphene oxide 2D monolayer of carbon atoms that presents a remarkable reactive surface due to the presence of oxygen functional groups. Although its electrical conductivity is quite inferior relatively to pristine graphene, this nanomaterial presents good mechanical and optical properties, enhanced water dispersibility, and multiple anchoring points for biomolecules and polymers, for which is being extensively explored for biomedical applications.

Hydrogel 3D polymeric network capable of using its intrinsic water retaining capacities to efficiently boost the regeneration of injured tissues. This smart/biomimetic material structurally similar to the extracellular matrix of many tissues can be often processed under relatively mild conditions and delivered in a minimally invasive manner. Moreover, it permits fabrication by 3D printing technologies to better reproduce native tissues architecture.

Intestinal dysbiosis Loss of microbial balance in the normal gut microbiota due to quantitative or qualitative changes in its composition, function, and/or distribution.

Laceration Type of spinal cord injury that involves a penetrating disruption of the spinal cord surface by bone fragments, knifes, or projectiles.

Magnetic field Vector field that describes the measurable forces between moving electric charges (electric currents) and magnetically ordered materials.

Magnetically ordered material Material with permanent magnetic moments whose orientations, relative to those of their close neighbors, are stable at a given temperature.

Magnetoelastic resonance Any mechanical system exhibits, when submitted to a periodical in time unidirectional stress, the so-called mechanical resonances, states at which the elastic energy provided to the system is absorbed with maximum efficiency, thus resulting in maximum strains along the stress direction. The magnetoelastic coupling linked to the magnetoelastic and magnetostrictive effects can be used to excite such a resonance by applying a time-varying magnetic field.

Magnetoelastic sensors Magnetic materials, from the point of view of sensorization, that react to variations in the different fields present in their environment with detectable changes in their magnetization state, interact with either charge or spin currents, and support controllable excitations such as the spin waves whose basic parameters can be also modified by the system environment. The two effects that underlie the functionalities of these sensors are magnetoelasticity and magnetostriction.

Magnetoelasticity Phenomenology linked to the modification of the state of magnetically ordered materials induced by the external application of mechanical stresses.

Magnetostriction Phenomenology related to the variation of the dimensions of a magnetically ordered material resulting from the application of a magnetic field.

Manganese enhanced magnetic resonance imaging Type of functional magnetic resonance imaging that relies on the strong paramagnetism of Mn^{2+}, which can be transported along the axons via microtubules permitting to describe neuroanatomical structures, to monitor neural activity and to evaluate axonal transport rates. At present, it is mainly used for studies of anatomy and cellular structure, tracing neural connections and brain function monitoring. Although it provides better resolution and image quality when compared with other types of functional magnetic resonance imaging, it has important toxicity problems.

Massive compression Very severe spinal cord injury in which the cord is macerated or pulpified to a varying degree and the epicenter of the lesion ends up in a cavitation full of connective tissue and nerve roots invasion. The tissue response in this case may share similarities with contusion, but also with the fibrosis observed in laceration injuries.

Mesenchymal scar Intraspinal piece of tissue consisting of fibroblasts, meningeal cells and deposits of collagen, laminin and inhibitory chondroitin sulfate proteoglycans filling an injury.

Nanoemulsions Transparent dispersions of oil droplets in water (or the inverse) stabilized by surfactants and smaller than nanomicelles in size. They allow the encapsulation of hydrophobic drugs in the dispersed oil phase and a tight control of drug release and have been used as drug delivery vectors to the central nervous system through intranasal administration, thus bypassing the blood–brain barrier. Solid lipid nanoparticles are a particular type of nanoemulsions with the hydrophobic phase in solid state.

Nanomicelles Less employed organic formulations, similar to liposomes in biodegradability but composed of amphiphilic molecules, other than phospholipids, forming a massive hydrophobic core and a hydrophilic shell that provides stability in circulation and the opportunity to deliver high payloads of hydrophobic drugs.

Nervous system Complex set of specialized structures in a multicellular organism whose mission is to control and regulate the function of all the components of the body, coordinating their actions and the relation of the whole organism with the external ambient. The nervous system is divided into two major components: the peripheral nervous system and the central nervous system. The first is composed of all the nerves starting at the central nervous system and extending throughout the body. Nerves are bundles of peripheral nerve fibers that form pathways from the central nervous system to the rest of the body and vice versa. On its turn, the central nervous system is formed by the brain and the spinal cord.

Neuron Basic functional and structural unit of the nervous system responsible for transmitting and guiding neural signals. Although with different shapes and sizes, a typical neuron comprises a cell body, where the nucleus is located, variable number of dendrites from which the signals are received and a longer filament called axon conducting the information and transmitting it to neighboring cells (e.g., muscle cells, neurons, glands).

Neuronanotechnology Emerging treatment approach in neuroscience. Nanoengineered materials offer the potential to interact with biological systems at the subcellular scale, with exceptional levels of control over physiological processes. Regenerative neural interfaces, once miniaturized and enriched by ad hoc designed nanoscale features, will better probe biological systems by mimicking bio-environments most acceptable for neural cells, facilitating biocompatibility and axon regrowth and regeneration.

Orthoses Devices to provide assistance and support for neurological or orthopedical pathologies. They contribute to the stabilization, immobilization, and correct posture and help to maintain their functions and prevent or reduce pain. There are traditionally two main types of orthoses: rigid orthoses, made of stiff materials to prevent voluntary and involuntary motions of the joints, and soft orthoses, composed of materials to achieve more compliant and adaptable performances, though not providing anti-gravitatory support. Recently, a new type of orthoses has been created, dynamic wearable orthosis, manufactured with metallic alloys with non-linear mechanical features and able to promote morphological changes in response to dynamic forces.

Post-traumatic syringomyelia Active cysts with pressure inside that forces them to expand rostro-caudally and may cause enormous damage, with progressive loss of function.

Prosthesis Artificial device that is implanted in a living organism to replace a body part and provide a minimum contribution to the correct functionality of the organ/system in which it is incorporated.

Regenerative neurosurgery Set of neurosurgical techniques aimed at restoring lost neurological functions. It includes both traditional surgical interventions in the acute period of brain and spinal cord traumas, intended to stabilize the injury and reduce secondary damage, and new, less invasive interventions to place intraparenchymal electrodes for electrostimulation and apply cellular therapies/immunotherapies into injury sites.

Resonance Phenomenon characteristic of mechanical and electromagnetic systems submitted to periodic excitations (isolated or combined stress and electromagnetic fields), whereby the system absorbs the excitation energy with maximum efficiency and experiences maximum amplitudes of the parameter through which it responds to excitation.

Robotic devices Sophisticated and automated electromechanical robotic devices developed in order to provide a more efficient way of rehabilitation. These robotic devices are able to increase the performance of an unable-bodied wearer by using an external source of energy that will be transformed into motion.

Scaffold 3D biomedical platform engineered from biomaterials with the purpose of resembling the function and shape of a specific extracellular matrix and, subsequently, inducing biomimetic cell responses. Depending on the tissue engineering strategy, a scaffold can be implanted without cells to be colonized in vivo or it can work as drug and/or cell vehicle for promoting a more controllable regeneration of injured tissues.

Schwann cell Cell with a vital role in maintaining the peripheral nervous system. It is responsible for the production of the myelin casing around neuronal axons and is equivalent to oligodendrocytes, which occur in the central nervous system and respond quickly to injury and aid axon regeneration.

Sensor Any type of device whose state can be modified by any of the fields describing its environment in a way that can be associated to a transient or steady event in a second system of the electric and/or electronic types.

Silicon-based materials Materials fabricated by macromolecular organosilicons that offer wide clinical application for their use as implants, electrodes, and tissue scaffolds. However, their dimensions and mechanical properties do not perfectly match those of the central nervous system tissue resulting in poor tissue integration.

Silicone Material formed by alternated chains of silicon and oxygen atoms, which are enrolled themselves. Its chemical inertness, waterproofing, resistance to oxidation, and stability against high and low temperatures make silicone a preferred material to manufacture orthotic and prosthetic devices.

Solid cord injury Type of spinal cord injury that seems normal macroscopically—without softening, discoloration, or cavity formation—but shows different features of damage and pathology under microscopic examination.

Spinal cord Neural tubular structure located within the vertebral column that integrates the central nervous system together with the brain, being responsible for connecting the encephalon to the peripheral nervous system. Due to its central role in transmitting sensory information to the brain and motor commands to the rest of the body, any disruption of the spinal cord neural system results in lifelong tragic consequences.

Tissue engineering Interdisciplinary research field that applies the principles of biomedical engineering, cell biology, clinical practice, and materials science to develop clinical approaches capable of improving or replacing the functionality of a damaged tissue/organ. A tissue engineering strategy often involves the synergistic combination of specific biomolecules, cells, and scaffolds toward the mimic of a targeted cell/tissue microenvironment (e.g., bone, cardiac, neural, skin).

Transcranial magnetic stimulation Noninvasive procedure that uses magnetic fields to stimulate nerve cells in the brain and improve symptoms of cerebral damage, either to decrease excitability in the healthy hemisphere or to increase cortical excitability in the injured one.

Traumatic brain injury Disruption of normal brain functions caused by a traumatic event that involves the head and damages the brain.

Virtual reality Computer-generated simulation of a 2D or 3D environment with which a person can physically interact by means of a set of specially designed electronic devices.

Index

Printed by Books on Demand, Germany